Selected Titles in This

294 **Robert Coquereaux, Ariel García, Roberto Trinchero, Editors,** Quantum symmetries in theoretical physics and mathematics, 2002

293 **Donald M. Davis, Jack Morava, Goro Nishida, W. Stephen Wilson, and Nobuaki Yagita, Editors,** Recent progress in homotopy theory, 2002

292 **A. Chenciner, R. Cushman, C. Robinson, and Z. Xia, Editors,** Celestial Mechanics, 2002

291 **Bruce C. Berndt and Ken Ono, Editors,** q-series with applications to combinatorics, number theory, and physics, 2001

290 **Michel L. Lapidus and Machiel van Frankenhuysen, Editors,** Dynamical, spectral, and arithmetic zeta functions, 2001

289 **Salvador Pérez-Esteva and Carlos Villegas-Blas, Editors,** Second summer school in analysis and mathematical physics: Topics in analysis: Harmonic, complex, nonlinear and quantization, 2001

288 **Marisa Fernández and Joseph A. Wolf, Editors,** Global differential geometry: The mathematical legacy of Alfred Gray, 2001

287 **Marlos A. G. Viana and Donald St. P. Richards, Editors,** Algebraic methods in statistics and probability, 2001

286 **Edward L. Green, Serkan Hoşten, Reinhard C. Laubenbacher, and Victoria Ann Powers, Editors,** Symbolic computation: Solving equations in algebra, geometry, and engineering, 2001

285 **Joshua A. Leslie and Thierry P. Robart, Editors,** The geometrical study of differential equations, 2001

284 **Gaston M. N'Guérékata and Asamoah Nkwanta, Editors,** Council for African American researchers in the mathematical sciences: Volume IV, 2001

283 **Paul A. Milewski, Leslie M. Smith, Fabian Waleffe, and Esteban G. Tabak, Editors,** Advances in wave interaction and turbulence, 2001

282 **Arlan Ramsay and Jean Renault, Editors,** Groupoids in analysis, geometry, and physics, 2001

281 **Vadim Olshevsky, Editor,** Structured matrices in mathematics, computer science, and engineering II, 2001

280 **Vadim Olshevsky, Editor,** Structured matrices in mathematics, computer science, and engineering I, 2001

279 **Alejandro Adem, Gunnar Carlsson, and Ralph Cohen, Editors,** Topology, geometry, and algebra: Interactions and new directions, 2001

278 **Eric Todd Quinto, Leon Ehrenpreis, Adel Faridani, Fulton Gonzalez, and Eric Grinberg, Editors,** Radon transforms and tomography, 2001

277 **Luca Capogna and Loredana Lanzani, Editors,** Harmonic analysis and boundary value problems, 2001

276 **Emma Previato, Editor,** Advances in algebraic geometry motivated by physics, 2001

275 **Alfred G. Noël, Earl Barnes, and Sonya A. F. Stephens, Editors,** Council for African American researchers in the mathematical sciences: Volume III, 2001

274 **Ken-ichi Maruyama and John W. Rutter, Editors,** Groups of homotopy self-equivalences and related topics, 2001

273 **A. V. Kelarev, R. Göbel, K. M. Rangaswamy, P. Schultz, and C. Vinsonhaler, Editors,** Abelian groups, rings and modules, 2001

272 **Eva Bayer-Fluckiger, David Lewis, and Andrew Ranicki, Editors,** Quadratic forms and their applications, 2000

271 **J. P. C. Greenlees, Robert R. Bruner, and Nicholas Kuhn, Editors,** Homotopy methods in algebraic topology, 2001

(Continued in the back of this publication)

Quantum Symmetries
in Theoretical Physics
and Mathematics

CONTEMPORARY MATHEMATICS

294

Quantum Symmetries in Theoretical Physics and Mathematics

Proceedings of the Bariloche School
January 10–21, 2000
Bariloche, Patagonia, Argentina

Robert Coquereaux
Ariel García
Roberto Trinchero
Editors

American Mathematical Society
Providence, Rhode Island

Editorial Board

Dennis DeTurck, managing editor

Andreas Blass   Andy R. Magid   Michael Vogelius

This volume contains the proceedings of a conference on Quantum Symmetries in Theoretical Physics and Mathematics, held in Bariloche, Patagonia, Argentina, on January 10–21, 2000.

2000 *Mathematics Subject Classification.* Primary 16W30, 17B37, 20G42, 81R50, 46L87, 46L37, 81T40, 82B20, 81Txx, 18G60.

All photographs included in the volume are used with the permission of Ariel O. Garcia.

Library of Congress Cataloging-in-Publication Data
Quantum symmetries in theoretical physics and mathematics : proceedings of the Bariloche school, January 10–21, 2000, Bariloche, Patagonia, Argentina / Robert Coquereaux, Ariel Garcia, Roberto Trinchero, editors.
    p. cm. — (Contemporary mathematics, ISSN 0271-4132 ; 294)
In English, with one contribution in French.
Includes bibliographical references.
ISBN 0-8218-2655-7 (alk. paper)
    1. Symmetry (Physics)—Congresses. 2. Quantum groups—Congresses. 3 Geometric quantization—Congresses. 4. Mathematical physics—Congresses. I. Coquereaux, Robert. II. Garcia, Ariel, 1970– III. Trinchero, Roberto, 1956– IV. Contemporary mathematics (American Mathematical Society) ; v. 294.

QC174.17.S9 Q35   2002
530.14′3—dc21
                                                                                                                                      2002024651

Copying and reprinting. Material in this book may be reproduced by any means for educational and scientific purposes without fee or permission with the exception of reproduction by services that collect fees for delivery of documents and provided that the customary acknowledgment of the source is given. This consent does not extend to other kinds of copying for general distribution, for advertising or promotional purposes, or for resale. Requests for permission for commercial use of material should be addressed to the Acquisitions Department, American Mathematical Society, 201 Charles Street, Providence, Rhode Island 02904-2294, USA. Requests can also be made by e-mail to `reprint-permission@ams.org`.

Excluded from these provisions is material in articles for which the author holds copyright. In such cases, requests for permission to use or reprint should be addressed directly to the author(s). (Copyright ownership is indicated in the notice in the lower right-hand corner of the first page of each article.)

    © 2002 by the American Mathematical Society. All rights reserved.
    The American Mathematical Society retains all rights
    except those granted to the United States Government.
    Printed in the United States of America.

    ∞ The paper used in this book is acid-free and falls within the guidelines
    established to ensure permanence and durability.
    Visit the AMS home page at URL: `http://www.ams.org/`

    10 9 8 7 6 5 4 3 2 1   07 06 05 04 03 02

Contents

Foreword	ix
List of participants	xii
Symétries quantiques en physique théorique et en mathématiques	xxi
Quantum symmetries in theoretical physics and mathematics	xxv
About finite dimensional Hopf algebras NICOLÁS ANDRUSKIEWITSCH	1
Lectures on differentials, generalized differentials and on some examples related to theoretical physics MICHEL DUBOIS-VIOLETTE	59
Modular invariants from subfactors JENS BÖCKENHAUER AND DAVID EVANS	95
The classification of subgroups of quantum $SU(N)$ ADRIAN OCNEANU	133
Uses of quantum spaces O. OGIEVETSKY	161
CFT, BCFT, ADE and all that J.-B. ZUBER	233

Foreword

This volume contains the written counterpart of several lectures that were given in Bariloche, Argentina, January 10–21, 2000.

This school on "Quantum symmetries in theoretical physics and mathematics" was probably the first mathematical meeting of the new century. It took place in Patagonia, at the beginning of the austral summer; it was therefore a summer school for many participants (but a winter school for the others!).

The organizers of the school want to thank all those who came and contributed to the success of the school: students, lecturers, secretaries ... and special thanks to Prof. J. Stasheff who could not attend, but helped us to obtain partial funding from the NSF. The meeting received major support from the CIMPA (France), NSF (USA) and ANPCyT (Argentina), but also from the ICTP (Trieste, Italy), the CLAF (Brazil) and the CNRS (France).

Last but not least, we want to express our gratitude towards our respective laboratories, namely the Instituto Balseiro, Centro Atómico Bariloche (CAB), and the Centre de Physique Théorique (CPT, Luminy, Marseille). Special thanks go to Madame M. Rossignol, from the administrative staff of CPT, who came to Bariloche and took care of money matters (fellowships, reimbursments, etc.) and to Gladys Campagno from CAB.

Two reports will be found at the beginning of this volume, one in English and one in French. The first was sent to NSF and the other to CIMPA, just after the school.

Both documents list the titles of the lectures given during the school (some more details are given in the French report). Here is the list of the main lecturers:

- N. Andruskiewitsch (Argentina),
- M. Dubois-Violette (France),
- D. Evans (Great Britain),
- A. Ocneanu (USA),
- O. Ogievetsky (France),
- N. Reshetikhin (USA),
- M. Rosso (France),
- A. Varchenko (USA),
- S. L. Woronowicz (Poland),
- J. B. Zuber (France).

Several participants also delivered talks:

- E. Batista - Topological geons in $2 + 1$ dimensions.
- E. Buffenoir - Harmonic analysis on complex quantum groups.
- H. Figueroa - On the antipode of Kreimer's Hopf algebra.

- P. Hajac - Quantum-group equivariant connections and Chern-Connes pairing for quantum-group quotients.
- W. Pusz - Quantum calculus on a classical space.
- E. Ragoucy - Connecting double Yangians, elliptic algebras and quantum groups.
- S. Sache - A q-deformed Poincaré algebra in $2+1$ dimensions and its representation theory.
- M. Socolovsky - On the topology of the symmetry group of the standard model.

The present volume contains the lectures given by N. Andruskiewitsch, M. Dubois-Violette, D. Evans, A. Ocneanu, O. Ogievetsky, and J. B. Zuber.

Gathering the following written material took much time and effort from the editors (not mentioning the time spent by the contributors themselves, to whom we address our warmest thanks), but we hope that it was worth doing it.

Nahuel Huapi Lake, Bariloche, Argentina

List of participants

Name, First Name	Email	Institution, Country
Abadie, Beatriz	abadie@cmat.edu.uy	Montevideo, URUGUAY
Abadie, Fernando	abadie@cmat.edu.uy	Montevideo, URUGUAY
Andruskiewitsch, Nicolás	andrus@famaf.unc.edu.ar	Córdoba, ARGENTINA
Armas, Pablo	parmas@dm.uba.ar	Buenos Aires, ARGENTINA
Asaeda, Marta	asaeda@math.psu.edu	Penn State, U.S.A.
Barberis, María Laura	barberis@fis.uncor.edu	Córdoba, ARGENTINA
Barkallil, Abdelilah	abarkall@ulb.ac.be	Bruxelles, BELGIUM
Batista, Eliezer	ebatista@mtm.ufsc.br	UFSC, Florianopolis, BRASIL
Bellich, Humberto	belich@cbpf.br	CBPF, Rio de Janeiro, BRASIL
Bonechi, Francesco	francesco.bonechi@fi.infn.it	Firenze, ITALY
Buffenoir, Eric	buffenoi@lpm.univ-montp2.fr	Montpellier, FRANCE
Casini, Horacio	casini@cab.cnea.gov.ar	CAB-IB, ARGENTINA
Coquereaux, Robert	coque@cpt.univ-mrs.fr	CPT, Marseille, FRANCE
Cuba Castillo, Guillermo	gcubac@cbpf.br	CBPF, Rio de Janeiro, BRASIL
DiLuiggi, Constanza	cdilui@dm.uba.ar	Buenos Aires, ARGENTINA
Doria, Celso Melchiades	cmdoria@mtm.ufsc.br	UFSC, Florianopolis, BRASIL
Dotti, Gustavo	gdotti@fis.uncor.edu	Córdoba, ARGENTINA
Dubois-Violette, Michel	patricia@qcd.th.u-psud.fr	Orsay, Paris, FRANCE
Evans, David	evansde@cardiff.ac.uk	Cardiff, U.K.
Fassarella, Lucio	fassarel@cbpf.br	CBPF, Rio de Janeiro, BRASIL
Fernandez, Victoria	victoria@venus.fisica.unlp.edu.ar	La Plata, ARGENTINA
Figueroa, Héctor	figo@cariari.ucr.ac.cr	COSTA RICA
García, Ariel	ariel@cab.cnea.gov.ar	CAB-IB, ARGENTINA
Graña, Matías	matiasg@dm.uba.ar	Buenos Aires, ARGENTINA
Guerra, Stefano	guerra@socrates.mat.unimi.it	Milano, ITALY
Hajac, Piotr	pmh@fm.sissa.it	Trieste, ITALY
Hassouni, Yassine	y-hassou@fsr.ac.ma	Rabat, MOROCCO
Heckenberger, Istvan	heckenbe@mathematik.uni-leipzig.de	Leipzig, GERMANY
Huerta, Marina	huerta@cabtep2.cnea.gov.ar	CAB-IB, ARGENTINA
Iglesias, Alberto	iglesias@grad.physics.sunysb.edu	Córdoba, ARGENTINA
Iglesias, Rodrigo	iglesias@mate.uncor.edu	U.S.A.
Iucci, Aníbal	aiucci@hotmail.com	La Plata, ARGENTINA
Jancsa, Patricia	jancsa@mate.uncor.edu	Córdoba, ARGENTINA
Juyumaya, Jesús	juyumaya@ictp.trieste.it	Trieste, ITALY
Laca, Marcelo	marcelo@frey.newcastle.edu.au	Newcastle, AUSTRALIA
Montani, Hugo	montani@cab.cnea.gov.ar	CAB-IB, ARGENTINA
Moura-Melo, Winder	winder@cbpf.br	CBPF, Rio de Janeiro, BRASIL
Muraki, Naofumi	muraki@iwate-pu.ac.jp	Iwate, JAPAN

LIST OF PARTICIPANTS

Name, First Name	Email	Institution, Country
Naón, Carlos	naon@venus.fisica.unlp.edu.ar	La Plata, ARGENTINA
Natale, Sonia	natale@mate.uncor.edu	Córdoba, ARGENTINA
Ocneanu, Adrian	adrian@math.psu.edu	Penn State, U.S.A.
Ogievetsky, Oleg	oleg@cpt.univ-mrs.fr	CPT, Marseille, FRANCE
Onofre, Rojas	or@atria.if.uff.br	UFF, Rio de Janeiro, BRASIL
Pusz, Wieslaw	wieslaw.pusz@fuw.edu.pl	Warsaw, POLAND
Ragoucy, Eric	ragoucy@lapp.in2p3.fr	LAPP, Annecy, FRANCE
Reshetikhin, Nicolai	reshetik@math.berkeley.edu	Berkeley, U.S.A.
Reula, Oscar	reula@fis.uncor.edu	Córdoba, ARGENTINA
Reznikoff, Sarah	sarah@math.berkeley.edu	Berkeley, U.S.A.
Rosso, Marc	rosso@math.u-strasbg.fr	Strasbourg, FRANCE
Sachse, Sebastian	sachse@ime.usp.br	IME, São Paulo, BRASIL
Salles, Mario Otavio	salles@ime.usp.br	IME, São Paulo, BRASIL
Schieber, Gil	schieber@if.ufrj.br	CPT, Marseille, FRANCE
Simring, Eric	simring@uclink4.berkeley.edu	Berkeley, U.S.A.
Socolovsky, Miguel	socolov@sisisa.podernet.com.mx	UNAM, Mexico, MEXICO
Szymanski, Wojciech	wojciech@frey.newcastle.edu.au	Newcastle, AUSTRALIA
Tahrie, El Hassan	tahrie@sciences.univ-oujda.ac.ma	Oujda, MOROCCO
Toppan, Francisco	toppan@cbpf.br	CBPF, Rio de Janeiro, BRASIL
Trinchero, Roberto	trincher@cab.cnea.gov.ar	CAB-IB, ARGENTINA
Varchenko, Alexandre	av@math.unc.edu	Chapell Hill, UNC, U.S.A.
Villanueva, David	david@ime.usp.br	IME, São Paulo, BRASIL
Woronowicz, Lech	slworono@fuw.edu.pl	Warsaw, POLAND
Zuber, Jean-Bernard	zuber@spht.saclay.cea.fr	Saclay, Paris, FRANCE

Secretaries:

Campagno, Gladys		CAB, Bariloche, ARGENTINA
Rossignol, Michèle		CPT, Marseille, FRANCE

A. Ocneanu and J.-B. Zuber

M. Rosso and N. Reshetikhin

A. Varchenko and S. L. Woronowicz

N. Andruskiewitsch and M. Dubois-Violette

D. Evans and A. Ocneanu

O. Ogievetsky and J.-B. Zuber

Symétries quantiques en physique théorique et en mathématiques

Bariloche, Patagonie, Argentine
10-21 Janvier 2000

Présentation générale

L'école de Bariloche qui s'est tenue en Argentine, en Janvier 2000 a réuni 68 participants, physiciens théoriciens ou mathématiciens (environ moitié-moitié) originaires de 15 pays différents. La majorité des stagiaires (40) sont venus de pays d'Amérique Latine ; il faut signaler tout d'abord l'Argentine, bien sûr (22 participants dont 6 locaux), ainsi que le Brésil (12 participants) mais plusieurs scientifiques sont également venus du Chili, du Mexique, de Costa Rica et de l'Uruguay. Les autres participants se répartissent entre différents pays d'Europe (16 personnes, dont 8 français), les Etats Unis (4), l'Australie (2), le Japon (2) et le Maroc (3). Signalons également la participation, tant au niveau des professeurs que des stagiaires, d'un bon nombre de scientifiques originaires – tout au moins au niveau familial – de pays d'Europe de l'Est, mais travaillant le plus souvent, maintenant, dans d'autres pays (USA, France etc.) La majorité des participants avait un age situé entre 30 et 45 ans, ce qui nous conduit à une moyenne d'age assez jeune. Signalons aussi que 5 des stagiaires étaient des femmes, ce qui représente une participation féminine très sensiblement supérieure à celle généralement obtenue lors d'écoles ou de conférences consacrées aux sciences physiques ou mathématiques.

Le projet de cette école a été présenté en premier lieu au Conseil Scientifique du CIMPA (Centre International de Mathématiques Pures et Appliquées) et a reçu l'approbation de ce dernier ainsi qu'une promesse de financement. Il s'agit donc d'une "école CIMPA". D'autres financements ont été, par la suite, obtenus par les organisateurs. L'école a été subventionnée essentiellement par le CIMPA (France), la NSF (USA) et l'ANPCyT (Argentine), et, à un niveau moindre, par l'ICTP (Trieste, Italie), le CLAF (Brésil), et le CNRS (France).

Il faut mentionner également l'aide logistique que les organisateurs ont reçu de leurs laboratoires respectifs, à savoir le Centre de Physique Théorique de Marseille (qui a pu dépêcher une secrétaire, Mme M. Rossignol), l'Instituto Balseiro et le Centro Atómico Bariloche. A noter aussi l'aide du professeur Jim Stasheff (University of North Carolina) en ce qui concernce l'obtention et la gestion des fonds NSF. L'ensemble des crédits obtenus par les directeurs de l'ecole a permis de financer les voyages de 46 personnes (8 orateurs et 38 stagiaires) et de financer également le séjour de tous les participants sans exception. Notons que les capacités d'accueil du Centro Atómico Bariloche ont grandement permis de minimiser les coûts en hôtellerie pure (5 participants seulement ont été logés dans un hôtel).

Le thème de l'école "Symétries quantiques en physique théorique et en mathématiques" est un sujet d'actualité, aussi bien en mathématiques qu'en physique. On sait que la découverte récente de toutes une classes d'algèbres de Hopf ("groupes quantiques") a relancé, en mathématiques pures, toute une gamme d'activités de recherche : tout d'abord en algèbre, bien entendu, mais aussi dans des thématiques voisines (topologie des variétés, mécanique, théorie des noeuds, systèmes dynamiques, algèbres d'opérateurs, théorie des nombres etc.). En physique, les groupes quantiques sont apparus tout d'abord dans l'étude des systèmes intégrables (méthode du scattering inverse) mais ils ont alors commencé à être utilisés dans de nombreuses autres discplines : physique mésoscopique, théorie statistique, transitions de phase, théories conformes, théorie quantique des champs, théorie des cordes, particules élémentaires etc.

Les organisateurs ont fait appel à des spécialistes internationaux et leur ont demandé de donner un cours jettant les bases de la théorie des groupes quantiques tout en réservant une partie de leur temps pour présenter quelques résultats récents. L'école de Bariloche étant effectivement une école, les différents orateurs ont été priés de consacrer une partie importante de leur temps de parole à un travail pédagogique d'exposition. Les professeurs ont su apporter un éclairage tout à fait personnel aux différentes notions qu'ils ont présentées. Les organisateurs sont également heureux de noter que la plupart des orateurs ont profité de cette rencontre de Bariloche pour présenter, á l'issue de leurs cours respectifs, plusieurs résultats originaux, souvent encore non publiés.

Les différents professeurs ont évidemment choisi des points de départ différents pour présenter les notions de symétries quantiques, points de départ qui reflètent leur préoccupation propres : certains ont centré leur cours sur une approche basée sur l'analyse (au sens mathématique du terme), d'autres sur la géométrie, d'autres ont préféré une approche purement algébrique, d'autres enfin sont partis des modèles physiques (théories conformes en particulier) pour introduire le sujet.

Contenu des cours

La liste des orateurs était la suivante :
N. Andruskiewitsch (Argentine), M. Dubois Violette (France), D. Evans (Grande Bretagne) , A. Ocneanu (USA), O. Ogievetsky (France) N. Reshetikhin (USA), M. Rosso (France), A. Varchenko (USA), S. L. Woronowicz (Pologne) et J. B. Zuber (France).

L'école a duré deux semaines et chacun des orateurs a donné un total de 5 heures de cours réparti sur trois cours distincts.

Le descriptif suivant est un résumé, un peu plus technique, du contenu des différents cours.

— N. Andruskiewitsch : Algèbres de Hopf de dimension finie.

Résumé : Tour d'horizon de ce qu'on sait de la classification et des propriétés des algèbres de Hopf de dimension finie. Une construction particulière faite à l'aide de la donnée d'un module de Yetter-Drinfeld a été particulièrement étudiée. De nombreux exemples ont été donnés.

- M. Dubois Violette : Propriétés homologiques et cohomologiques des algèbres non commutatives (en particulier les algèbres de Hopf).

    Résumé : Après avoir rappelé un certain nombre de faits connus concernant les complexes différentiels (en particulier la cohomologie de Hochschild d'une algèbre non commutative à valeur dans un bimodule), l'orateur a étudié la généralisation de ces concepts au cas $d^N = 0$ ($N > 2$). Ces cohomologies généralisées ont été alors utilisées pour l'étude d'espaces de tenseurs irréductibles possédant des symétries décrites par des diagrammes d'Young quelconques (le cas usuel se limitant aux tenseurs completement antisymétriques).

- D. Evans : Interpretation des théories de champs conformes en terme de sous-facteurs et d'algèbres de Von Neumann.

    Résumé : Les propriétés d'invariance modulaire des théories de champs conformes ont été interprétées en termes de théorie des opérateurs : systèmes de bimodules, facteurs et sous facteurs.

- A. Ocneanu : Classification des symétries quantiques de type $SL(3)$.

    Résumé : L'orateur a d'abord décrit un travail datant d'environ cinq ans "Path on Dynkin diagrams, ADE, quantum group symmetries, Platonic solids and singularities" où les symétries de l'espace des chemins sur les diagrammes de Dynkin avaient été classifiés et reliées d'une part aux représentations de groupes quantiques aux racines de l'unité, et, d'autre part, aux théories de champs conformes possédant une symétrie quantique de type $SL_q(2)$. Par la suite, le professeur Ocneanu a présenté (pour la première fois) la classification concernant le cas $SL(3)$. Cette classification retrouve, tout en la précisant et en la complétant, les résultats obtenus, grâce à des motivations physiques, par Di Francesco et Zuber. Une liste de graphes généralisant les diagrammes de Dynkin ADE a été donnée. Cette liste apparait (poster couleur fourni par le Pr. Ocneanu) sur la page web dédiée à l'école de Bariloche. Voir http ://www.cpt.univ-mrs.fr/~coque/Bariloche2000.html

- O. Ogievetsky : Matrices R perturbatives et non perturbatives.

    Résumé : Le cours a débuté avec un exposé décrivant la méthode d'obtention des matrices R quantiques par quantification des matrices r classiques et a continué par une discussion des triplets de Belavin-Drinfeld. Les propriétés des espaces quantiques (algèbres quadratiques) associées à la décomposition spectrale des matrices R (tordues par transposition tensorielle) ont alors été étudiées. Pour terminer, l'orateur s'est attaché à une discussion du cas où les groupes quantiques ne peuvent pas être obtenus par déformation (ou par twist) des algèbres enveloppantes d'algèbres de Lie. Une classification des algèbres de Hopf, ni commutatives, ni co-commutatives, mais possédant la même série de Poincaré que celle de l'algèbre classique des fonctions sur le groupe $GL(3)$ a été obtenue par l'auteur et exposée lors de cette école pour la première fois. L'orateur a également abordé les sujets suivants : formes réelles des algèbres de Hopf et algèbres q-multilinéaires.

- N. Reshetikhin : Systèmes intégrables classiques et quantiques.

    Résumé : Après une introduction générale sur le thème des variétés symplectiques et des variétés de Poisson, l'orateur a présenté la méthode de quantification par déformation formelle, la méthode utilisant des idéaux de déformation ainsi que les cas d'intégrabilité au sens de Liouville (et une généralisation à un cas dit dégénéré). Il a montré comment les systèmes intégrables classiques sont reliés aux feuilles symplectiques des groupes de Lie-Poisson et comment cette construction se généralise au cas des systèmes intégrables quantiques.

- M. Rosso : Point de vue combinatoire sur les groupes quantiques.

  Résumé : L'orateur a tout d'abord décrit la structure de l'algèbre (de Hopf) des battages de carte sur un espace vectoriel et le concept de "mots de Lyndon". Il a alors construit un analogue quantique de l'algèbre des battages de cartes (algebra of quantum shuffles) et montré comment récupérer, à partir de cette notion, la sous algèbre de Borel de toutes les algèbres enveloppantes quantiques obtenues usuellement par déformation. Un sous-produit de cette construction concerne l'obtention de formules totalement explicites pour les matrices R universelles.

- A. Varchenko : Groupes quantiques dynamiques.

  Résumé : Toute équation K-Z (Knizhnik-Zamolodchikov) généralisée peut être reliée à la donnée d'une surface de Riemann avec un certain nombre de points marqués. Dans le cas de la sphère, les conditions d'integrabilité correspondantes conduisent à l'equation de Yang-Baxter "classique", dont la quantification conduit à l'equation de Yang-Baxter et aux groupes quantiques (algèbres de Hopf quadratiques quasi-triangulaires). Dans le cas du tore, l'équation obtenue, "équation classique de Yang-Baxter dynamique", conduit, après quantification, à l'équation de Neveu-Felder et aux groupes quantiques dynamiques. Les groupes quantiques elliptiques peuvent alors être obtenus naturellemnt et étudiés. Après une description de ce qui précède, l'orateur a montré L'existence de relations exsitant avec l'étude des fonctions spéciales (q-Lame, Mac Donald etc).

- S.L. Woronowicz : Groupes quantiques non compacts et opérateurs unitaires multiplicatifs.

  Résumé : L'orateur a montré que toute l'information contenue dans un groupe quantique de matrices (en particulier celle reliée à sa structure de C^* algèbre) pouvait être codée dans un opérateur particulier "l'opérateur de Skandalis", lequel doit en particulier satisfaire une identité pentagonale. L'exemple du groupe quantique "az+b" a été étudié en détails.

- J.B. Zuber : CFT, BCFT, ADE and all that.

  Résumé : Le cours de J.B. Zuber avait pour but de présenter certaines des particularités qu'on rencontre en physique dans l'étude des théories de champs conformes de dimension 2. Les mots clefs de son cours sont les graphes (ceux de Dynkin, et certaines généralisations de ces derniers). L'étude des fonctions de partition invariantes modulaires, conduisent, dans le cas des théories conformes affines de type SL(2), à une classification ADE déjà obtenue il y a plusieurs années. Dans le cas de $SL(3)$ une classification est connue mais les graphes correspondants et leur interprétation mathématique est encore "Terra Incognita" (voir cependant le cours d'A. Ocneanu à cette école). Une relation entre ces classifications et l'étude des théories de champs conformes en présence d'une frontière a été également décrite.

Quantum symmetries in theoretical physics and mathematics

Bariloche, Patagonia, Argentina
January 10-21, 2000

Final Report NSF Grant INT-9979435

The workshop took place at the Instituto Balseiro and Centro Atómico Bariloche from January 10 to 21, 2000. There were 68 participants from 15 countries, the majority (41) from Latin America, 16 from Europe, 4 from the US, 3 from Marocco, 2 from Australia and 2 from Japan. There were five women, well above average for a meeting of this sort. The scientific range of the participants ran from graduate students to post-docs to young researchers on up to leading theoreticians.

The workshop received major support from the CIMPA (France), the NSF and the ANPCyT (Argentina), and also from the ICTP (Trieste, Italy), the CLAF (Brazil) and the CNRS (France). This provided accomodations for all participants and travel support for 8 lecturers and 38 other participants. The facilities at the Centro Atómico Bariloche provided accomodations at bargain rates.

The theme, "Quantum symmetries in theoretical physics and mathematics", is at the cutting edge of research in both mathematics and physics. The discovery some 15 years ago of quantum groups (a special class of Hopf algebras) has led to an ongoing torrent of activity in algebra but also in topology (manifold and knot theories), mechanics, dynamical systems, operator theory and even number theory. In physics, quantum groups first appeared in integrable systems but now appear also in statistical mechanics, conformal field theory, quantum field theory and more.

The organizers invited international specialists to give a course each on the basics of relevant aspects of quantum group theory leading up to their recent results. The workshop was thus in major part a 'school' providing exposition for the training of the next generation of researchers. The various lecturers provided significantly different points of view: from analysis, geometry, algebra and physical models, especially conformal.

The lecturers were N. Andruskiewitsch (Argentina), M. Dubois Violette (France), D. Evans (UK), A. Ocneanu (USA), O. Ogievetsky (France), N. Reshetikin (USA), M. Rosso (France), A. Varchenko (USA), S. Woronowicz (Poland), and J. B. Zuber (France).

Each gave a series of lectures, 5 hours in all, as follows:

- N. Andruskiewitsch: Finite dimensional Hopf algebras.

- M. Dubois Violette: Homological and cohomological properties of non-commutative algebras.
- D. Evans: Interpretation of conformal field theories in terms of sub-factors of von Neumann algebras.
- A. Ocneanu: Classification of quantum symmetries of type $SL(3)$.
- O. Ogievetsky: Perturbative and non-perturbative R-matrices.
- N. Reshetikin: Classical and quantum integrable systems.
- M. Rosso: The combinatorial point of view on quantum groups.
- A. Varchenko: Dynamical quantum groups.
- S. Woronowicz: Non-compact quantum groups and multiplicative unitary operators.
- J.B. Zuber: CFT, BCFT, ADE and all that.

About finite dimensional Hopf algebras

Nicolás Andruskiewitsch

Dedicado a mis padres, Ana Bauchiero e Igor Andruskiewitsch.

Contents

1. Introduction — 1
2. Examples — 4
3. General results — 11
4. The semisimple case — 16
5. The pointed case — 22
6. The general case — 45
7. Forms — 48
8. Appendix. Questions — 50
References — 52

1. Introduction

These Notes are the written suport to the Lectures I gave at the school "Quantum Symmetries in Theoretical Physics and Mathematics", held in Bariloche, January 2000. Both the Lectures and the Notes intend to survey the state of the art on the classification of finite dimensional Hopf algebras over a field. In despite of many interesting results of the last years, our knowledge of the structure of Hopf algebras is still in a primary stage. The impact of quantum groups in the area has not been realized yet in its full significance. I hope these Notes will convey the interest to a range of questions related to the classification of finite dimensional Hopf algebras.

This report is aimed at results on classification or structure of finite dimensional Hopf algebras. Applications to low dimensional topology and conformal field theory, and related topics like modular categories, operator algebras, various generalizations of the notion of Hopf algebra (face algebras, weak Hopf algebras, etc.), and many other aspects will not be touched.

1991 *Mathematics Subject Classification.* Primary 16W30, 17B37.
Key words and phrases. Hopf algebras, quantum groups.
Partially supported by CONICET, CONICOR, SeCYT (UNC) and FAMAF (República Argentina).

1.1. The first results on classification of Hopf algebras were obtained already by Hopf, Leray and Borel; their work was extended by Milnor and Moore [**MM**, Section 7], Cartier [**Ca**, Th. 2] and Kostant [**Sw**, Th. 8.1.5 and 13.0.1], [**Ko**, Th. 3.3]. The following result is known as the Cartier-Kostant-Milnor-Moore theorem.

THEOREM 1.1. *A cocommutative Hopf algebra over an algebraically closed field* **k** *of characteristic 0 is a semidirect product of a group algebra and the enveloping algebra of a Lie algebra. In particular, a finite dimensional cocommutative Hopf algebra over* **k** *is a group algebra.* □

It is also known that a commutative Hopf algebra over **k** is the algebra of rational functions on a pro-algebraic group; see also [**Ga**], [**Ser3**]. This is no longer true over algebraically closed fields of positive characteristic: restricted enveloping algebras of p-Lie algebras are finite dimensional cocommutative Hopf algebras but not group algebras. The duals of these finite dimensional Hopf algebras are Hopf kernels of Frobenius homomorphisms on algebraic groups. Notice that the classification (or even a characterization) of all cocommutative finite dimensional Hopf algebras over an algebraically closed field of positive characteristic is not known. See [**FV**] and references therein.

1.2. The resemblance of the theory of Hopf algebras with group theory, almost a tautology at the level of cocommutative Hopf algebras, was boldly marked by the discovery of quantum groups. Even this name was chosen by Drinfeld to emphasize the ideological bridge between the two theories.[1] However, despite of the fact that concepts of group theory served as inspiration and guide to recent research – an explicit example being Kaplansky's conjectures – many elementary facts about finite groups are hard to translate to finite dimensional Hopf algebras, or even false at least in the first naive approach. For instance, the Nichols-Zöller freeness theorem is a counterpart of Lagrange's theorem: a finite dimensional Hopf algebra is free over any Hopf subalgebra. However the known proof is not straightforward and relies on the Krull-Remak-Schmidt theorem. Sylow's theorems are no longer true (just look at the dual of a group algebra of a simple group). The class equation expressing the order of a finite group as the sum of the cardinals of its conjugacy classes can be somehow recovered, but without reference to isotropy subgroups.

On the other hand, our stock of examples of finite dimensional Hopf algebras is not as wide as one would like. We collect the main techniques of construction of Hopf algebras in Section 2.

1.3. The theory of Hopf algebras has received a deep impact with the introduction of quantum groups [**Dr1**]. On one hand, a whole range of new techniques and concepts – the double, quasitriangular Hopf algebras, braided categories, twisting, ... , just to mention the most used – came into use and form now part of the standard baggage of a Hopf algebraist. On the other hand, the fundamental examples of quantized enveloping algebras [**Dr1**], [**J**], their duals and twisted variations thereof, were intensively studied. In particular, a very important class of non-trivial Hopf algebras was discovered by Lusztig: these are called Frobenius-Lusztig kernels since they appear as Hopf kernels of a quantum Frobenius homomorphism. They can be thought of as liftings to characteristic 0 of Frobenius kernels, and as such, they play an important rôle in representation theory [**AJS**].

---

[1] One should also mention that, in this spirit, formal groups are antecessors of quantum groups.

1.4. The classification of finite dimensional Hopf algebras over an algebraically closed field **k** of characteristic 0 follows two definitely different tracks: the semisimple case and the non-semisimple case. Indeed, there are several equivalent characterizations of semisimple Hopf algebras. In positive characteristic, the correct setting for those characterizations is that of semisimple and cosemisimple Hopf algebras. By a beautiful result of Etingof and Gelaki [**EG3**], problems in this setting can be reduced to similar problems in characteristic 0. See Section 3. A nice survey on classification of semisimple Hopf algebras is [**Mo2**]; we will discuss in Section 4 some more recent results which are not included there.

1.5. Substantial results, though not definitive, on the classification of non-semisimple Hopf algebras are known only for the class of pointed Hopf algebras over an algebraically closed field of characteristic 0. In this case, the lifting principle, due to H. Schneider and the author [**AS2**], provides the adequate framework for this problem and allows Lie theory to enter into the picture through quantum groups. The goal of the series of papers [**AS2, AS3, AS4, AS5**] is to show that Frobenius-Lusztig kernels, or Lusztig's small quantum groups, exhaust a natural class of finite dimensional Hopf algebras. See also [**AS7**] for the infinite dimensional case. An account is given in Section 5; we refer to the survey paper [**AS6**] for a more detailed exposition.

1.6. There are few results on classification of all Hopf algebras of a prescribed dimension; we collect them in Section 6. Section 7 contains some remarks on forms of Hopf algebras. Due to the lack of space, some other interesting topics are not touched; we do not comment on work of Etingof and Gelaki on classification of triangular Hopf algebras, see for instance [**EG7, Ge3**]. We touch Kaplansky's conjectures only in the course of the exposition; for a systematic overview of the actual state of them, see [**Sr2**].

1.7. Acknowledgements. The development of the point of view on the material presented here benefited from uncountable conversations along the years with Hans Schneider and my students Sonia Natale and Matías Graña. I am very grateful to them for sharing with me their insights. I also profited from conversations with A. Abella, M. Beattie, S. Caeneppel, S. Dăscălescu, W. Ferrer Santos, S. Gelaki, Y. Kashina, A. Masuoka, S. Montgomery, Y. Sommerhäuser, P. Schauenburg, Y. Zhu.

I also thank the Directors of the School, and Editors of this volume, R. Coquereaux and R. Trinchero, for the splendid organization of the School, and for the invitation to deliver the Lectures and write the Notes.

I have tried to give appropriate credit to all the papers I am aware of, but even with the help of the contemporary electronical techniques, some results or authors might have remained unrecognized in this paper; I offer my sincere apologies.

1.8. Conventions and notations. We shall work over a field **k**; suitable hypotheses on **k** will be fixed accordingly. Our references for the theory of Hopf algebras are [**Sw**], [**Mo1**], [**Sch3**], [**Ma3**], [**DaNR**]. The notation for Hopf algebras is standard: Δ, $\mathcal{S}$, ε, denote respectively the comultiplication, the antipode, the counit. We use Sweedler's notation but drop almost always the summation symbol. That is, $\Delta(x) = \sum_{(x)} x_{(1)} \otimes x_{(2)} = x_{(1)} \otimes x_{(2)}$.

Similarly, if C is a coalgebra and V is a left comodule with structure map $\delta : V \to C \otimes V$, then $\delta(x) = \sum_{(x)} x_{(-1)} \otimes x_{(0)} = x_{(-1)} \otimes x_{(0)}$.

If A is a Hopf algebra, the well-known (left) adjoint representation ad of A on itself is given by $\operatorname{ad} x(y) = x_{(1)} y \mathcal{S}(x_{(2)})$. There are other versions of adjoint representations that we will not consider.

If C is a coalgebra, we denote by $G(C)$ the set of group-like elements of C, cf. (2.1) below. If $g, h \in G(C)$, then we let $\mathcal{P}_{g,h}(C) = \{x \in C \mid \Delta(x) = g \otimes x + x \otimes h\}$ denote the space of all (g, h)-primitive elements. If A is a Hopf algebra then $\mathcal{P}_{1,1}(A) =: \mathcal{P}(A)$ is the space of primitive elements. In this case, assume that $\Gamma \hookrightarrow G(A)$ is a finite abelian subgroup; then Γ acts on each $\mathcal{P}_{g,h}(A)$ by conjugation, and if χ is a character of Γ, then we define $\mathcal{P}_{g,h}(A)^\chi = \{a \in \mathcal{P}_{g,h}(A) \mid uau^{-1} = \chi(u)a,$ for all $u \in \Gamma\}$.

If C is a coalgebra and A is an algebra, then $\operatorname{Hom}(C, A)$ bears an associative algebra structure with the so-called convolution product; this is given by $f * g(c) = f(c_{(1)})g(c_{(2)})$. Whenever a morphism is said to be invertible, this refers to the convolution product, unless explicitly stated. For instance, the axioms on the antipode say that $\mathcal{S}$ is the (convolution) inverse of the identity.

2. Examples

In this Section, we list several known ways to obtain finite dimensional Hopf algebras. Here **k** is an arbitrary field.

2.1. Duals. If H is a finite dimensional Hopf algebra, then its linear dual H^* is a again a Hopf algebra, by transposition of all the structure maps (so that the multiplication of H^* is the transpose of the comultiplication of H and so on).

2.2. From groups. Let G be a finite group. The algebra $\mathbf{k}^G$ of functions from G to $\mathbf{k}$ is a semisimple Hopf algebra with pointwise product and comultiplication $\Delta(\phi)(g, h) = \phi(gh)$, $\phi \in \mathbf{k}^G$, $g, h \in G$, where we are identifying $\mathbf{k}^G \otimes \mathbf{k}^G \simeq \mathbf{k}^{G \times G}$. The antipode is $\mathcal{S}(\phi)(g) = \phi(g^{-1})$ and the counit is $\varepsilon(\phi) = \phi(1)$. For $g \in G$, let $\delta_g \in \mathbf{k}^G$ be the function which is 0 outside g and takes the value 1 in g. The dual Hopf algebra is the group algebra $\mathbf{k}G$; if $\{e_g\}$ is the dual basis of $\{\delta_g\}$, then $e_g e_h = e_{gh}$ and

$$(2.1) \qquad \Delta(e_g) = e_g \otimes e_g.$$

An element of a Hopf algebra whose coproduct is given by (2.1) is called a group-like. The elements e_g can be thought as Dirac measures on the discrete topological space G. It is well-known that $\mathbf{k}^G$ is always semisimple, and that the group algebra $\mathbf{k}G$ is semisimple provided that the characteristic of $\mathbf{k}$ does not divide the order of G (Maschke's theorem).

2.3. By extension. It is natural to think of building Hopf algebras from smaller ones, just as in group theory. This was treated by many people in different settings: by G. I. Kac for his finite dimensional C^*-Hopf algebras [**Ka1**]; by Singer for connected Hopf algebras, in the so-called abelian case [**Si**]; by Majid and Majid-Soibelman [**Ma1**], [**MaS**]; by Hofstetter [**Ho**]; and also in [**By1**], [**By2**], [**Sch3**], [**Mk1**], [**AnDe**], [**An1**]. See also [**Mk8**]. The notion of extension also appears in [**Lu3**] without an explicit axiomatization. Briefly, the situation is the following.

(a). A sequence of Hopf algebra maps $1 \to A \xrightarrow{\iota} C \xrightarrow{\pi} B \to 1$, where 1 denotes the Hopf algebra **k**, is *exact* if
1. ι is injective. Identify then A with its image.
2. π is surjective.
3. $\ker \pi = CA^+$. (A^+ is the augmentation ideal, *i. e.* the kernel of the counit).
4. $A = \{x \in C : (\pi \otimes \mathrm{id})\Delta(x) = 1 \otimes x\}$.

Then the equality $\pi\iota = \varepsilon$ follows either from (3) or from (4). In such case, C is called an extension of the Hopf algebras A and B.

An injective morphism of Hopf algebras $A \xrightarrow{\iota} C$ is normal if $\iota(A^+)C = C\iota(A^+)$. In such case, the unique possible B completing the exact sequence is the "Hopf cokernel" C/CA^+. Now condition (4) above can be dropped if $A \xrightarrow{\iota} C$ is faithfully flat and A is stable by the adjoint action of C; and B does indeed complete the exact sequence. Examples of inclusions of Hopf algebras which are not faithfully flat were given by Schauenburg: they are related to the lack of bijectivity of the antipode [**Scg3**]. Indeed, if H is a Hopf algebra whose antipode S is not surjective then H is *not* faithfully flat over $S(H)$. Examples of such Hopf algebras are known. In the finite dimensional case, any Hopf algebra is free over its Hopf subalgebras and this problem does not arise.

A similar analysis proceeds for a conormal, surjective, morphism of Hopf algebras $C \xrightarrow{\pi} B$; but the rôle of the "faithfully flat" requirement is played now by "faithfully coflat", see for instance [**AnDe**, Proposition 1.2.11].

In conclusion: if C is a finite dimensional Hopf algebra and A is a normal Hopf subalgebra of C, then the quotient coalgebra $B := C/CA^+$ is a quotient Hopf algebra of C and it fits into an exact sequence as above.

(b). If G is a group, A is a normal abelian subgroup and $N := G/A$, then N acts on A and G can be reconstructed from N, the N-module A and a 2-cocycle $\sigma : N \times N \to A$, unique up to its class in $H^2(N, A)$. Conversely, given a group N, an N-module A and a 2-cocycle $\sigma : N \times N \to A$ we can build a group G (whose underlying set is $A \times N$) with A as a normal subgroup and such that $N \simeq G/A$.

A similar, but much more involved, situation can be considered for Hopf algebras. Given Hopf algebras A and B, we can build a Hopf algebra structure on the vector space $A \otimes B$ from the following additional data: a "weak action" $\rightharpoonup: B \otimes A \to A$, a "2-cocycle" $\sigma : B \otimes B \to A$, a weak coaction $B \to A \otimes B$ and a "dual 2-cocycle" $A \to B \otimes B$, fulfilling a list of compatibility conditions. Conversely, given an extension as above, one can find a weak action, a weak coaction, a cocycle an a dual cocycle which are compatible and so that C can be reconstructed from A, B and these data, whenever the extension is *cleft*. Here cleft means that there exists an invertible comodule section of π. This section is somehow the analogue of the section of sets $N \to G$ at the group level, which always exists. But in some other settings, like algebraic groups, the existence of such section in the category is not always true and similar considerations are carried out.

In any case, an extension of finite dimensional Hopf algebras is always cleft [**Sch1**]. In conclusion: if C is a finite dimensional Hopf algebra and A is a normal Hopf subalgebra of C, then C can be reconstructed from A, the quotient Hopf algebra $B = C/CA^+$ and data as above.

(c). If A is commutative and B is cocommutative, the weak action and the weak coaction are an actual action and an actual coaction, respectively. Once they are fixed, the 2-cocycle and the dual 2-cocycle have a cohomological meaning, their sum being a 2-cocycle in the total complex associated to a double complex. This is usually called the *abelian case*. To compute then how many extensions C (up to isomorphisms) can be constructed from A, B, a fixed action and a fixed coaction, amounts to compute the order of the second cohomology group $H^2(A, B)$. A very useful devise for this computation is the so-called Kac exact sequence (in honor of G. I. Kac, who introduced it); see for instance [**Mk8**].

(d). In several instances, properties of C can be deduced from similar properties of A and B, and vice versa. For example, C is semisimple if and only if both A and B are; use [**BM**].

(e). Let us say that a finite dimensional Hopf algebra is *simple* if it has no proper normal Hopf subalgebras. It is natural to ask for "genuine" examples of simple Hopf algebras. Clearly, if G is a simple group then both $\mathbf{k}G$ and $\mathbf{k}^G$ are simple Hopf algebras. For more examples, see the discussion in the next Subsection.

On the other hand, a routine dimension argument shows that any finite dimensional Hopf algebra can be obtained as an iterated extension of simple Hopf algebras. We can then attach to any finite dimensional Hopf algebra the sequence of its simple subfactors in *some* presentation as iterated extension. We do not know the answer to the following:

QUESTION 2.1. *Does the Jordan-Hölder theorem hold for finite dimensional Hopf algebras? That is, is the sequence of simple subfactors alluded to above unique up to permutation?*

For some explicit non-trivial examples of Hopf algebras that can be obtained by extensions, see [**AN2, Mk6, Mk7, Mk8, Mul2**].

2.4. By twisting. Let A be a Hopf algebra and $F \in A \otimes A$ be an invertible element. Let $\Delta_F := F\Delta F^{-1} : A \to A \otimes A$; it is again an algebra map. It is coassociative if and only if

$$(2.2) \qquad (1 \otimes F)(\mathrm{id} \otimes \Delta)(F) = (F \otimes 1)(\Delta \otimes \mathrm{id})(F)U,$$

where U is an element of $A \otimes A \otimes A$ that centralizes $(\Delta \otimes \mathrm{id})\Delta(A)$. It is counital if and only if

$$(2.3) \qquad (\mathrm{id} \otimes \varepsilon)(F) = (\varepsilon \otimes \mathrm{id})(F) \text{ is central in } A.$$

If (2.2) and (2.3) hold, then A_F (the same algebra, but with comultiplication Δ_F) is again a Hopf algebra. We shall say in this case that F is a *pseudo-cocycle* and that A_F is obtained from A via twisting by F. When $U = 1$ and $(\mathrm{id} \otimes \varepsilon)(F) = 1$, F is a cocycle in a suitable sense. As the multiplication does not change, if A is semisimple then A_F is semisimple. This construction is originally due to Drinfeld [**Dr3**] and Reshetikhin [**Re**]; it was applied to group algebras in [**Nk**], [**Mov**], [**EG2**]. See also [**EV, Mk10, HS, IK1, Sek, V**].

Let us assume for the rest of this Subsection that **k** is algebraically closed and has characteristic 0. As a first application of the twisting, examples of non-trivial twistings of group algebras of simple groups are given:

THEOREM 2.2. [**Nk**]. *There exists $F \in \mathbf{k}\mathbb{A}_5 \otimes \mathbf{k}\mathbb{A}_5$ such that $(\mathbf{k}\mathbb{A}_5)_F$ is a simple non-commutative, non-cocommutative Hopf algebra.* □

Here $\mathbb{A}_5$ is the alternating group of order 60, the smallest non-abelian simple group. This Theorem was generalized to all non-abelian simple groups [**Hof**]. See also [**Bi**] for more examples. As a matter of fact, all possible twisting of group algebras by a cocycle are described in [**EG7**].

QUESTION 2.3. Up to our knowledge, the only known simple semisimple Hopf algebras are group algebras, their twistings and duals of them. Are there more?

One could begin by the following:

QUESTION 2.4. What are all the simple semisimple Hopf algebras of dimension 60?

One says that a finite dimensional Hopf algebra is *trivial* if it is isomorphic to a group algebra or to a dual group algebra. Further, one has the following natural definition.

DEFINITION 2.5. [**MonW**] A finite dimensional Hopf algebra is *semi-solvable* if it can be obtained by successive extensions from group algebras or duals of group algebras.

Question 2.3 is very close to the following.

QUESTION 2.6. Consider the subcategory of the category of all semisimple Hopf algebras which contains the group algebras and is closed by taking duals, by extensions and by twistings. Is this subcategory equal to the category of all semisimple Hopf algebras?

2.5. Twisting the multiplication. There is a dual version of the twisting operation, which amounts to a twist of the multiplication. Let A be a Hopf algebra and let $\sigma : A \times A \to \mathbf{k}$ be an invertible 2-cocycle, that is

$$\sigma(x_{(1)}, y_{(1)})\sigma(x_{(2)}y_{(2)}, z) = \sigma(y_{(1)}, z_{(1)})\sigma(x, y_{(2)}z_{(2)}),$$
$$\sigma(1,1) = 1,$$

for all $x, y, z \in A$. Then A_σ – the same A but with the multiplication $._\sigma$ below – is again a Hopf algebra, where

$$x._\sigma y = \sigma(x_{(1)}, y_{(1)})x_{(2)}y_{(2)}\sigma^{-1}(x_{(3)}, y_{(3)}).$$

REMARK 2.7. If A is a finite dimensional Hopf algebra and $\sigma : A \times A \to \mathbf{k}$ is an invertible 2-cocycle, then $A_\sigma \simeq \left((A^*)^F\right)^*$ where $F \in A^* \otimes A^*$ corresponds to σ by the canonical isomorphism.

REMARK 2.8. There is a very convenient setting where this last construction can be performed. Let U, H be Hopf algebras. Let $\tau : U \otimes H \to \mathbf{k}$ be a bilinear map such that for all $u, v \in U$, $a, b \in H$

- $\tau(uv, a) = \tau(u, a_{(1)})\tau(v, a_{(2)})$,

- $\tau(u, ab) = \tau(u_{(1)}, b)\tau(u_{(2)}, a)$,

- $\tau(1, a) = \varepsilon(a)$,

- $\tau(u, 1) = \varepsilon(u)$.

Let A be the tensor product Hopf algebra $A = U \otimes H$ and let $\sigma : A \otimes A \to \mathbf{k}$ be the bilinear map

$$\sigma(u \otimes a, v \otimes b) = \varepsilon(u)\tau(v, a)\varepsilon(b), \quad u, v \in U, \quad a, b \in H.$$

Then τ is convolution invertible with inverse given by $\tau^{-1}(v, a) = \tau(\mathcal{S}v, a) = \tau(v, \mathcal{S}^{-1}a)$; σ is an invertible 2-cocycle – with inverse

$$\sigma^{-1}(u \otimes a, v \otimes b) = \varepsilon(u)\tau^{-1}(v, a)\varepsilon(b), \quad u, v \in U, \quad a, b \in H;$$

and consequently A_σ is a Hopf algebra. Note that U and H are naturally identified with Hopf subalgebras of A_σ; and the multiplication induces a linear isomorphism $U \otimes H \simeq A_\sigma$. This means that U and H form a *matched pair* and that A_σ is the *double crossproduct* of U and H. See [**DoT**]; and also [**Ma3**, Chapter 7] for a detailed exposition and examples.

REMARK 2.9. There is a simple way to describe such maps τ. In the notation of Remark 2.8, assume that H is finite dimensional and let $\varphi : U \to (H^*)^{\text{cop}}$ be a Hopf algebra homomorphism. Then $\tau : U \otimes H \to \mathbf{k}$, $\tau(v, a) = \varphi(v)(a)$, is invertible – with inverse given by $\tau^{-1}(v, a) = \varphi(\mathcal{S}v)(a) = \varphi(v)(\mathcal{S}^{-1}a)$, and satisfies the requirements above. Conversely, given such a τ there is a unique such φ.

2.6. The double. Let H be a finite dimensional Hopf algebra. Let $U = (H^*)^{\text{cop}}$; by Remark 2.9 applied to $\varphi = \text{id}$, we obtain a linear map τ which gives rise in turn to a cocycle σ, as explained in Remark 2.8. The Hopf algebra $((H^*)^{\text{cop}} \otimes H)_\sigma$ is called the *Drinfeld double*, or simply the double of H, and is denoted by $D(H)$. Introduced by Drinfeld in the seminal paper [**Dr1**], it is one of the most important constructions in Hopf algebra theory. We refer to [**Mo1**], [**Ma3**], for detailed expositions. We content ourselves by listing some of the fundamental properties of the double:

- [**Dr1**]. $(D(H), \mathcal{R})$ is a quasitriangular Hopf algebra, where $\mathcal{R} \in H \otimes H^* \subset D(H) \otimes D(H)$ is the canonical element.

- [**Dr1**]. If (H, R) is quasitriangular, then R induces a Hopf algebra projection $\pi : D(H) \to H$, with Hopf kernel $(H^*)^{\text{cop}}$.

- [**Dr2**]. If (H, R) is quasitriangular, and $R = \sum_i R_i \otimes R^i$, then the *Drinfeld element* of H is $u = \sum_i \mathcal{S}(R^i)R_i$. It is invertible and satisfies
$$\mathcal{S}^2(x) = uxu^{-1}, \quad x \in H, \quad \Delta(u) = (u \otimes u)(R_{21}R)^{-1} = (R_{21}R)^{-1}(u \otimes u),$$
where $R_{21} = \sum_i R^i \otimes R_i$.

- [**ReS**]. The double of the double $D(D(H))$ is isomorphic to $(D(H) \otimes D(H))_\sigma$, for a suitable cocycle σ.

I do not know if the double of an arbitrary finite dimensional Hopf algebra can always be presented as a non-trivial extension, as is the case for quasitriangular Hopf algebras.

The following Proposition is useful in classification problems. We identify $D(H)^*$ with $H \otimes H^*$ as vector spaces.

PROPOSITION 2.10. *(i)*. [**R4**]. *The group $G(D(H))$ is isomorphic to $G(H^*) \times G(H)$, via multiplication.*
(ii). [**R4**]. *The elements of the group $G(D(H)^*)$ are of the form $g \times \eta$, where $\eta \in G(H^*)$, $g \in G(H)$, and $\eta \times g$ is in the center of $D(H)$.*
(iii). [**Sch5**]. *If H is factorizable, the group $G(H^*)$ is isomorphic to $G(H) \cap Z(H)$.* □

2.7. By bosonization. This is a more sophisticated kind of extension. It was discovered by Radford and interpreted in braided-categorical terms by Majid, see [**Ma2**], [**R2**]. It is a distinguished feature of Hopf algebra theory with no analogue in group theory; it is a very useful tool in classification problems, specially in the pointed case.

Let A and H be Hopf algebras with bijective antipodes, and assume there are Hopf algebra maps $\pi : A \to H$ and $\iota : H \to A$ such that $\pi\iota = \text{id}_H$; so that π is surjective and ι is injective. By analogy with elementary group theory, one seeks to reconstruct A from H and the kernel of π as a semidirect product. The rôle of the kernel of π is played in this situation by the algebra of coinvariants

(2.4) $$R := A^{\text{co }\pi} = \{a \in A : (\text{id} \otimes \pi)\Delta(a) = a \otimes 1\}.$$

In contrast with the group case, this is *not* a usual Hopf algebra; but it is a *braided* Hopf algebra.

To explain what this means, let us first recall some definitions. Given a Hopf algebra H, a *Yetter-Drinfeld module* over H is a vector space V provided with

- a structure of left H-module: $. : H \otimes V \to V$;

- and a structure of left H-comodule: $\delta : V \to H \otimes V$, such that

- the following compatibility condition is satisfied:

(2.5) $$\delta(h.v) = h_{(1)}v_{(-1)}\mathcal{S}(h_{(3)}) \otimes h_{(2)}.v_{(0)}.$$

The category of all Yetter-Drinfeld modules, with morphisms the maps of modules and comodules, is denoted by $^H_H\mathcal{YD}$. (If $H = \mathbf{k}\Gamma$ and the base field is clearly fixed, we simply write $^\Gamma_\Gamma\mathcal{YD}$). It is a braided category; that is, if $M, N \in {}^H_H\mathcal{YD}$, then:

- The tensor product $M \otimes N$ and the dual M^*, with the natural module and comodule structures, are again Yetter-Drinfeld modules.

- There exists a natural isomorphism $c : M \otimes N \to N \otimes M$ given explicitly by

(2.6) $$c(m \otimes n) = m_{(-1)}.n \otimes m_{(0)}$$

which satisfies appropriate axioms, see for instance [**Mo1**]. When H is finite dimensional, a Yetter-Drinfeld module over H is nothing but a left module over the Drinfeld double $D(H)$ of H; and the braiding is just the one defined by the universal R-matrix of $D(H)$. (Here we mean one of the versions of the double, corresponding to left-left Yetter-Drinfeld modules).

REMARK 2.11. A natural way to introduce Yetter-Drinfeld modules is through Hopf bimodules. A *Hopf bimodule* over a Hopf algebra H is simultaneously a bimodule and a bicomodule (so it has left and right actions, and left and right coactions); all this structure is required to satisfy natural compatibility conditions. Given a Hopf bimodule M, the space of right coinvariants $M^{co\,H} = \{m \in M : \delta_r(m) = 1 \otimes m\}$ bears a natural structure of a Yetter-Drinfeld module. One gets in this way an equivalence between the two categories. See [**Ni, Wo2**]; this was rediscovered several times, see *e. g.* [**Scg1, AnDe, Ro1, BD**].

We can then consider braided Hopf algebras in the braided category ${}^H_H\mathcal{YD}$. We recall that R is a braided Hopf algebra in ${}^H_H\mathcal{YD}$ if:

- R is an algebra; the product and the unit are morphisms in ${}^H_H\mathcal{YD}$.

- R is a coalgebra; the coproduct and the counit are morphisms in ${}^H_H\mathcal{YD}$.

- The coproduct Δ_R is an algebra map; but here we consider in $R \otimes R$ not the usual product but the one defined by
$$m_{R \otimes R} = (m_R \otimes m_R)(\mathrm{id} \otimes c \otimes \mathrm{id}) : R \otimes R \otimes R \otimes R \to R \otimes R.$$
(This multiplication differs from the usual tensor product multiplication in the appearance of c instead of the usual flip).

- The identity $\mathrm{id} : R \to R$ has a convolution-inverse $\mathcal{S}_R$ in $\mathrm{End}\,R$, the antipode.

The archetypical example of a braided Hopf algebra in ${}^H_H\mathcal{YD}$ is the algebra of coinvariants R as in (2.4). Namely, R is a subalgebra of A and it is also stable under the adjoint action of A; the coaction of H on R, and the comultiplication of R are given by
$$\delta = (\pi \otimes \mathrm{id})\Delta, \qquad \Delta_R(r) = r_{(1)}\iota\pi\mathcal{S}(r_{(2)}) \otimes r_{(3)}.$$

Conversely, let R be a braided Hopf algebra in ${}^H_H\mathcal{YD}$. Then the *bosonization* or biproduct of R by H is a usual Hopf algebra $A = R\#H$, with underlying vector space $R \otimes H$, whose multiplication and comultiplication are given by

(2.7) $$(r\#h)(s\#f) = r(h_{(1)}.s)\#h_{(2)}f,$$

(2.8) $$\Delta(r\#h) = r^{(1)}\#(r^{(2)})_{(-1)}h_{(1)} \otimes (r^{(2)})_{(0)}\#h_{(2)}.$$

The maps $\pi : A \to H$ and $\iota : H \to A$, $\pi(r\#h) = \epsilon(r)h$, $\iota(h) = 1\#h$, are Hopf algebra homomorphisms; we have $R = \{a \in A : (\mathrm{id} \otimes \pi)\Delta(a) = a \otimes 1\}$. Therefore, these constructions are inverse to each other.

We would like to have new examples of Hopf algebras constructed by bosonization. Let R be a braided Hopf algebra in ${}^H_H\mathcal{YD}$ and let $A = R\#H$. First, A is semisimple if and only if both R and H are (again by [**BM**]). There are some non-trivial examples of semisimple Hopf algebras which can be constructed as bosonization [**Ge1, AN2, Sr1**]. Unfortunately, all these examples can also be described as extensions.

DEFINITION 2.12. [**AS2**]. A Hopf algebra H is *very simple* if it is simple, and it can not be presented as a bosonization in a non-trivial way.

QUESTION 2.13. Does there exist a semisimple Hopf algebra which is simple but not very simple? That is, a non-trivial bosonization but not an extension.

A positive answer to this Question would also answer Question 2.3, by the negative.

EXAMPLE 2.14. Let $N > 1$ be a natural number and let $R = \mathbf{k}[x]/(x^N)$. Assume that char $\mathbf{k} = 0$; then there is no algebra map from R to $R \otimes R$ (with the usual tensor product multiplication) which sends x to $x \otimes 1 + 1 \otimes x$.

Now, let Γ be a group, $g \in Z(\Gamma)$ and χ a one-dimensional representation of Γ such that $\chi(g) =: q$ is a root of 1 of order N. We consider $R \in {}_\Gamma^\Gamma \mathcal{YD}$ by $h.x^j = \chi(h)^j x^j$, $h \in \Gamma$, $\delta(x^j) = g^j \otimes x^j$. Then $c(x^j \otimes x^t) = q^{jt} x^t \otimes x^j$. Using the quantum binomial formula, one can see without effort that there is an algebra map $\Delta_R : R \to R \otimes R$ (with the "braided" tensor product multiplication) sending x to $x \otimes 1 + 1 \otimes x$; and that this is indeed a braided Hopf algebra.

If g has order N and Γ is generated by g, then the bosonization $T(q) = R \# \mathbf{k}\Gamma$ is called a Taft algebra; one has $T(q) \simeq \mathbf{k}\langle g, x | q^N = 1, x^N = 0, gx = qxg\rangle$ with g group-like and $\Delta(x) = x \otimes 1 + g \otimes x$. It is also not difficult to see that $T(q) \simeq T(q)^*$ as Hopf algebras, and that the proper Hopf subalgebras of $T(q)$ are contained in $\mathbf{k}\langle g \rangle$. It follows at once that $T(q)$ is simple (but not very simple). For a general criterion of simplicity of pointed Hopf algebras, see [**AS0**].

If R is a braided Hopf algebra in ${}_H^H\mathcal{YD}$ then there is also a braided adjoint representation ad_c of R on itself given by

$$\text{ad}_c x(y) = \mu(\mu \otimes \mathcal{S})(\text{id} \otimes c)(\Delta \otimes \text{id})(x \otimes y),$$

where μ is the multiplication and $c \in \text{End}(R \otimes R)$ is the braiding. Note that if $x \in \mathcal{P}(R)$ then the braided adjoint representation of x is just

(2.9) $$\text{ad}_c x(y) = \mu(\text{id} - c)(x \otimes y) =: [x,y]_c.$$

The element $[x, y]_c$ defined by the second equality for any x and y, regardless of whether x is primitive, will be called a braided commutator.

When $A = R \# H$, then for all $b, d \in R$,

(2.10) $$\text{ad}_{(b\#1)}(d\#1) = (\text{ad}_c b(d)) \# 1.$$

3. General results

In this Section, we review some basic fundamental results that allow to organize the study of classification problems of finite dimensional Hopf algebras. The field $\mathbf{k}$ is general, except when explicitly stated.

3.1. Freeness results. Let H be a Hopf algebra. Let $R \subseteq H$ be a subalgebra such that $\Delta(R) \subseteq H \otimes R$; that is, R is a left coideal subalgebra. A (left) relative (H, R)-Hopf module is a left R-module M provided with a left comodule structure $\delta : M \to H \otimes M$ such that

$$\delta(r.m) = r_{(1)} m_{(-1)} \otimes r_{(2)}.m_{(0)},$$

for all $m \in M$, $r \in R$. The category of left relative (H, R)-Hopf modules is denoted by ${}_R^H\mathcal{M}$. Here is one of the fundamental results that opened the recent theory of finite dimensional Hopf algebras.

THEOREM 3.1. [**NiZ**]. *If B is a Hopf subalgebra of a finite dimensional Hopf algebra A then any left relative (A, B)-Hopf module is a free left B-module.* □

An example of relative (A, B)-Hopf module is A itself, with left multiplication of B and the comultiplication of A. Clearly, a subspace C of A is a relative (A, B)-Hopf submodule if $BC \subseteq C$ and $\Delta(C) \subseteq A \otimes C$. The following applications of Theorem 3.1 illustrate better its importance.

COROLLARY 3.2. *Let A be a finite dimensional Hopf algebra.*

1. [**NiZ**]. *If B is a Hopf subalgebra of A, then A is free as B-module with left multiplication. In particular, $\dim B$ divides $\dim A$.*

2. *(Masuoka). If C is a subcoalgebra of A and S is a subgroup of $G(A)$ such that $gC \subseteq C$ for all $g \in S$, then $|S|$ divides $\dim C$.*

3. *(Masuoka). If $\pi : A \to H$ is a Hopf algebra epimorphism and B is a Hopf subalgebra of A such that $B \subseteq A^{\operatorname{co}\pi}$, then $\dim B$ divides $\dim A^{\operatorname{co}\pi}$; cf. (2.4).*

□

In the case of group algebras, part (i) of the Corollary amounts to Lagrange's theorem. In the infinite dimensional case, part (i) of the Corollary is false; a counterexample is given in [**OS2**]. A substitute would be the following statement: "any inclusion of Hopf algebras is faithfully flat"; but this is also false as shown by Schauenburg [**Scg3**].

For a Hopf subalgebra B of a finite dimensional Hopf algebra A, the integer

$$[A : B] := \frac{\dim A}{\dim B}$$

will be called the *index* of B in A. If $B = \mathbf{k}G$ is a group algebra, we just write $[A : G] := [A : B]$.

Recently, the Nichols-Zöller theorem was generalized to the braided case.

THEOREM 3.3. [**Scf, Tk1**]. *Let H be a Hopf algebra with bijective antipode and let A be a finite dimensional braided Hopf algebra in ${}^H_H\mathcal{YD}$. If B is a braided Hopf subalgebra of A, then A is free as B-module with left multiplication. In particular, $\dim B$ divides $\dim A$.* □

Notice that Theorem 3.3 follows directly from Corollary 3.2 in the case when H is also finite dimensional [**AS1**]. A non-categorical variation of this result will be discussed later, see Theorem 5.42.

3.2. Integrals and applications. Let A be a Hopf algebra. We say that $\lambda \in A^*$ is a *right integral on A* if

$$\lambda\beta = \lambda\langle\beta, 1\rangle,$$

for all $\beta \in A^*$; or equivalently if $\langle\lambda, x\rangle 1 = \langle\lambda, x_{(1)}\rangle x_{(2)}$, for all $x \in A$. Similarly left integrals on A are defined. Furthermore, we say that $\Lambda \in A$ is a *left integral in A* if

$$x\Lambda = \langle\varepsilon, x\rangle\Lambda,$$

for all $x \in A$. We collect several facts about integrals:

THEOREM 3.4. *Let A be a Hopf algebra.*

1. [**LaS**]. *If A is finite dimensional, then the space of left integrals in A is one-dimensional.*

2. *If the space of left integrals in A is non-zero, then A is finite dimensional. This implies:*

3. [**Sw**, Ex. 1-4, pp. 107]. *If A is semisimple[2], then it is finite dimensional.*

4. [**Sul**]. *The space of left integrals on A has dimension ≤ 1.*

$\square$

If A is a Hopf algebra with a non-zero integral *on* it, then A is called *co-Frobenius*. It need not be finite dimensional. For instance, cosemisimple Hopf algebras (like compact quantum groups) are co-Frobenius. Co-Frobenius Hopf algebras are a matter of current interest. See [**DaNT**] and references therein.

We fix now a finite dimensional Hopf algebra A, a non-zero right integral $\lambda \in A^*$ and a a non-zero left integral $\Lambda \in A$. Notice that neither λ is necessarily a left integral in A^*, nor Λ is necessarily a right integral in A. The failures are respectively measured by the so-called modular (or distinguished group-like) elements $a \in A$, $\alpha \in A^*$. They are determined by

$$\beta\lambda = \lambda\langle\beta, a\rangle, \qquad \Lambda x = \langle\alpha, x\rangle\Lambda.$$

We say that A is unimodular if Λ is a right integral in A; that is, if $\alpha = \varepsilon$. It is known that:

1. Semisimple Hopf algebras are unimodular.

2. Factorizable (finite dimensional) Hopf algebras are unimodular [**R3**]. In particular, a Drinfeld double is unimodular (a direct proof is given in [**R4**, Th. 4]), and hence any finite dimensional Hopf algebra can be embedded into a unimodular one.

The modular elements are related to the antipode by the following important formula discovered by Radford:

THEOREM 3.5. [**R1**]. $\mathcal{S}^4(x) = \operatorname{ad}\alpha(\operatorname{ad} a(x)) = \operatorname{ad} a(\operatorname{ad}\alpha(x))$. $\square$

See [**Sch3**] for an elegant conceptual proof of Radford's formula.

COROLLARY 3.6. [**R1**]. *The order of the antipode of a finite dimensional Hopf algebra is finite.*

Indeed, it is not difficult to see that the actions of the modular elements commute with each other.

---

[2] In this paper, we understand by "semisimple ring" one whose category of modules is semisimple. Some authors use "artinian semisimple", and reserve "semisimple" for a ring whose Jacobson radical is zero.

If $A = \mathbf{k}G$ is the group algebra of a finite group G, then $\Lambda = \sum_{g \in G} g$ is a (right and left) integral. Note that this is a Haar measure on the finite group G; this justifies the name of *integral* given above. The well known Maschke theorem asserts that $\mathbf{k}G$ is semisimple iff the characteristic of $\mathbf{k}$ does not divide the order $|G|$ of G. As $|G| = \sum_{g \in G} <\epsilon, g> = <\epsilon, \sum_{g \in G} g>$, Maschke theorem is equivalent to the statement "$\mathbf{k}G$ is semisimple iff $<\epsilon, \Lambda> \neq 0$". In this way, it can be generalized to finite dimensional Hopf algebras:

THEOREM 3.7. (**Maschke Theorem for Hopf algebras**). *The following statements are equivalent.*

1. *A is semisimple.*
2. *$<\epsilon, \Lambda> \neq 0$.*

□

The square of the antipode $\mathcal{S}$ is intimately related to the semisimplicity of A:

THEOREM 3.8. [**OS1**]. *Let $\lambda \in A^*$ (resp. $\Lambda \in A$) be integrals as above. We normalize them by $\lambda(\Lambda) = 1$. Then $\operatorname{tr} \mathcal{S}^2 = \varepsilon(\Lambda)\lambda(1)$. Therefore the following are equivalent:*

1. $\operatorname{tr} \mathcal{S}^2 \neq 0$.
2. *A and A^* are semisimple.*

□

Here is a useful consequence of the preceding ideas.

LEMMA 3.9. [**Z1**]. *Assume that the characteristic of $\mathbf{k}$ is zero. Let H be a Hopf algebra of odd dimension. If both H and H^* are unimodular, then H is semisimple. In particular, if both $G(H)$ and $G(H^*)$ are trivial, then H is semisimple.*

PROOF. By Radford's formula 3.5, $\mathcal{S}^4 = \operatorname{id}$. If H were not be semisimple, then $\operatorname{tr} \mathcal{S}^2 = 0$ by Theorem 3.8. Let $H_\pm$ be the eigenspace of $\mathcal{S}^2$ of eigenvalue ± 1. Then $\dim H_+ + \dim H_- = \dim H$, $\dim H_+ - \dim H_- = 0$; hence $\dim H$ is even, a contradiction. □

Here is a finer result about the relation between the square of the antipode and the semisimplicity of the Hopf algebra. It answers a conjecture of Kaplanksky (1975).

THEOREM 3.10. [**LaR1, LaR2**]. *If the characteristic of $\mathbf{k}$ is zero, the following conditions on a finite dimensional Hopf algebra A are equivalent:*

1. *A is semisimple.*
2. *A is cosemisimple, that is, A^* is semisimple.*
3. *A is involutory, i. e. $\mathcal{S}^2 = \operatorname{id}$.*

□

It follows at once from Theorem 3.10 that Hopf subalgebras and quotient Hopf algebras of a semisimple Hopf algebra are also semisimple.

In positive characteristic, the preceding Theorem is false: group algebras are involutory, but not always semisimple. Conditions 1 and 2 imply that the characteristic of $\mathbf{k}$ does not divide the dimension of A [**LaR2**, Th. 2]; also, if we suppose that $\dim A \neq 0$, then condition 3 implies conditions 1 and 2. It seems that the right analogue is that of semisimple and cosemisimple Hopf algebras. This is enhanced by the following beautiful result. For the rest of this Section, **k** is an algebraically closed field of characteristic $p > 0$.

We need some notation. The idea of Theorem 3.11 below is that one can lift the category of semisimple and cosemisimple Hopf algebras over **k** to some analogous category but in characteristic 0. We need a field of characteristic 0; for this, one considers the ring $\mathcal{O}$ of Witt vectors of **k** and its field of fractions K. It is known that char $K = 0$, $\mathcal{O}$ is a discrete valuation ring, say with maximal ideal $\mathfrak{M}$, and $\mathcal{O}/\mathfrak{M} \simeq \mathbf{k}$.

THEOREM 3.11. [**EG3**, Section 2]. *Let A be a semisimple and cosemisimple Hopf algebra over **k** of dimension N. Then there exists a unique, up to isomorphism, Hopf algebra $\overline{A}$ over $\mathcal{O}$, free of rank N, such that $\overline{A}/\mathfrak{M}\overline{A} \simeq A$ as Hopf algebras. The Hopf algebra $A_0 := \overline{A} \otimes_{\mathcal{O}} K$ satisfies the following properties:*

- *It is semisimple and cosemisimple.*

- *The dimensions of its irreducible modules and comodules are the same as those of A.*

- *If A is quasitriangular, then the universal R-matrix also lifts to an universal R-matrix for A_0, which is then also quasitriangular. If A is triangular, then A_0 also is.*

□

This Lifting Theorem allows to deduce many results in the positive characteristic case, from the characteristic 0 case. Notably, here is an answer to another conjecture of Kaplansky.

THEOREM 3.12. [**EG3**]. *The following conditions on a finite dimensional Hopf algebra A are equivalent:*

1. *A is semisimple and cosemisimple.*
2. *$\dim A \neq 0$ and $\mathcal{S}^2 = \mathrm{id}$.*

□

It is worth mentioning that there exists only a finite number of isomorphism classes of semisimple and cosemisimple Hopf algebras of fixed dimension n. This was conjectured by Kaplansky and proved by Ștefan [**St2**]. An alternative proof was found in [**Sch4**]; this proof was later rediscovered independently in [**EG3**]. In both papers [**Sch4**] and [**EG3**], a new and much simpler proof of the following theorem of Radford is presented: The number of Hopf algebra automorphisms of a semisimple and cosemisimple Hopf algebra is finite.

3.3. The exponent. In this subsection, **k** is any field. In analogy with elementary group theory, one can consider the following notion.

DEFINITION 3.13. [**EG5, Kas1**] The *exponent* of a Hopf algebra H is

$$\exp H := \min\{N \in \mathbb{N} : \mu_N \circ (\mathrm{id} \otimes \mathcal{S}^{-2} \otimes \cdots \otimes \mathcal{S}^{-2N+2})\Delta_N = \varepsilon.1\} \in \mathbb{N} \cup \infty.$$

Here μ_N and Δ_N are iterations of the multiplication and comultiplication, respectively.

In [**EG5**] several properties and alternative characterizations of the exponent are proved. For example, the exponent of a Hopf algebra H equals the order of the Drinfeld element of the Drinfeld double $D(H)$. Notably, inspired by Vafa's formula from Conformal Field Theory, they show:

THEOREM 3.14. [**EG5**] *If H is a semisimple and cosemisimple Hopf algebra over $\mathbf{k}$, then $\exp H$ divides $(\dim H)^3$.* □

This gives support to the following natural conjecture.

CONJECTURE 3.15. [**Kas1**] *If H is a semisimple and cosemisimple Hopf algebra over $\mathbf{k}$, then $\exp H$ divides $\dim H$.*

The validity of the Conjecture was verified by Kashina in most of the known cases. A further motivation for this conjecture is given by

PROPOSITION 3.16. [**Kas1**, Th. 6] *If H is a semisimple and cosemisimple Hopf algebra over $\mathbf{k}$, and R is a a semisimple and cosemisimple Hopf algebra in ${}^H_H\mathcal{YD}$, then the order of the square of the antipode of R divides $\exp H$.* □

More questions about the exponent can be found in [**EG5**].

4. The semisimple case

In this Section we survey known results on classification of semisimple Hopf algebras. The base field $\mathbf{k}$ is now algebraically closed; it has characteristic 0, except when stated otherwise.

4.1. The Class Equation. A first fundamental result is the so-called Class Equation, found by G. I. Kac in 1972 [**Ka2**] and rediscovered by Y. Zhu 20 years later [**Z1**]. To state it, we need some preliminaries that have interest on their own.

Let H be a finite dimensional Hopf algebra and let V be a finite dimensional left H-module, corresponding to a representation $\rho : H \to \operatorname{End} V$. We define the *character* of V as the functional $\chi_V \in H^*$ given by

$$\chi_V(h) = \operatorname{tr}_V(\rho(h)), \qquad h \in H.$$

Here are some elementary properties of the characters: if V and W are finite dimensional H-modules, then

- $\chi_U = \chi_V + \chi_W$ for any extension U of V by W, in particular $\chi_{V \oplus W} = \chi_V + \chi_W$;
- $\chi_{V \otimes W} = \chi_V \cdot \chi_W$;
- $\chi_{V^*} = \mathcal{S}(\chi_V)$.

Hence, the abelian subgroup of H^* generated by the characters of the finite dimensional modules of H is finitely generated, and is a subring of H^*. It will be denoted by $R_{\mathbb{Z}}(H)$. If T is any ring, we set $R_T(H) := R_{\mathbb{Z}}(H) \otimes T$.

Now assume that the Hopf algebra H is actually semisimple. It can be shown that

- $R_{\mathbb{Z}}(H)$ is a free abelian group of finite rank. Concretely, if $V_1, \ldots, V_s$ form a set of representatives of isomorphism classes of irreducible modules, then $\chi_{V_1}, \ldots, \chi_{V_s}$ form a $\mathbb{Z}$-basis of $R_{\mathbb{Z}}(H)$.

- $R_{\mathbf{k}}(H)$ is isomorphic to the **k**-subalgebra of H^* spanned by the characters. In particular, $\chi_{V_1}, \ldots, \chi_{V_s}$ form a **k**-basis of $R_{\mathbf{k}}(H)$.

- $R_{\mathbf{k}}(H)$ is a semisimple subalgebra of H^*.

More information about the character ring of a Hopf algebra can be found in [**CoZ, Co, Lo2, NiR, Wi2**].

THEOREM 4.1. [**Ka2, Z1**] *(The Class Equation for semisimple Hopf algebras). Let H be a semisimple Hopf algebra. Let $\lambda \in H^*$ be an integral, such that $<\lambda, 1> = 1$, and let $\lambda = e_1, e_2, \ldots, e_n$, be a complete set of orthogonal primitive idempotents in $R_{\mathbf{k}}(H)$. Then*
$$\dim H = \sum_{i=1}^{n} \dim(e_i H^*),$$
with $\dim(e_1 H^*) = 1$, *and* $\dim(e_i H^*) \,/\, \dim H$, $\forall 1 \leq i \leq n$. □

The original proof of Kac and Zhu used some algebraic number theory. A shorter proof was given in [**Lo1**] reducing the use of algebraic number theory to a minimum.

EXAMPLE 4.2. Let $H = \mathbf{k}G$ be a group algebra. Then the algebra of characters $R(G) \subset \mathbf{k}^G$ is the algebra of class functions, whose idempotents are the characteristic functions of the conjugacy classes. In this case, the Class Equation expresses the elementary fact that the cardinals of the conjugacy classes divide the order of G. Hence the name of Theorem 4.1.

EXAMPLE 4.3. Let $H = \mathbf{k}^G$ be a the dual group algebra. Now the algebra of characters $R(G)$ is all the dual $\mathbf{k}G$. The orthogonal primitive idempotents in $R(\mathbf{k}^G)$ are the orthogonal primitive idempotents in $\mathbf{k}G$. In this case, the Class Equation says that the dimensions of the irreducible G-modules divide the order of G (Frobenius theorem).

Let us now state some applications of the Class Equation. The following Theorem was obtained by Masuoka, finding the right setting for previous results of Kac and Zhu.

THEOREM 4.4. [**Mk3**]. *Let H be a semisimple Hopf algebra such that $\dim H = p^m$, with p a prime, and $m \geq 1$. Then H contains a central grouplike $g \neq 1$.*

PROOF. (Sketch). The Class Equation implies the existence of a *non-trivial* orthogonal primitive idempotent e in $R_{\mathbf{k}}(H)$ such that $\dim(eH^*) = 1$. But e corresponds to a central group-like in H, as desired; see for instance [**Sch3**, 4.14]. □

The following result was conjectured by Kaplansky in 1975 and proved by Zhu; similar ideas were used by G. I. Kac to obtain an analogous result in the framework of C^*-algebras [**Ka2**].

THEOREM 4.5. [**Ka2, Z1**]. *Let p be a prime number. Recall that the characteristic of **k** is 0. A Hopf algebra of dimension p is necessarily semisimple and isomorphic to the group algebra of $\mathbb{Z}/(p)$.*

PROOF. We can assume that $p > 2$, the case $p = 2$ being not difficult. By Theorem 4.4, it is enough to show that H is semisimple. Assume that H is not semisimple. By Lemma 3.9, then either $G(H)$ or $G(H^*)$ has to be non-trivial. But this would imply, by Corollary 3.2, that H itself is a group algebra, contradicting the assumption. □

Theorem 3.11 allows to extend Theorem 4.5 to characteristic $q \neq p$.

THEOREM 4.6. *Let p be a prime number and assume that char $\mathbf{k} = q$. Let H be a Hopf algebra of dimension p.*

*(i) [**EG3**, Theorem 3.4]. If H is semisimple and cosemisimple, then $q \neq p$ and H is isomorphic to the group algebra of $\mathbb{Z}/(p)$.*

*(ii) [**EG3**, Corollary 3.4]. If $q > p$, then H is necessarily semisimple and cosemisimple.* □

QUESTION 4.7. What are all the Hopf algebras of dimension p, in the remaining characteristics, in particular in characteristic p?

We assume again that the base field has characteristic 0. Here is another application of Theorem 4.4.

COROLLARY 4.8. [**Mk3**]. *Let p be a prime number. Semisimple Hopf algebras of order p^2 are group algebras, in particular, commutative and cocommutative.* □

This follows from Theorem 4.4 by an argument on extensions. A harder analysis of extensions gives the classification also for p^3, p an odd prime. (For $p = 2$, this was also done by Masuoka, and the list is slightly different).

THEOREM 4.9. [**Mk2**]. *There are $p + 8$ semisimple Hopf algebras of order p^3:*
1. *Three group algebras of abelian groups.*
2. *Two group algebras of non-abelian groups, and their duals.*
3. *Finally, $p+1$ Hopf algebras which are neither commutative nor cocommutative. They are extensions of the group algebra $\mathbf{k}(\mathbb{Z}/(p) \times \mathbb{Z}/(p))$ by $\mathbf{k}(\mathbb{Z}/(p))$, and are self-dual.*

□

In principle, one could go further by classifying extensions. In this direction, Kashina classified all semisimple Hopf algebras of order 16 [**Kas2**]. As in the group case, the computations become harder and harder; Kashina's paper witnesses this. To deal with other dimensions, one needs more sophisticated tools. We will see some of them in the following Subsection. Let us state before another nice application of the Class Equation.

THEOREM 4.10. *Let K be a Hopf subalgebra of a semisimple Hopf algebra H.*
(a). Assume that there exists a prime number r such that the index $[H : K] = r^n$, $n \geq 1$, and $r^n < s$, for all prime number s dividing the dimension of H, $s \neq r$. Then H^* admits a non-trivial central group-like element.
(b). Assume furthermore that $n = 1$. Then K is normal in H. □

Part (a) is [**Na1**, Th. 2.2.1]; a small change in the proof of *loc. cit.* is required to obtain (b). For $r = 2$, (b) was obtained by Masuoka, see *e. g.* [**Mk4**]. Statement (b) appeared without proof already in [**KM**] (I am grateful to Y. Kashina for communicating this to me). An alternative proof of (b), using Theorem 4.12 below, was independently found by Bakhturin, Montgomery and Gelaki.

4.2. The Frobenius property. The validity of the following Conjecture, stated by Kaplansky in 1975, would have deep consequences in the classification of semisimple Hopf algebras.

CONJECTURE 4.11. *(Kaplansky). If H is a semisimple Hopf algebra, then the sizes of the matrices occuring in any full matrix constituent of H divide the dimension of H.*

In general, the conjecture is still open. A semisimple Hopf algebra satisfying the conclusion of the Conjecture will be said to have *the Frobenius property*. A semisimple Hopf algebra H is known to have the Frobenius property in the following cases:

- If H is a group algebra. This is a classical result by Frobenius; here is where the name comes from.

- If H has a form over a ring of algebraic integers. This was proved by Larson in 1971 [**La**]. Another proof was offered in [**AN1**].

- If $R(H)$ is central in H^* [**ZS**, Theorem 8]. (The result is more specific).

- If H is semi-solvable [**MonW**], *cf.* Definition 2.5.

- If H has a two-dimensional irreducible representation, then $\dim H$ is even [**NiR**].

- If H has a nontrivial self-dual simple module, then $\dim H$ is even. As a consequence, if H has a simple module of even dimension, then $\dim H$ is even [**KaSZ**].

- If H is cotriangular, *i. e.* if H^* is triangular [**EG6**].

But the most striking instance of Hopf algebras with the Frobenius property is given by the following result of Etingof and Gelaki.

THEOREM 4.12. [**EG1**]. *If H is a semisimple Hopf algebra and V is an irreducible module over the Drinfeld double $D(H)$, then the dimension of V divides the dimension of H.* □

COROLLARY 4.13. [**EG1**]. *If H is a quasitriangular semisimple Hopf algebra then it has the Frobenius property.*

PROOF. The universal R matrix provides a surjective Hopf algebra map from $D(H)$ to H; so that any irreducible representation of H is also an irreducible representation of $D(H)$. □

The original proof of Theorem 4.12 used the Verlinde formula from modular categories. A second proof was offered shortly after in [**TZ**], based on [**Z2**] which uses in turn the Class Equation. A third proof of Theorem 4.12 appears in [**Sch5**]; it also uses the Class Equation, combined with a nice generalization of a result of Drinfeld [**Dr2**]. In addition, it is shown in [**Sch5**] that the Verlinde formula in the case of Hopf algebras can be derived from the Class Equation. These ideas suggest that the Class Equation and the Verlinde formula are deeply intertwined.

4.3. Semisimple Hopf algebras of low dimension.
A first application of Theorem 4.12 to classification results is the following.

THEOREM 4.14. [**EG4, GeW, Mk4**]. *Let $p \neq q$ be prime numbers. A semisimple Hopf algebra of dimension pq is necessarily commutative or cocommutative.* □

This was proved in the case $2p$ in [**Mk4**] without Theorem 4.12; in [**GeW**], it was shown for Hopf algebras of Frobenius type. The general case was derived in [**EG4**] from [**GeW**] and Theorem 4.12; but the derivation is not straightforward and needs extra arguments. Alternative proofs of Theorem 4.14 were offered in [**Sr1, Na1**]; both use Theorem 4.12. The proof in [**Na1**] is based on the following result, inspired in turn by the proof of [**EG4**, Lemma 2].

THEOREM 4.15. [**Na1**]. *Let $p \neq q$ be prime numbers, and n, m non-negative integers. Let H be a semisimple Hopf algebra of dimension $p^n q^m$. Assume H is of Frobenius type. If $p > q^b$, where*

$$b := \max\{j : \exists \text{ an irreducible character of } H \text{ of degree } q^j\},$$

then H has a non-trivial one-dimensional representation. □

COROLLARY 4.16. [**Na1**]. *Suppose H is any semisimple Hopf algebra of dimension $p^a q^b$. If $p > q^b$, then $G(D(H)^*)$ is non-trivial.* □

The meaning of the Corollary is clarified by Proposition 2.10; one has indeed a short exact sequence

(4.1) $$1 \to \mathbf{k}G(D(H)^*) \xrightarrow{\iota} D(H) \xrightarrow{\pi} K \to 1,$$

for *any* finite dimensional Hopf algebra H (not necessarily semisimple). This sequence was considered with profit in [**Na1, Na4**]. Note that K is of Frobenius type, when H is semisimple.

The following question was raised by S. Montgomery, in a slightly different formulation.

QUESTION 4.17. *What is the analogue of Burnside's $p^a q^b$-Theorem for semisimple Hopf algebras?*

The only known results in this direction are in the case of dimension pq^2, besides those already reviewed.

THEOREM 4.18. *Let $p \neq q$ be prime numbers. Let H be a semisimple Hopf algebra of dimension pq^2.*

(i). [**Na1**]. *If H is not simple, then H is known. Specifically, either H is trivial or belongs to one of three families of extensions: $\mathcal{A}_l$, $0 \leq l \leq q-1$, $p = 1$ mod q (constructed in [**Ge1**]); $\mathcal{B}_{\lambda_j}$, $0 \leq j \leq \frac{p-1}{2}$ $q = 1$ mod p; or $\mathcal{B}^*_{\lambda_j}$ (constructed in [**Mk5**] for $p = 2$). See [**Na1**] for details.*

(ii). [**Na1, Na2, Na5**]. *Suppose that both H and H^* are of Frobenius type. Then either H or H^* has a non-trivial central group-like. Therefore, if Kaplansky's conjecture 4.11 is true, then all possible H are known.*

(iii). *Even without knowing the validity of Kaplansky's conjecture, the classification is known in some cases. For instance, either H or H^* has a non-trivial central group-like under the following restrictions:*

- $p = 2, 3$ or $p^2 < q$ [**Na1**]; or

- $p > q^4$ and $p \neq 1 \mod q$ [**Na1**]; or

- $\dim H \leq 100$ [**Na5**].

$\square$

The proof uses most of the machinery already explained; one of the main new tools is a systematic consideration of the exact sequence (4.1).

The following question was raised by S. Montgomery.

QUESTION 4.19. Classify all semisimple Hopf algebras H such that $\dim H < 60$.

The actual status of this Problem is the following:

- It is open for dimensions 24, 30, 32, 36, 40, 42, 48, 54, 56, and known for the rest of the dimensions.

- For the dimensions where the classification is known, there are non-trivial examples in some of the dimensions; all of them are extensions.

- For the dimensions where the classification is *not* known, there are non-trivial examples in all the dimensions but 30; all of them are extensions.

The proof, for the cases where the answer is known, follows from Theorems 4.5, 4.9, 4.14, 4.18 and the main result of [**Kas2**]. We should notice that some low cases were previously known, *e.g.* 12 was done in [**F**], 18 appears in [**Mk5**]; and that to apply Theorem 4.18 one needs in some cases to check that a Hopf algebra of the desired dimension is of Frobenius type, but this follows from a counting argument in low dimension. The classification of semisimple Hopf algebras of dimension 16 [**Kas2**] was recently simplified in [**KMM**] using another new tool: the Frobenius-Schur indicator, introduced in [**LiM**]. We should also mention that the classification of Kac algebras of dim $pq^2 < 60$ and other interesting results are obtained in [**IK2**] via completely different methods, arising from the theory of subfactors.

It is the personal opinion of this author that one should concentrate efforts in determining whether the Hopf algebras with the remaining dimensions are semi-solvable, rather than in computing explicitly all the extensions. For, it is enough for many applications to know that a Hopf algebra is semi-solvable, in analogy with the similar situation for finite groups. And on the contrary, if it is discovered that there are examples of those dimensions which are not semi-solvable, then these examples would be really new.

5. The pointed case

We have seen that the semisimple case is essentially different to the non-semisimple case, *cf.* Theorems 3.8 and 3.10. We can replace "semisimple" by "cosemisimple" in the preceding sentence, by Theorem 3.10, if the field **k** has characteristic 0; this fits better in the following framework.

We first suppose that **k** is any field. Let us say that a coalgebra C is *simple* if any non-zero subcoalgebra D is equal to C. If C is simple then it is finite dimensional, and indeed it is dual to a simple finite dimensional algebra. When **k** is algebraically closed, it is then dual to a matrix algebra.

Given any coalgebra C, we can consider the sum of all its simple subcoalgebras:

$$C_0 := \sum_{D \subseteq C,\, D \text{ simple}} D.$$

Then C_0, itself a subcoalgebra of C, is a direct sum of simple subcoalgebras; it is called the *coradical* of C. Furthermore, we can define recursively a filtration of C by

$$C_{n+1} := \{x \in C : \Delta(x) \in C_n \otimes C + C \otimes C_0\}.$$

Then all the C_n's are subcoalgebras of C, so that $C_0 \subseteq C_1 \subseteq \ldots C_n \subseteq C_{n+1} \ldots$; and the filtration is exhaustive, *i. e.* $C = \bigcup_{n \geq 0} C_n$. It is called the *coradical filtration* of C.

When C is finite dimensional, so that it is the dual coalgebra of an algebra A, then $C_n = (\operatorname{Jac} A^{n+1})^\perp$, where Jac denotes the Jacobson radical.

It is natural to separate the investigation of finite dimensional Hopf algebras according to the shape of its coradical, and the relative position in the full Hopf algebra. One extreme case is when the Hopf algebras coincide with their coradical; this is the cosemisimple case. The opposite case is when the coradical is as simple as possible, but not the full Hopf algebra.

DEFINITION 5.1. A coalgebra (not necessarily finite dimensional) C is called *pointed* if any simple subcoalgebra of C has dimension one.

This can be phrased alternatively as "any simple comodule has dimension one"; or as "the coradical is cocommutative", when **k** is algebraically closed. If H is a pointed Hopf algebra, then H_0 is the group algebra of $G(H)$.

A first information about the coradical filtration of a pointed coalgebra is given by the Theorem of Taft and Wilson.

THEOREM 5.2. [**TaW**]; see [**Mo1**, Theorem 5.4.1]. *Let C be a pointed coalgebra. Then:*

- *If $n \geq 1$, the n-th term of the coradical filtration can be decomposed as*

$$C_n = \sum_{g,h \in G(C)} C_n(g,h), \text{ where}$$

$C_n(g,h) = \{x \in C : \Delta(x) = x \otimes h + g \otimes x + u, \text{ for some } u \in C_{n-1} \otimes C_{n-1}\}.$

- *The first term of the coradical filtration can be expressed as*

$$C_1 = kG(C) + (\oplus_{g,h \in G(C)} \mathcal{P}_{g,h}).$$

□

This Section is devoted to finite dimensional pointed Hopf algebras. For shortness, we shall say "pointed" for pointed non-cosemisimple. From now on, the field **k** is supposed algebraically closed and of characteristic 0.

5.1. Summary. We now give an account of results on classification of pointed Hopf algebras. Let p be a prime number.

Assume first that $p = 2$. All pointed Hopf algebras with coradical $\mathbb{Z}/(2)$ were classified in [**Ni**]; there is exactly one isomorphism class in each dimension 2^n. By a different method, the same result was obtained later in [**CD2**]. Pointed Hopf algebras of dimension 16, resp. 32, were classified in [**CDR**]; respectively in [**Gñ1**]. See also [**AnDa**].

Assume now that $p > 2$. The only pointed Hopf algebras of dimension p^2 are the Taft algebras. This follows at once from the Theorem of Taft and Wilson and Corollary 3.2 (i). This was known to Nichols [**Ni**] and was rediscovered in independently by Chin, Ştefan and the present author.

Pointed Hopf algebras of dimension p^3 were classified in [**AS2, CD1, StvO**] by different methods. Pointed Hopf algebras of dimension p^4 were classified in [**AS3**], using results of [**AS2**]. As a matter of fact, there exist infinitely many isomorphism classes of pointed Hopf algebras of dimension p^4; hence, one of the ten conjectures of Kaplansky is not true. This was shown independently in [**AS2, BDG, Ge2**]. However, it was shown that all the Hopf algebras in these families are twists of each other [**Mk9**].

The classification of pointed Hopf algebras of dimension p^5 follows from [**Gñ2**] combined with [**AS5**]. The Lifting Method allows to obtain many other classification results; with these ideas, for instance, it is also possible to classify pointed Hopf algebras of dimension pq^2, q another prime [**AN3**].

An important classification result is given in [**AS5**], where the list of all finite dimensional pointed Hopf algebras H such that $G(H)$ is an abelian group of odd prime exponent, is presented. See Theorem 5.38.

There are classification results of Hopf algebras with special properties. Minimal triangular pointed Hopf algebras were classified in [**Ge3**]. Finite dimensional pointed Hopf algebras H such that the index $[H : G(H)]$ is either p, p^2 or p^3 are classified in [**Gñ3**] (this result has antecedents in [**AS2**, Th. 0.2], [**AS3**, Prop. 7.4], [**D**]). Finite dimensional pointed Hopf algebras H with $[H : \mathbf{k}G(H)] < 32$ are classified in [**Gñ4**]. Many other interesting results can be found in [**AS2, AS3, AS4, AS6, AS5, Gñ3, Gñ4**]; their formulation is too technical to be included here. Some of them will be evoked below.

5.2. Methods. We shall devote the next Subsection to the "Lifting method" [**AS2**]; here we review another methods.

The Theorem of Taft and Wilson has been a very valuable tool in the study of pointed Hopf algebras; it is applied combined with variations of the Proposition 5.3 below; the proposition appears in [**AN3**] but the idea of the proof goes back to [**Ni, AS1, St1**]. We need first some notation.

Let M and N be non-negative integers such that $2 \leq M$ divides N and let $\xi \in \mathbf{k}^\times$ be a primitive M-th. root of unity. Consider the algebra $K_\mu(N, \xi)$, generated by elements x and g with relations

$$x^M = \mu(1 - g^M), \quad g^N = 1, \quad gx = \xi xg,$$

where $\mu = 0$, if $M = N$, and $\mu \in \{0, 1\}$, if $M \neq N$. The formulas

$$\Delta(g) = g \otimes g, \quad \Delta(x) = 1 \otimes x + x \otimes g,$$
$$\epsilon(x) = 0, \quad \epsilon(g) = 1,$$
$$\mathcal{S}(g) = g^{-1}, \quad \mathcal{S}(x) = -xg^{-1},$$

determine a Hopf algebra structure in $K_\mu(N, \xi)$. It follows from [**AS2**, Theorem 5.5] that the dimension of $K_\mu(N, \xi)$ is MN. If $M = N$, then $K_\mu(N, \xi) \simeq T(\xi)$, where $T(\xi)$ is the Taft algebra corresponding to ξ.

PROPOSITION 5.3. *Let H be a finite dimensional Hopf algebra. Suppose that $\mathbf{k}(g - h) \neq \mathcal{P}_{g,h}$, for some $g, h \in G(H)$. Then H contains a Hopf subalgebra K isomorphic to $K_\mu(N, \xi)$, for some root of unity $\xi \in \mathbf{k}$, and some $\mu \in \{0, 1\}$. In particular, if $\dim H$ is free of squares, then H does not contain non-trivial skew primitive elements.* □

Note that $\mathbf{k}(g - h) \neq \mathcal{P}_{g,h}$ implies that H is not cosemisimple.

In the pioneering paper [**Ni**], the author introduces many interesting concepts, including the "bialgebras of type one" and the now called "Nichols algebras" (see Subsection 5.3 below). He approaches a general pointed Hopf algebra by a limit procedure from bialgebras of type one. He discusses several examples; most of them were recently interpreted in [**Gñ4**].

In the papers [**BDG, CD1, CD2, CDR, D**] the authors construct pointed Hopf algebras as iterated Ore extensions; then they determine the isomorphism classes and, in suitable situations, obtain classification results. This is a very simple and elegant method but it does not seem to be applicable to complicated situations.

In the paper [**StvO**], the authors study a projection $H_1 \to H_0$ and interpret it in terms of coalgebra cohomology. This idea is related to the problem of lifting discussed below. They are able to obtain the classification of pointed Hopf algebras of dimension p^3, as said above.

5.3. The lifting method.

5.3.1. *Overview.* The lifting method [**AS2, AS3**] seems to be the most powerful method to understand pointed Hopf algebras up to now. Let us first roughly overview the method in words. We shall then discuss the main tools of the method and illustrate it with some examples. For more details, see [**AS2, AS3, AS4, AS6, AS5, Gñ3, Gñ4**]. The method works in principle for more general Hopf algebras than pointed; it is enough to assume that the coradical is a Hopf subalgebra.

Let A be a Hopf algebra whose coradical $H = A_0$ is a Hopf subalgebra. We attach to A several invariants; a very sensible one is an algebra R which is not a usual Hopf algebra, but a braided Hopf algebra in the category ${}^H_H\mathcal{YD}$. If A is finite dimensional, then R also is; in many cases (and conjecturally, always), one is able to classify such an R in combinatorial terms. The last step is to recover A from R and H; this is the so-called lifting step. The outcome is, loosely speaking, that finite dimensional pointed Hopf algebras are variations of Frobenius-Lusztig kernels, the finite quantum groups constructed by Lusztig.

Let us describe the invariants of A that we shall consider. Let

(5.1) $$\operatorname{gr} A = \oplus_{n \geq 0} \operatorname{gr} A(n)$$

be the graded vector space associated to the coradical filtration, where $\operatorname{gr} A(n) = A_n/A_{n-1}$, $n > 0$, and $\operatorname{gr} A(0) = A_0 = H$. In general, $\operatorname{gr} A$ is a graded coalgebra [**Sw**, Chapter 11]; but in this situation, it is a graded Hopf algebra because the coradical is a Hopf subalgebra [**Mo1**, 5.2.8]. The graded projection $\pi : \operatorname{gr} A \to \operatorname{gr} A(0) = H$ is a Hopf algebra map and a retraction of the inclusion $\iota : \operatorname{gr} A(0) \to \operatorname{gr} A$. We can then apply the general remarks of Subsection 2.7. Let

$$R = \{a \in \operatorname{gr} A : (\operatorname{id} \otimes \pi)\Delta(a) = a \otimes 1\}$$

be the algebra of coinvariants of π; R is a braided Hopf algebra in the category ${}^H_H\mathcal{YD}$ of Yetter-Drinfeld modules over H and $\operatorname{gr} A$ can be reconstructed from R and H as a bosonization: $\operatorname{gr} A \simeq R \# H$.

In the present case, the braided Hopf algebra has some extra properties. First, R inherits the gradation from $\operatorname{gr} A$: $R = \oplus_{n \geq 0} R(n)$, where $R(n) = \operatorname{gr} A(n) \cap R$, and R is a *graded* braided Hopf algebra; also, $R(0) = \mathbf{k}1$. Second, since we began from the coradical filtration of A, we can conclude that the coradical filtration of R coincides with the filtration given by the gradation:

$$R_n = \oplus_{m \leq n} R(n).$$

The important consequence of this is that $R(1) = P(R)$, the space of primitive elements of R. The reason of this importance will be explained in the next Subsection.

5.3.2. *Nichols algebras.* One of the important tools in the Lifting Method is the notion of Nichols algebras. Nichols algebras can be presented in several different ways, all of them relatively technical. We choose one of these ways and then state the main property of Nichols algebras relating them to the Lifting Method.

DEFINITION 5.4. A braided vector space (V, c) is a finite dimensional vector space V provided with an isomorphism $c : V \otimes V \to V \otimes V$ which is a solution of the braid equation, that is

$$(c \otimes \operatorname{id})(\operatorname{id} \otimes c)(c \otimes \operatorname{id}) = (\operatorname{id} \otimes c)(c \otimes \operatorname{id})(\operatorname{id} \otimes c).$$

A Yetter-Drinfeld module is always a braided vector space; however, a braided vector space could be realized as a Yetter-Drinfeld module over a Hopf algebra with bijective antipode only when the braiding is *rigid*, see for instance [**Tk2**]; and if so, it can be realized in many different ways.

Let (V, c) be a braided vector space. The braiding $c : V \otimes V \to V \otimes V$ induces representations of the braid groups $\mathbb{B}_n \to \operatorname{Aut}(V^{\otimes n})$ by $\sigma_i \mapsto \operatorname{id} \otimes c \otimes \operatorname{id}$; here σ_i are the standard generators of the braid group $\mathbb{B}_n$, and c acts on the tensor product of the i-th. and $(i+1)$-st. copies of V. On the other hand, there is a well-known set-theoretical section s from the symmetric group $\mathbb{S}_n$ to $\mathbb{B}_n$, usually called the Matsumoto section. See [**Bou**]. We consider the quantum symmetrizer

$$\mathfrak{S}_n := \sum_{\sigma \in \mathbb{S}_n} s(\sigma)$$

as an element of the group algebra $k\mathbb{B}_n$, and by abuse of notation, as an endomorphism of $V^{\otimes n} = T^n(V)$.

DEFINITION 5.5. The *quantum symmetric algebra* of braided vector space (V, c) is the quotient $\mathfrak{B}(V) := T(V)/\mathcal{J}$, where $\mathcal{J} = \oplus_{n \geq 0} \ker \mathfrak{S}_n$.

If $V \in {}^H_H\mathcal{YD}$, where H is a Hopf algebra with bijective antipode, then $\mathfrak{B}(V)$ is actually a braided Hopf algebra in ${}^H_H\mathcal{YD}$ called the *Nichols algebra* of V over H.

REMARK 5.6. Nichols algebras apeared first in [**Ni**], as the invariant part of "bialgebras of type one". Woronowicz rediscovered them in his approach to "quantum differential calculus" [**Wo2**]. Lusztig used them, in a different language, to present quantum groups in an invariant way: indeed, the algebras $\mathfrak{f}$ in [**Lu3**] (defined by the non-degeneracy of certain invariant bilinear form) are Nichols algebras of Yetter-Drinfeld modules arising from generalized Cartan matrices. See also [**Ro1, Ro2, Scg2, Rz, BD**]. Also, the positive parts of the "small quantum groups" of [**Lu1**], [**Lu2**]- which we shall call *Frobenius-Lusztig kernels*- are also Nichols algebras [**Mul1, Ro1**]. The presentation via non-degeneracy of the invariant bilinear form always holds [**AnG**].

We now explain the important properties characterizing Nichols algebras; this also shows its relation with the lifting method.

THEOREM 5.7. (Nichols). *Let $V \in {}^H_H\mathcal{YD}$. A braided graded Hopf algebra $T = \oplus_{n \geq 0} T(n)$ with $T(0) = \mathbf{k}1$ is isomorphic to the Nichols algebra of V if and only if*
- $V \simeq T(1)$ *in* ${}^H_H\mathcal{YD}$,
- $P(T) = T(1)$, *and*
- T *is generated as an algebra by $T(1)$.*

PROOF. See [**AnG**, Prop. 3.2.12 or Th. 3.2.29]. □

Let us return to the situation of the preceding Subsection, and the corresponding notation A, H, $\operatorname{gr} A$, R. It follows from Theorem 5.7 that the subalgebra R' of R generated by $R(1)$ is isomorphic to the Nichols algebra of V: $R' \simeq \mathfrak{B}(V)$. To summarize, we give a list of invariants of our initial Hopf algebra A:
- The graded braided Hopf algebra R; it is called the *diagram* of A.

- The braided vector space (V, c), where $V := R(1) = P(R)$ and $c : V \otimes V \to V \otimes V$ is the braiding in ${}^H_H\mathcal{YD}$. It will be called the *infinitesimal braiding* of A.

- The dimension of $V = P(R)$, called the *rank* of A, or of R.

- The subalgebra R' of R generated by $R(1)$: $R' \simeq \mathfrak{B}(V)$.

5.3.3. *Description of the method.* Now we can describe the *lifting method* to deal with Hopf algebras whose coradical is a Hopf subalgebra. This method can be used to solve different problems, but we shall restrict our attention to questions of finite dimensionality. Let us fix a finite dimensional cosemisimple Hopf algebra H. To determine all finite dimensional Hopf algebras A with $A_0 \simeq H$ as Hopf algebras, then we have to address the following steps.

(a). Determine when $\mathfrak{B}(V)$ is finite dimensional, for all braided vector spaces (V, c) in a suitable class.

(b). For those V as in (a), find in how many ways, if any, they can be realized as Yetter-Drinfeld modules over H.

For instance, if $R = \mathbf{k}[x]/(x^N)$ as in Example 2.14 with a fixed q determining the braiding, then it can be realized over $H = \mathbf{k}\Gamma$ whenever there exist $g \in Z(\Gamma)$ and χ a one-dimensional representation of Γ such that $\chi(g) =: q$.

(c). For $\mathfrak{B}(V)$ as in (a), compute all Hopf algebras A such that $\operatorname{gr} A \simeq \mathfrak{B}(V) \# H$ ("lifting").

(d). Investigate whether any *finite dimensional* graded braided Hopf algebra $R = \oplus_{n \geq 0} R(n)$ in $^H_H\mathcal{YD}$ satisfying $R(0) = \mathbf{k}1$ and $P(R) = R(1)$, is generated by its primitive elements, *i. e.* is a Nichols algebra.

In the next Subsections, we shall discuss the meaning and difficulty of these steps.

5.3.4. *The pointed case.* We shall assume in the rest of this Section that $H = \mathbf{k}\Gamma$ is a group algebra. We shall consider suitable classes of braided vector spaces; to justify this, let us recall the classification of all finite dimensional Yetter-Drinfeld modules over a finite group.

Let us first treat the case of a finite abelian group Γ. If $V \in {}^\Gamma_\Gamma\mathcal{YD}$ is finite dimensional, then the action of Γ is diagonalizable. Therefore,

(5.2) $$V = \oplus_{g \in \Gamma, \chi \in \widehat{\Gamma}} V_g^\chi,$$

where $V^\chi = \{v \in V : h.v = \chi(h)v, \forall h \in \Gamma\}$, $V_g = \{v \in V : \delta(v) = g \otimes v\}$ and $V_g^\chi = V^\chi \cap V_g$. Conversely, any vector space with a decomposition (5.2) is a Yetter-Drinfeld module over Γ. The braiding is given by $c(v \otimes w) = \chi(g)w \otimes v$, $v \in V_g$, $w \in V^\chi$. In other words, there exists a basis $x_1, \ldots, x_\theta$ of V and $g(1), \ldots, g(\theta) \in \Gamma$, $\chi(1), \ldots, \chi(\theta) \in \widehat{\Gamma}$ such that $x_i \in V_{g_i}^{\chi_i}$, $1 \leq i \leq \theta$; and then $c(x_i \otimes x_j) = \chi_j(g_i)x_j \otimes x_i$, $1 \leq i, j \leq \theta$.

Let us now treat the case of a finite non-abelian group Γ. The irreducible Hopf bimodules over Γ were classified in [**DiPR, Ci**]. By Remark 2.11, the classification of irreducible Yetter-Drinfeld modules follows; this classification was also obtained in [**Wi1**]. In the notation of [**AnG**], we have:

Let $g \in \Gamma$ and let $\rho : \Gamma^g \to \operatorname{End} V$ be an irreducible representation of the isotropy subgroup Γ^g, that is, the centralizer of g in Γ. Let W be the space of the induced representation of ρ. If one fixes a set of elements $h_1, \ldots, h_s$ such that $(h_i g h_i^{-1})_{1 \leq i \leq s}$ is a numeration of the conjugacy class $\mathcal{O}_g$, then as a vector space

$$W = \oplus_{1 \leq i \leq s} h_i \otimes V.$$

There is a Yetter-Drinfeld module structure on W given by

(5.3) $$\delta(h_i \otimes v) = h_i g h_i^{-1} \otimes h_i \otimes v,$$
(5.4) $$x.(h_i \otimes v) = h_j \otimes \rho(t)(v), \quad \text{if } xh_i = h_j t, t \in \Gamma^g.$$

Indeed, $xh_igh_i^{-1}x^{-1}$ is of the form $h_jgh_j^{-1}$ for a unique j in $\{1,\ldots,s\}$, and then there exists a unique $t \in \Gamma^g$ such that $xh_i = h_jt$.

This Yetter-Drinfeld module will be denoted $M(g,\rho)$; it is not difficult to see that it is irreducible.

If we fix a collection $(g)_{g\in\mathcal{C}}$ of representatives of conjugacy classes of Γ, and for each of them a class $\widehat{\Gamma^g}$ of representatives of irreducible representations, then the Yetter-Drinfeld modules $M(g,\rho)$, $g \in \mathcal{C}$, $\rho \in \widehat{\Gamma^g}$, are pairwise non-isomorphic; and these are all the irreducible Yetter-Drinfeld modules over Γ.

DEFINITION 5.8. We shall say that a braided vector space (V,c) is of *group type* if there exists a basis $x_1,\ldots,x_\theta$ of V such that

(5.5) $$c(x_i \otimes x_j) = g_i(x_j) \otimes x_i;$$

necessarily $g_i \in GL(V)$. Notice that $V \in {}^G_G\mathcal{YD}$, where G is the subgroup of $GL(V)$ generated by $g_1,\ldots,g_\theta$.

Furthermore, we shall say that (V,c) is of *finite group type* (resp., of *abelian group type*) if G is finite (resp. abelian).

We shall say that (V,c) is of *diagonal type* if V has a basis $x_1,\ldots,x_\theta$ such that

(5.6) $$c(x_i \otimes x_j) = q_{ij}(x_j \otimes x_i),$$

for some q_{ij} in $\mathbf{k}$. (Observe that any collection q_{ij} defines a solution of the braid equation by (5.6)).

The suitable class related to pointed Hopf algebras is that of braided vector spaces of group type. Let us first concentrate in the case when Γ is a finite abelian group; so, we focus on braided vector spaces of finite abelian group type; clearly, they are of diagonal type. The first fundamental question in the classification program of finite dimensional pointed Hopf algebras with abelian coradical is then the following:

QUESTION 5.9. Given a matrix $(q_{ij})_{1\leq i,j\leq\theta}$ whose entries are roots of 1, when is $\mathfrak{B}(V)$ finite dimensional, where V is a vector space with basis $x_1,\ldots,x_\theta$ and braiding (5.6)? If this is so, compute $\dim \mathfrak{B}(V)$, and give a "nice" presentation by generators and relations.

The meaning of "nice" will be clarified later. We shall give several criteria answering partially Question 5.9.

Let us keep the notation q_{ij}, V as above. We shall denote $N_i = \operatorname{ord} q_{ii}$. If W is a vector subspace of V spanned by some subset of $x_1,\ldots,x_\theta$, then W is a braided vector space itself, and $\mathfrak{B}(W)$ can be identified with the subalgebra of $\mathfrak{B}(V)$ generated by W (use Theorem 5.7). Moreover, $\dim \mathfrak{B}(W)$ divides $\dim \mathfrak{B}(V)$ by Theorem 3.3.

If $V = \mathbf{k}x$ is a one-dimensional braided vector space, say with $c(x \otimes x) = qx \otimes x$, then $\mathfrak{B}(V) \simeq \mathbf{k}[X]$, the polynomial algebra in one variable, when $q = 1$; or else $\mathfrak{B}(V) \simeq \mathbf{k}[X]/(X^N)$ where N is the order of q, when $N > 1$. The first criterion follows easily:

REMARK 5.10. If $\dim \mathfrak{B}(V)$ is finite, then

(5.7) $$q_{ii} \neq 1, \quad 1 \leq i \leq \theta.$$

The analysis of the following family of examples gives already several interesting consequences.

DEFINITION 5.11. Assume that (5.7) holds. We shall say that $\mathfrak{B}(V)$ is a *quantum linear space* if

(5.8) $$q_{ij}q_{ji} = 1, \quad 1 \leq i \neq j \leq \theta.$$

Let us see how the Lifting Method works in the case of quantum linear spaces.

LEMMA 5.12. [**AS2**] *Let V be a braided vector space of diagonal type, with braiding given by a matrix $(q_{ij})_{1 \leq i,j \leq \theta}$ whose entries are roots of 1. Assume that (5.7) holds.*

(a). $\dim \mathfrak{B}(V) \geq \prod_{1 \leq i \leq \theta} N_i$; the equality holds if and only if $\mathfrak{B}(V)$ is a quantum linear space.

(b). If $\mathfrak{B}(V)$ is a quantum linear space, then it is isomorphic to the algebra presented by generators $x_1, \ldots, x_\theta$ with relations

(5.9) $x_i^{N_i} = 0, \quad 1 \leq i \leq \theta,$

(5.10) $x_i x_j = q_{ij} x_j x_i, \quad 1 \leq i < j \leq \theta.$

$\square$

We now illustrate step (b) of the Lifting Method in the setting of quantum linear spaces. Let Γ be a finite abelian group. Let $\theta(\Gamma)$ be the greatest integer θ such that Γ admits a quantum linear space of rank θ. The computation of $\theta(\Gamma)$ and the classification of quantum linear spaces over Γ, are subtle combinatorial questions. For instance, $\theta(\Gamma) = 2$ if Γ is a cyclic p-group, where p is an odd prime; and $\theta(\Gamma \times \widetilde{\Gamma}) = \theta(\Gamma) + \theta(\widetilde{\Gamma})$ if the orders of Γ and $\widetilde{\Gamma}$ are relatively prime. See [**AS2**]. However, $\theta(\mathbb{Z}/2) = \infty$; hence $\theta(\Gamma) = \infty$ for any finite abelian group Γ of even order.

We shall describe now all possible finite dimensional pointed Hopf algebras with abelian coradical whose diagram is a quantum linear space.

We fix a decomposition $\Gamma = \langle y_1 \rangle \oplus \cdots \oplus \langle y_\sigma \rangle$ and we denote by M_ℓ the order of y_ℓ, $1 \leq \ell \leq \sigma$.

A compatible datum of quantum linear space $\mathcal{D}$ for Γ consists of families $g_1, \ldots, g_\theta \in \Gamma$, $\chi_1, \ldots, \chi_\theta \in \widehat{\Gamma}$, $\mu_1, \ldots, \mu_\theta \in \{0, 1\}$, $\lambda_{ij} \in \mathbf{k}$, $1 \leq i < j \leq \theta$ such that (5.7), (5.8) hold for $q_{ij} := \chi_j(g_i)$ and

(5.11) μ_i is arbitrary if $g_i^{N_i} \neq 1$ and $\chi_i^{N_i} = 1$ but 0 otherwise;

(5.12) λ_{ij} is arbitrary if $g_i g_j \neq 1$ and $\chi_i \chi_j = 1$ but 0 otherwise.

THEOREM 5.13. [**AS2**, Th. 5.5]. *Let A be a pointed finite dimensional Hopf algebra with coradical $H = k(\Gamma)$, where Γ is an abelian group as above. We assume that the diagram of A is a quantum linear space.*

Then there exists a compatible datum of quantum linear space $\mathcal{D}$ such that A is isomorphic as Hopf algebra to the algebra presented by generators h_ℓ, $1 \leq \ell \leq \sigma$, and a_i, $1 \leq i \leq \theta$ with defining relations

(5.13) $h_\ell^{M_\ell} = 1$, $1 \leq \ell \leq \sigma$;

(5.14) $h_\ell h_t = h_t h_\ell$, $1 \leq t < \ell \leq \sigma$;

(5.15) $a_i h_\ell = \chi_i^{-1}(y_\ell) h_\ell a_i$, $1 \leq \ell \leq \sigma$, $1 \leq i \leq \theta$;

(5.16) $a_i^{N_i} = \mu_i \left(1 - g_i^{N_i}\right)$, $1 \leq i \leq \theta$;

(5.17) $a_j a_i = \chi_i(g_j) a_i a_j + \lambda_{ij}(1 - g_i g_j)$, $1 \leq i < j \leq \theta$;

and where the Hopf algebra structure is determined by

(5.18) $\Delta(h_\ell) = h_\ell \otimes h_\ell$, $1 \leq \ell \leq \sigma$;

(5.19) $\Delta(a_i) = a_i \otimes 1 + g_i \otimes a_i$, $1 \leq i \leq \theta$.

Conversely, given a compatible datum of quantum linear space $\mathcal{D}$, the algebra $\mathcal{A}(\mathcal{D})$ presented by generators h_ℓ, $1 \leq \ell \leq \sigma$, and a_i, $1 \leq i \leq \theta$ with defining relations (5.13), (5.14), (5.15), (5.16) *and* (5.17) *has a unique Hopf algebra structure determined by* (5.18) *and* (5.19). *It is pointed; $G(\mathcal{A}(\mathcal{D})) \simeq \Gamma$ and*

$$\dim \mathcal{A}(\mathcal{D}) = |\Gamma| \prod_{1 \leq i \leq \theta} N_i.$$

□

REMARK 5.14. The meaning of this formulas is as follows: (5.13), (5.14) and (5.18) are responsible of the group Γ. The families $g_1, \ldots, g_\theta \in \Gamma$, $\chi_1, \ldots, \chi_\theta \in \widehat{\Gamma}$ define a Yetter-Drinfeld module structure over Γ on the braided vector space (V, c) with basis $x_1, \ldots, x_\theta$, by

$$\delta(x_i) = g_i \otimes x_i, \qquad h.x_i = \chi_i(h) x_i.$$

Then (5.15) and (5.19) reflect (2.7) and (2.8). Finally, (5.16) and (5.17) are the "lifting" of the formulas (5.9) and (5.10). This illustrates what we mean by "nice" relations; (5.9) and (5.10) are "nice" because it is possible to lift them.

REMARK 5.15. The definition of compatible datum of quantum linear space and the preceding Theorem can be extended to non-abelian finite groups by requiring the elements $g_1, \ldots, g_\theta$ to be central.

REMARK 5.16. Given two compatible data $\mathcal{D}$ and $\mathcal{D}'$ for groups Γ and Γ' respectively, a Hopf algebra isomorphism $A(\mathcal{D}) \to A(\mathcal{D}')$ induces, first, an isomorphism of the groups Γ and Γ'; and second, sends skew-primitive elements to skew-primitive elements. The presence of the relations (5.16) and (5.17) imposes further conditions; this gives rise to constructions of infinite families of non-isomorphic pointed Hopf algebras of the same dimension, in particular of dimension p^4 [**AS2**]. For more on the isomorphism problem, see [**AS3, AS4, B**].

REMARK 5.17. The basic hypothesis of Theorem 5.13 is that the diagram of A is a quantum linear space. This hypothesis is very restrictive and contains the fact that R is generated in degree 1; with tricky but elementary arguments we can also show that R is generated in degree one in some restricted situations, e. g. when $\dim R(1) = 1$. To go on, we need more powerful ideas, that will be explained in the next Subsection.

5.3.5. *Braidings of Cartan type.* Let (V, c) be a braided vector space of diagonal type, with a basis $x_1, \ldots, x_\theta$ such that (5.6) holds, where the q_{ij}'s are roots of 1. Clearly, the matrix (q_{ij}) determines such a braided vector space (up to change of basis).

DEFINITION 5.18. We shall say that (V, c) is *of Cartan type* if for all i, j, $q_{ii} \neq 1$ and there exists $a_{ij} \in \mathbb{Z}$ such that

(5.20) $$q_{ij} q_{ji} = q_{ii}^{a_{ij}}.$$

The integers a_{ij} are uniquely determined when chosen in the following way:

(5.21) $a_{ii} = 2$;

(5.22) $-\operatorname{ord} q_{ii} < a_{ij} \leq 0$, $i \neq j$.

It follows that (a_{ij}) is a generalized Cartan matrix (GCM) in the sense of the book [**K**]. We shall adapt the terminology from generalized Cartan matrices and Dynkin diagrams to braidings of Cartan type. For instance, we shall say that (V, c) is of *finite Cartan type* if it is of Cartan type and the corresponding GCM is actually of finite type, i. e. a Cartan matrix associated to a finite dimensional semisimple Lie algebra. We shall say that a Yetter-Drinfeld module V is *of Cartan type* (resp., connected, ...) if the matrix (q_{ij}) as above is of Cartan type (resp., connected, ...).

Let (a_{ij}) be a generalized Cartan matrix. Let $\mathcal{X}$ be the set of connected components of the Dynkin diagram corresponding to it. For each $I \in \mathcal{X}$, we let $\mathfrak{g}_I$ be the Kac-Moody Lie algebra corresponding to the generalized Cartan matrix $(a_{ij})_{i,j \in I}$ and $\mathfrak{n}_I$ be the Lie subalgebra of $\mathfrak{g}_I$ spanned by all its positive root vectors. We denote by Φ_I, resp. Φ_I^+, the set of all roots, resp. all positive roots of $\mathfrak{g}_I$. We omit the subindex I when $I = \{1, \ldots, \theta\}$.

The fundamental examples of braidings of Cartan type are given as follows. Let $(a_{ij})_{1 \leq i,j \leq \theta}$ be a generalized Cartan matrix; assume that it is *symmetrizable*, i. e. that there exists positive integers $d_1, \ldots, d_\theta$ such that $d_i a_{ij} = d_j a_{ji}$ for all i, j. Let $q \neq 1$ be a root of unity of order N and let $q_{ij} := q^{d_i a_{ij}}$, $1 \leq i, j \leq \theta$. We assume that N is relatively prime to the entries of the Cartan matrix; in particular, it is odd. Let $(\mathbb{V}, \mathfrak{c})$ denote the corresponding braided vector space with a basis $X_1, \ldots, X_\theta$.

To state the following important Theorem, we need first to discuss a delicate notion. Recall the definition of braided commutator (2.9). When $(a_{ij})_{1 \leq i,j \leq \theta}$ is a finite Cartan matrix, Lusztig defined root vectors $X_\alpha \in \mathfrak{B}(\mathbb{V})$, $\alpha \in \Phi^+$ [**Lu2**]. The definition is not unique and depends on the choice of a decomposition of the longest element of the Weyl group. Moreover, one can see from [**Lu3**] that, up to a non-zero scalar, each root vector can be written as an iterated braided commutator in some sequence $X_{\ell_1}, \ldots, X_{\ell_a}$ of simple root vectors such as

$[[X_{\ell_1}, [X_{\ell_2}, X_{\ell_3}]_c]_c, [X_{\ell_4}, X_{\ell_5}]_c]_c$. See also [**Ri**]. We now fix for each $\alpha \in \Phi^+$ such a representation of X_α as an iterated braided commutator. For a general braided vector space (V, c) of finite Cartan type, we define root vectors x_α in the tensor algebra $T(V)$, $\alpha \in \Phi^+$, as the same formal iteration of braided commutators in the elements $x_1, \ldots, x_\theta$ instead of $X_1, \ldots, X_\theta$ but with respect to the braiding c given by the general matrix (q_{ij}).

THEOREM 5.19. [**Lu1, Lu2, Lu3, Ro1, Mul1**]. *The algebra $\mathfrak{B}(\mathbb{V})$ is finite dimensional if and only if (a_{ij}) is a finite Cartan matrix.*

If this happens, then $\mathfrak{B}(\mathbb{V})$ can be presented by generators X_i, $1 \leq i \leq \theta$, and relations

$$(5.23) \qquad \mathrm{ad}_c(X_i)^{1-a_{ij}}(X_j) = 0, \qquad i \neq j,$$

$$(5.24) \qquad X_\alpha^N = 0, \qquad \alpha \in \Phi^+.$$

Moreover, the following elements constitute a basis of $\mathfrak{B}(\mathbb{V})$:

$$X_{\beta_1}^{h_1} X_{\beta_2}^{h_2} \ldots X_{\beta_P}^{h_P}, \qquad 0 \leq h_j \leq N - 1, \quad 1 \leq j \leq P.$$

Here $\beta_1, \ldots, \beta_P$ is a well-chosen numeration of the set of positive roots Φ^+. In particular,

$$\dim \mathfrak{B}(\mathbb{V}) = N^{\dim \mathfrak{n}}.$$

□

The basis and dimension statements of this Theorem are due to Lusztig, in the setting of the now called Frobenius-Lusztig kernels. Rosso and Müller showed that Frobenius-Lusztig kernels are Nichols algebras.

REMARK 5.20. The definition of braided vector space of Cartan type is also valid for braided vector spaces of diagonal type with general q_{ij} (not necessarily roots of 1). Also, there is a braided vector space $(\mathbb{V}, c)$ attached to q general, (not necessarily a root of 1); when q is not a root of 1, $\mathfrak{B}(\mathbb{V})$ is the +-part of the quantized enveloping algebra of the Lie algebra $\mathfrak{g}$ [**Lu3, Ro1, Ro2**].

Theorem 5.19 motivates the following definition.

DEFINITION 5.21. Let (V, c) be a braided vector space of Cartan type with associated Cartan matrix (a_{ij}) as in (5.21), (5.22). We say that (V, c) is of *FL-type* if there exists positive integers $d_1, \ldots, d_\theta$ such that for all i, j,

(5.25) $d_i a_{ij} = d_j a_{ji}$ hence (a_{ij}) is symmetrizable;

(5.26) There exists $q \in \mathbf{k}^\times$ such that $q_{ij} = q^{d_i a_{ij}}$.

Furthermore, we shall say that (V, c) is *locally of FL-type* if for any subset $I \subset \{1, \ldots, \theta\}$ of cardinal 2, the submatrix $(q_{ij})_{i,j \in I}$ gives a braiding of FL-type.

The following result follows from Theorem 5.19 in combination with the twisting operation, see Subsection 2.4.

THEOREM 5.22. [**AS3, AS5**]. *Let (V, c) be a braided vector space of Cartan type. We also assume that q_{ij} has odd order for all i, j.*

(i). Assume that (V,c) is locally of FL-type and that, for all i, the order of q_{ii} is relatively prime to 3 whenever $a_{ij} = -3$ for some j, and is different from 3, 5, 7, 11, 13, 17. If $\mathfrak{B}(V)$ is finite dimensional, then (V,c) is of finite Cartan type.

(ii). If (V,c) is of finite Cartan type, then $\mathfrak{B}(V)$ is finite dimensional, and if moreover 3 does not divide the order of q_{ii} for all i in a connected component of the Dynkin diagram of type G_2, then

$$\dim \mathfrak{B}(V) = \prod_{I \in \mathcal{X}} N_I^{\dim \mathfrak{n}_I},$$

where $N_I = \mathrm{ord}(q_{ii})$ for all $i \in I$, $I \in \mathcal{X}$. The Nichols algebra $\mathfrak{B}(V)$ is presented by generators x_i, $1 \leq i \leq \theta$, and relations

(5.27) $$\mathrm{ad}_c(x_i)^{1-a_{ij}}(x_j) = 0, \qquad i \neq j,$$

(5.28) $$x_\alpha^{N_I} = 0, \qquad \alpha \in \Phi_I^+, I \in \mathcal{X}.$$

Moreover, the following elements constitute a basis of $\mathfrak{B}(V)$:

$$x_{\beta_1}^{h_1} x_{\beta_2}^{h_2} \ldots x_{\beta_P}^{h_P}, \qquad 0 \leq h_j \leq N_I - 1, \text{ if } \beta_j \in \Phi_I^+, \quad 1 \leq j \leq P.$$

Recall that $\beta_1, \ldots, \beta_P$ is a well-chosen numeration of the set of positive roots Φ^+. □

There is a large class of finite abelian groups where the preceding result allows to answer completely step (a).

COROLLARY 5.23. [**AS3**]. *Let p be an odd prime number, Γ a finite direct sum of copies of $\mathbb{Z}/(p)$ and V a finite dimensional Yetter-Drinfeld module over Γ. We assume that $q_{ii} \neq 1$ for all i. Then (V,c) is of Cartan type and*

(i). If V is of finite Cartan type, then $\mathfrak{B}(V)$ is finite dimensional, and

$$\dim \mathfrak{B}(V) = p^M,$$

where $M = \sum_{I \in \mathcal{X}} \dim \mathfrak{n}_I$ is the number of positive roots of the root system of (a_{ij}).

(ii). If $\mathfrak{B}(V)$ is finite dimensional and $p > 17$, then V is of finite Cartan type. □

5.3.6. *Yetter-Drinfeld modules of Cartan type over a fixed abelian group.* Let us now address step (b) of the lifting method in the setting of Theorem 5.22. This means, to determine which braidings of finite Cartan type actually appear over a general finite abelian group Γ. This reduces, for each fixed finite Cartan matrix $(a_{ij}) \in \mathbb{Z}^{\theta \times \theta}$, to find all the sequences $g(1), \ldots, g(\theta) \in \Gamma$, $\chi(1), \ldots, \chi(\theta) \in \widehat{\Gamma}$ such that

(5.29) $\langle \chi(i), g(i) \rangle \neq 1, \quad$ for all i;

(5.30) $\langle \chi(j), g(i) \rangle \langle \chi(i), g(j) \rangle = \langle \chi(i), g(i) \rangle^{a_{ij}}, \quad$ for all i,j.

This can be interpreted as a problem of computational number theory: to compute all the solutions of a system of quadratic congruences. See [**AS3**, Section 8] for a discussion. This author feels that a compact answer to such question is out of reach; however, for each fixed finite abelian group, the computation could be performed. In this sense, it is worth mentioning the following consequence of Theorem 5.19, found by M. Graña.

PROPOSITION 5.24. [**Gñ4**] *Let Γ be a finite group of odd order. Then there are only finitely many isomorphism classes of Yetter-Drinfeld modules V such that $\mathfrak{B}(V)$ has finite dimension.*

PROOF. Let us assume for simplicity that Γ is abelian. Assume that 3 is relatively prime to the order of Γ. Let V be a Yetter-Drinfeld module with a basis $x_1, \ldots, x_\theta$ with $x_i \in V_{g_i}^{\chi_i}$ for all i. Suppose that $\mathfrak{B}(V)$ has finite dimension. We claim the following more specific result:

There is no $i \neq j$ such that $g_i = g_j$ and $\chi_i = \chi_j$.

For, if such i and j exist, the Yetter-Drinfeld submodule W generated by x_i and x_j has diagonal braiding with matrix

$$\begin{pmatrix} q & q \\ q & q \end{pmatrix},$$

where $q = \chi_i(g_i)$. Let N be the order of q; we can assume that $q \neq 1$, and so $N > 3$. But this braiding is of Cartan type, actually of FL-type, with generalized Cartan matrix

$$\begin{pmatrix} 2 & 2-N \\ 2-N & 2 \end{pmatrix},$$

which is not of finite type; thus $\mathfrak{B}(W)$ has infinite dimension. If 3 divides the order of Γ and the order of q is 3, then one proves similarly that there is no i, j, k all different such that $g_i = g_j = g_k$ and $\chi_i = \chi_j = \chi_k$. The proof for Γ non-abelian follows an analogous argument. □

Let $\Gamma \simeq \mathbb{Z}/(p)$, where p is an odd prime number. Then the determination of all Yetter-Drinfeld modules over Γ with braiding of finite Cartan type is given in [**AS3**, Theorem 1.3]. We shall not repeat the list here, but instead we discuss for illustration the following case.

LEMMA 5.25. *There exists a Nichols algebra with Dynkin diagram G_2 if and only if $p \equiv 1 \mod 3$.*

PROOF. If V is a Yetter-Drinfeld module of dimension 2 with a braiding of Cartan type G_2, then there exists a generator u of Γ, $q \in \mathbf{k}^\times$ of order p and integers b, d such that

$$g(1) = u, \quad g(2) = u^b, \quad \langle \chi(1), u \rangle = q, \quad \langle \chi(2), u \rangle = q^d.$$

So $q_{11} = q$, $q_{22} = q^{bd}$. By definition of the Cartan matrix associated to a braiding, we have $p > 3$. We should have

$$\langle \chi(1), g(2) \rangle \langle \chi(2), g(1) \rangle = q_{11}^{-1} = q_{22}^{-3}.$$

This means $b + d \equiv -1 \equiv -3bd \mod p$. Thus $3b^2 + 3b + 1 \equiv 0 \mod p$ and looking at the discriminant of this equation we see that it has a solution if and only if -3 is a square $\mod p$. By the quadratic reciprocity law, this happens exactly when $p \equiv 1 \mod 3$. □

5.3.7. *Liftings.* We now discuss the step (c) of the lifting method for pointed Hopf algebras whose diagram is a Nichols algebra of Cartan type.

Let $\Gamma = \langle y_1 \rangle \oplus \cdots \oplus \langle y_\sigma \rangle$ be a finite abelian group; we denote by M_ℓ the order of y_ℓ, $1 \leq \ell \leq \sigma$. Let A be a pointed Hopf algebra with $G(A) \simeq \Gamma$ and whose diagram $R \simeq \mathfrak{B}(V)$, with V of finite Cartan type; say as usual with a basis $x_1, \ldots, x_\theta$, $x_i \in V_{g_i}^{\chi_i}$ for all i, and with Cartan matrix (a_{ij}). Then Theorem 5.22 and formulas (2.7) imply that $\operatorname{gr} A$ can be presented by generators h_ℓ, $1 \leq \ell \leq \sigma$, and x_i, $1 \leq i \leq \theta$ with defining relations (5.13), (5.14), (5.15) (with x_i instead of a_i), (5.27) and (5.28). We can then lift the generators h_ℓ to group-like elements with the same name (by abuse of notation), and the generators x_i to elements $a_i \in \mathcal{P}_{g_i,1}^{\chi_i}$; this choice guarantees that relations (5.13), (5.14) and (5.15) hold now in A. It is not difficult to see that the elements h_ℓ, $1 \leq \ell \leq \sigma$, and a_i, $1 \leq i \leq \theta$ are now generators of the algebra A. It remains to see what happens with the relations (5.27) and (5.28). We split this in three cases:

- lifting of the "quantum Serre relations" $x_j x_i - \chi_i(g_j) x_i x_j = 0$, when $i \neq j$ live in different components of the Dynkin diagram;

- lifting of the "quantum Serre relations" $\operatorname{ad}_c(x_i)^{1-a_{ij}}(x_j) = 0$, when $i \neq j$ live in the same component of the Dynkin diagram;

- lifting of the "power of root vectors relations" $x_\alpha^{N_I} = 0$, α a positive root.

In the first case, a straightforward computation shows that $a_j a_i - \chi_i(g_j) a_i a_j$ is again a skew-primitive. It can then be shown that

$$a_j a_i = \chi_i(g_j) a_i a_j + \lambda_{ij}(1 - g_i g_j),$$

where λ_{ij} satisfies (5.12). To deal with this, we introduce the following notion.

DEFINITION 5.26. We say that two vertices i and j *are linkable* (or that i *is linkable to* j) if

(5.31) $i \not\sim j$,

(5.32) $g_i g_j \neq 1$ and

(5.33) $\chi_i \chi_j = 1$.

Here $i \not\sim j$ means that i and j live in different connected components of the Dynkin diagram.

These are the first elementary properties related to this notion.

(5.34) If i is linkable to j, then $\chi_i(g_j)\chi_j(g_i) = 1$, $\chi_j(g_j) = \chi_i(g_i)^{-1}$

(5.35) If i and k, resp. j and ℓ, are linkable, then $a_{ij} = a_{k\ell}$, $a_{ji} = a_{\ell k}$.

(5.36) A vertex i can not be linkable to two different vertices j and h.

A *linking datum* is a collection $(\lambda_{ij})_{1 \leq i < j \leq \theta, i \not\sim j}$ of elements in $\mathbf{k}$ such that λ_{ij} is arbitrary if i and j are linkable but 0 otherwise. Given a linking datum, we say that two vertices i and j *are linked* if $\lambda_{ij} \neq 0$. We shall use this Definition in Theorem 5.38 below.

For the second case, we can offer the following:

THEOREM 5.27. [**AS5**, Theorem 6.8]. *Let $I \in \mathcal{X}$. Assume that $N_I \neq 3$. If I is of type B_n, C_n or F_4, resp. G_2, assume further that $N_I \neq 5$, resp. $N_I \neq 7$. Then the quantum Serre relations hold for all $i \neq j \in I$.* □

QUESTION 5.28. What happens in the low cases excluded in the hypothesis of Theorem 5.27?

The third case is the most complicated. First, if the root α is simple, say corresponding to a vertex i, then an easy computation shows that $a_i^{N_i}$ is again a skew-primitive. It can then be shown that

$$(5.37) \qquad a_i^{N_i} = \mu_i \left(1 - g_i^{N_i}\right),$$

where μ_i satisfies (5.11). Now, if the root α is *not* simple then $a_i^{N_i}$ is not necessarily a skew-primitive, but a skew-primitive "modulo root vectors of shorter lenght".

EXAMPLE 5.29. We consider the case when the Dynkin diagram has type A_2. That is, let $g_1, g_2 \in \Gamma$, $\chi_1, \chi_2 \in \widehat{\Gamma}$ be such that

$$(5.38) \qquad q = \chi_1(g_1) = \chi_2(g_2), \quad \chi_1(g_2)\chi_2(g_1) = q^{-1},$$

where $q \in \mathbf{k}$ is a primitive p-th root of unity, $p > 1$ an odd integer. Let $V := \mathbf{k}x_1 + \mathbf{k}x_2$ be a Yetter-Drinfeld module over Γ with $x_i \in V_{g_i}^{\chi_i}$, $i = 1, 2$. In $\mathfrak{B}(V)$, we define a non-simple root vector by

$$x_{1,2} = x_1 x_2 - \chi_2(g_1) x_2 x_1.$$

Then $\Delta(x_{1,2}) = g_1 g_2 \otimes x_{1,2} + x_{1,2} \otimes 1 + (1 - q^{-1})x_1 g_2 \otimes x_2$, and using the quantum Serre relations, one can prove that

$$\Delta(x_{1,2}^p) = g_1^p g_2^p \otimes x_{1,2}^p + x_{1,2}^p \otimes 1 + (q-1)^p \chi_1(g_2)^{\frac{p(p-1)}{2}} x_1^p g_2^p \otimes x_2^p.$$

Let A be a pointed Hopf algebra such that $G(A) \simeq \Gamma$ and the diagram R of A is isomorphic to $\mathfrak{B}(V)$. As before, we lift $x_1, x_2 \in \operatorname{gr} A$ to $a_1, a_2 \in A$ such that $x_i \in \mathcal{P}(A)_{g_i}^{\chi_i}$, $i = 1, 2$. As we said, there exist $\mu_1, \mu_2 \in \{0, 1\}$ satisfying (5.11) (with $N_1 = N_2 = p$) such that (5.37) holds.

Let $a_{1,2} = a_1 a_2 - \chi_2(g_1) a_2 a_1$. It is possible to show that there exists $\lambda \in \mathbf{k}$ such that

$$(5.39) \qquad a_{1,2}^p = \mu_1 \mu_2 (q-1)^p (1 - g_2^p) + \lambda(1 - g_1^p g_2^p),$$

where λ can be chosen in the following way:

(5.40) $\lambda = 0$, if $g_1^p g_2^p = 1$ or $\chi_1^p \chi_2^p \neq \varepsilon$.

THEOREM 5.30. [**AS4**]. *Let A be a pointed Hopf algebra such that $G(A) \simeq \Gamma$ and the diagram R of A is isomorphic to $\mathfrak{B}(V)$ with V as above of type A_2. Then A can be presented by generators h_ℓ, $1 \leq \ell \leq \sigma$, and a_i, $1 \leq i \leq 2$, subject to relations (5.13), (5.14), (5.15) (5.37) and (5.39).*

An algebra presented by generators h_ℓ, $1 \leq \ell \leq \sigma$, and a_i, $1 \leq i \leq 2$ and relations (5.13), (5.14), (5.15) (5.37) and (5.39), where $\mu_1, \mu_2 \in \{0,1\}$ satisfy (5.11) and $\lambda \in \mathbf{k}$ satisfy (5.40), is a pointed Hopf algebra of dimension $\operatorname{ord}\Gamma \, p^3$. □

That is, we have a complete answer when the type is A_2. The general case presents a delicate combinatorial question. It is solved for type A_n in [**AS6**].

5.3.8. *Generation in degree one.* Let us now discuss step (d) of the Lifting method. More generally, let H be any Hopf algebra with bijective antipode and let $R = \oplus_{n \geq 0} R(n)$ be a graded braided Hopf algebra in ${}^H_H\mathcal{YD}$.

QUESTION 5.31. Assume that
- $R(0) = \mathbf{k}1$ and
- $P(R) = R(1)$.

Is it always true that R is generated as an algebra by its primitive elements, *i. e.* is R a Nichols algebra? If the answer is "no", we also seek for additional conditions that would force a positive answer.

There is a dual version of this problem. Let $S = \oplus_{n \geq 0} S(n)$ be a graded braided Hopf algebra in ${}^H_H\mathcal{YD}$.

QUESTION 5.32. Assume that
- $S(0) = \mathbf{k}1$ and
- S is generated as an algebra by $S(1)$.

Is it always true that $P(S) = S(1)$, *i. e.* is S a Nichols algebra? If the answer is "no", we also seek for additional conditions that would force a positive answer.

Both questions are equivalent in the category of graded braided Hopf algebras, whose homogeneous components are finite dimensional. Indeed, if $R = \oplus_{n \geq 0} R(n)$ is such a Hopf algebra and $S = \oplus_{n \geq 0} S(n)$ is its graded dual, that is $S(n) = R(n)^*$ for all n, then the answer to Question 5.31 for R is yes if and only if the answer to Question 5.32 for S is yes. See *e. g.* [**AS3**, Lemma 5.5].

EXAMPLE 5.33. Question 5.32 has a negative answer over a field $\mathbf{F}$ of positive characteristic p. For, let $S = \mathbf{F}[X]$; this a (usual) Hopf algebra when X is required to be primitive. But X^{p^j} are also primitive for all j. Finite dimensional counterexamples are the quotients $\mathbf{F}[X]/(X^{p^j})$, $j \geq 2$.

More general counterexamples arise by considering the infinitesimal or Frobenius kernels of powers of the Frobenius homomorphism of an algebraic group; the preceding are just those related to the affine group of dimension one.

EXAMPLE 5.34. Question 5.32 has a negative answer over $\mathbf{k}$, but the counterexamples are infinite dimensional. For, let q be a primitive root of unity of odd order p. Let $H = \mathbf{k}\mathbb{Z}$; say g is a generator of $\mathbb{Z}$. Let $S = \mathbf{k}[X]$; this a braided Hopf algebra in ${}^H_H\mathcal{YD}$ with $\delta(X) = g \otimes X$, $g.X = qX$, and X is required to be primitive. But using the quantum binomial formula, X^p is also primitive.

More general counterexamples arise by considering the "positive" parts of quantized enveloping algebras at roots of one [**Lu1, Lu2**].

We are not aware of any counterexample to Question 5.31 over $\mathbf{k}$, when both H and R are finite dimensional. In the case when H is a group algebra, a positive answer to Question 5.31, or to Question 5.32, in this setting is equivalent to a positive answer to the following:

CONJECTURE 5.35. [**AS3**]. *Any pointed finite dimensional Hopf algebra over $\mathbf{k}$ is generated by group-like and skew-primitive elements.*

A strong indication that the conjecture is true is given by:

THEOREM 5.36. [**AS5**]. *Let A be a finite-dimensional pointed Hopf algebra with coradical $\mathbf{k}\Gamma$, and let R be the diagram of A, that is*

$$\operatorname{gr} \mathcal{A} \simeq R \# \mathbf{k}\Gamma,$$

and $R = \oplus_{n \geq 0} R(n)$ is a graded braided Hopf algebra in $_\Gamma^\Gamma \mathcal{YD}$ with $R(0) = \mathbf{k}1, R(1) = \mathcal{P}(R)$.

Assume that $R(1)$ is a Yetter-Drinfeld module of finite Cartan type with braiding $(q_{ij})_{1 \leq i,j \leq \theta}$. For all i, let $q_i = q_{ii}, N_i = \operatorname{ord}(q_i)$. Assume that $\operatorname{ord}(q_{ij})$ is odd and N_i is not divisible by 3 and > 7 for all $1 \leq i, j \leq \theta$.

1. *For any $1 \leq i \leq \theta$ contained in a connected component of type B_n, C_n or F_4 resp. G_2, assume that N_i is not divisible by 5 resp. by 5 or 7.*
2. *If i and j belong to different components, assume $q_i q_j = 1$ or $\operatorname{ord}(q_i q_j) = N_i$.*

Then R is generated as an algebra by $R(1)$, that is A is generated by skew-primitive and group-like elements. □

Let us discuss the idea of the proof of Theorem 5.36. As mentioned, one could focus the attention on the graded dual S of R; then one has a surjection of graded braided Hopf algebras $S \to \mathfrak{B}(V)$, where $V = S(1)$. But we know the defining relations of $\mathfrak{B}(V)$, since it is of finite Cartan type. So that we are reduced to show that these relations also hold in S. For instance, take a quantum Serre relation $\operatorname{ad}_c(x_i)^{1-a_{ij}}(x_j) = 0$, $i \neq j$; and consider the Yetter-Drinfeld submodule W of S generated by x_i and $\operatorname{ad}_c(x_i)^{1-a_{ij}}(x_j)$. The assumptions (1) and (2) of the Theorem guarantee that W also is of Cartan type, but not of finite Cartan type. Thus $\operatorname{ad}_c(x_i)^{1-a_{ij}}(x_j) = 0$ in S.

REMARK 5.37. We see that the main point in the proof of Theorem is to have control on Nichols algebras of rank 2. When this is granted, the proof can be extended to other settings; see [**Gñ1, Gñ4**]. There also versions of this Theorem in the infinite dimensional case, over $\mathbb{C}$ and for "positive" braidings; see [**AS7**].

Putting together the previous results, we get a complete answer in a significant case. Let p be an odd prime number. Let s be a natural number and let $\Gamma(s) = (\mathbb{Z}/(p))^s$.

THEOREM 5.38. [**AS5**]. *(a). Let $p > 17$. Let $\mathcal{A}$ be a pointed finite-dimensional Hopf algebra such that $G(\mathcal{A}) \simeq \Gamma(s)$. Then there exist*

- *a finite Cartan matrix $(a_{ij}) \in \mathbb{Z}^{\theta \times \theta}$ [**K**];*
- *elements $g_1, \ldots, g_\theta \in \Gamma(s)$, $\chi_1, \ldots, \chi_\theta \in \widehat{\Gamma(s)}$ such that*

(5.41) $$\langle \chi_i, g_i \rangle \neq 1, \qquad \text{for all } 1 \leq i \leq \theta,$$
(5.42) $$\langle \chi_j, g_i \rangle \langle \chi_i, g_j \rangle = \langle \chi_i, g_i \rangle^{a_{ij}}, \qquad \text{for all } 1 \leq i, j \leq \theta;$$

- *and a linking datum $(\lambda_{ij})_{1 \leq i < j \leq \theta, i \not\sim j}$, cf. Definition 5.26;*

such that $\mathcal{A}$ can be presented as algebra by generators $a_1, \ldots, a_\theta$, $y_1, \ldots, y_s$ and relations

(5.43) $\qquad\qquad\qquad y_h^p = 1, \qquad y_m y_h = y_h y_m, \qquad 1 \leq m, h \leq s,$

(5.44) $\qquad\qquad\qquad y_h a_j = \chi_j(y_h) a_j y_h, \qquad 1 \leq h \leq s, 1 \leq j \leq \theta,$

(5.45) $\qquad\qquad\qquad (\operatorname{ad} a_i)^{1-a_{ij}} a_j = 0, \qquad 1 \leq i \neq j \leq \theta, \quad i \sim j,$

(5.46) $\qquad\qquad\qquad a_i a_j - \chi_j(g_i) a_j a_i = \lambda_{ij}(1 - g_i g_j), \qquad 1 \leq i < j \leq \theta, \quad i \not\sim j;$

(5.47) $\qquad\qquad\qquad a_\alpha^p = 0, \qquad \alpha \in \Phi^+;$

and where the Hopf algebra structure is determined by

(5.48) $\quad \Delta y_h = y_h \otimes y_h, \qquad \Delta a_i = a_i \otimes 1 + g_i \otimes a_i, \qquad 1 \leq h \leq s, 1 \leq i \leq \theta.$

(b). *Conversely, let* $(a_{ij}) \in \mathbb{Z}^{\theta \times \theta}$ *be a finite Cartan matrix,* $g_1, \ldots, g_\theta \in \Gamma(s)$, $\chi_1, \ldots, \chi_\theta \in \widehat{\Gamma(s)}$ *such that* (5.41), (5.42) *hold and* (λ_{ij}) *a "linking datum" for* (a_{ij}), $g_1, \ldots, g_\theta$ *and* $\chi_1, \ldots, \chi_\theta$. *Assume that* $p > 3$ *if the Cartan matrix* (a_{ij}) *has a connected component of type* G_2. *Then the algebra* $\mathcal{A}$ *presented by generators* $a_1, \ldots, a_\theta$, $y_1, \ldots, y_s$ *and relations* (5.43), (5.44), (5.45), (5.46), (5.47) *has a unique Hopf algebra structure determined by* (5.48). $\mathcal{A}$ *is pointed,* $G(\mathcal{A}) \simeq \Gamma(s)$ *and* $\dim \mathcal{A} = p^{s+|\Phi^+|}$. $\qquad\square$

REMARK 5.39. As a corollary of the Theorem and its proof, we get the complete clasification of all finite dimensional pointed Hopf algebras with coradical of prime dimension p, $p \neq 5, 7$. This same result was obtained by Musson, using the Lifting method [**Mus**].

5.3.9. *Low dimension.* To proceed with the lifting method beyond the Cartan case, and also because of the reason explained in Remark 5.37, we need to answer the following particular case of Question 5.9.

QUESTION 5.40. Given a braided vector space V of diagonal type and dimension 2, decide when $\mathfrak{B}(V)$ is finite dimensional. If so, compute $\dim \mathfrak{B}(V)$, and give a "nice" presentation by generators and relations.

Examples of finite dimensional Nichols algebras $\mathfrak{B}(V)$ of rank 2 over cyclic groups of even order which are *not* of Cartan type were known already to Nichols [**Ni**, pp. 1540 ff.]. All of them are encompassed by the following Proposition, which gives a partial answer to Question 5.40.

We recall first the following computation in a Nichols algebra, see [**Ro2**]. If $i \neq j$, then
$$\operatorname{ad}_c(x_i)^r(x_j) = (r)!_{q_{ii}} \prod_{0 \leq k \leq r} \left(1 - q_{ii}^k q_{ij} q_{ji}\right) x_i^r x_j.$$

Given $q \in \mathbf{k}$, $q \neq 0$, we denote by $N(q)$ the order of q if q is a root of 1 different from 1; and ∞ otherwise.

PROPOSITION 5.41. [**Gñ4**]. *Let* (V, c) *be a braided vector space of diagonal type and dimension 2, with basis* $\{x_1, x_2\}$ *and matrix* (q_{ij}). *Let* $N_j = N(q_{jj})$ *for* $j = 1, 2$. *Let* $r + 1$ *be the nilpotency order of* ad_{x_2} *on* x_1. *For* $1 \leq i \leq r$ *let* $M_i = N(q_{11}(q_{12}q_{21})^i q_{22}^{i^2})$. *Then*
$$\dim \mathfrak{B}(V) \geq N_1 N_2 \prod_{1 \leq i \leq r} M_i.$$

Furthermore, suppose that the nilpotency order of ad_{x_1} on x_2 is 2. Then:
1. If $r = 1$ then the equality holds.
2. If $r = 2$, and $N(q_{11}) \neq 2$ or $N(q_{22}) \neq 3$ then the equality holds.
3. If $r = 2$, $N(q_{11}) = 2$ and $N(q_{22}) = 3$ then the equality holds if and only if $q_{12}q_{21} = -1$, or $q_{12}q_{21} = q_{22}$, or $q_{12}q_{21} = -q_{22}$.

$\square$

We shall say that the Nichols algebras in the second part of the Proposition have generalized type A_2 or B_2, by analogy with the Cartan case, since they have similar PBW bases.

An useful tool to deal with low index problems is the next Theorem. Let Γ be an arbitrary group. We denote by $P_V(t)$ the Hilbert series of a graded vector space V.

THEOREM 5.42. [Gñ3]. Let $V \in {}_\Gamma^\Gamma \mathcal{YD}$ be finite dimensional and let $W \subset V$ be $\mathbf{k}\Gamma$-subcomodule. Let $\Gamma' \subset \Gamma$ be the smallest subgroup such that $\delta(W) \subseteq \mathbf{k}\Gamma' \otimes W$. We assume that:
- W is stable under the action of Γ'. Thus $c(W \otimes W) = W \otimes W$, and we can consider $\mathfrak{B}(W)$.
- $V = W \oplus W'$, where W' is a $\mathbf{k}\Gamma$-subcomodule and a $\mathbf{k}\Gamma'$-submodule.
- $\mathfrak{B}(W)$ is finite dimensional.

Then there exists a graded subalgebra K of $\mathfrak{B}(V)$ such that $\mathfrak{B}(V) \simeq K \otimes \mathfrak{B}(W)$ as right $\mathfrak{B}(W)$-modules and left K-modules. In particular, $P_{\mathfrak{B}(V)}(t) = P_K(t)P_{\mathfrak{B}(W)}(t)$; hence $\dim \mathfrak{B}(W)$ divides $\dim \mathfrak{B}(V)$.

The proof of Theorem 5.42 uses "quantum differential operators"; the subalgebra K can be presented as the algebra of invariant elements of the quantum differential operators generated by W.

REMARK 5.43. An alternative proof of Theorem 5.42 is given in [MlS], for more general inclusions of braided Hopf algebras and without the finiteness assumption on the subalgebra. The proof offered in [MlS] shows that Theorem 5.42 is closer to Radford's biproducts than to Nichols-Zöller Theorem.

An application of Theorem 5.42 is the following.

THEOREM 5.44. [Gñ3]. Let A be a finite dimensional pointed Hopf algebra. Let p be a prime number.
1. If the index $[A : G(A)] = p$, then A is a lifting of a quantum line.
2. If the index $[A : G(A)] = p^2$, then A is a lifting of a quantum line or a quantum plane.
3. If the index $[A : G(A)] = p^3$, then A is a lifting of a quantum line, a quantum plane, or a Nichols algebra of type A_2.

In particular, A is generated by group-like and skew primitive elements. $\square$

Parts (1) and (2) of Theorem 5.44 were proved in [AS2] under the assumption $G(A)$ abelian; part (1) was proved in [AS3, D] under the assumption that p is the lowest prime dividing the order of $G(A)$.

Combining Theorem 5.44 with *ad-hoc* techniques, it is possible to show:

THEOREM 5.45. [**Gñ4**]. *Let A be a finite dimensional pointed Hopf algebra. If the index $[A : G(A)] < 32$, then A is a lifting of a quantum line, a quantum plane, a Nichols algebra of (generalized) type A_2 or B_2, or a Nichols algebra $\mathfrak{B}(\mathbf{V}(\mathbb{S}_3))$, see Example 5.54 below. In particular, A is generated by group-like and skew primitive elements.* □

COROLLARY 5.46. *Let A be a finite dimensional pointed Hopf algebra; assume that $\dim A < 100$. Then A is a lifting of a quantum line, a quantum plane, a Nichols algebra of (generalized) type A_2 or B_2, or a Nichols algebra $\mathfrak{B}(\mathbf{V}(\mathbb{S}_3))$. In particular, A is generated by group-like and skew primitive elements.*

PROOF. (Sketch). The only cases not covered by Theorem 5.45 are when $\dim R' \geq 32$; but in these cases, $G(A)$ should have order 2 or 3; and we know the answer anyway by [**Ni**], resp. Theorem 5.38. □

All these Hopf algebras can be explicitly presented, and their isomorphisms types can be explicitly given. The computation of the liftings can be actually done; the most involved case should be 64.

5.3.10. *Nichols algebras over non-abelian finite groups.* To classify pointed Hopf algebras with non-abelian group of group-likes by the Lifting Method, we need to deal first of all with step (a). This means, given a braided vector space V of group type, *cf.* Definition 5.8, to decide when V is finite dimensional and compute it explicitly. It is convenient to consider the following definition, for general braided vector spaces.

DEFINITION 5.47. Let (V, c) be a braided vector space. We say that (V, c) is *decomposable* if there exist proper vector subspaces V_1, V_2, of V such that $V = V_1 \oplus V_2$ and $c(V_i \otimes V_j)) \subseteq V_j \otimes V_i$, $1 \leq i, j \leq 2$. Otherwise, we say that (V, c) is *indecomposable* or a *fat point*.

In the same spirit, a *decomposition* of a braided vector space (V, c) is a direct sum $V = \oplus_{1 \leq i \leq r} V_i$, with $0 \neq V_i \neq V$, and $c(V_i \otimes V_j)) \subseteq V_j \otimes V_i$, $1 \leq i, j \leq r$. We shall denote $c_{ij} := c|_{V_i \otimes V_j} : V_i \otimes V_j \to V_j \otimes V_i$.

A decomposition $V = \oplus_{1 \leq i \leq r} V_i$ shall be called *irreducible* if (V_i, c_{ii}) is indecomposable, $1 \leq i \leq r$.

A braided vector space (V, c) of diagonal type is indecomposable if and only if $\dim V = 1$. This justifies the name of fat point.

We propose to split our problem in two parts:

- First, deal with fat points of group type.

- Next, deal with decomposable braided vector spaces of group type.

We would be happy if the second case could be interpreted as a "generalized Dynkin diagram" with fat points. Here is a first clue that this approach could be correct.

THEOREM 5.48. [**Gñ3**]. *Let $V = \oplus_{1 \leq i \leq r} V_i$ be a decomposition of a braided vector space (V, c). Assume that $\mathfrak{B}(V_i)$ is finite dimensional for all i. Then $\dim \mathfrak{B}(V) \geq \prod_{i=1}^{n} \dim \mathfrak{B}(V_i)$. Furthermore, the equality holds if and only if $c_{ij} = c_{ji}^{-1}$ $\forall i \neq j$.* □

REMARK 5.49. [**Gñ3**]. Let (V, c) be a braided vector space of group type. Assume that $\mathfrak{B}(V)$ is finite dimensional and let $V = \oplus_{1 \leq i \leq r} V_i$ be a decomposition of (V, c). For each $1 \leq i \leq r$, let $W_i \subseteq V_i$ be a (possibly zero) subspace of V_i satisfying the same hypothesis as in Theorem 5.42, with respect to V. Then $\prod_{i=1}^{r} P_{\mathfrak{B}(W_i)}(t)$ divides $P_{\mathfrak{B}(V)}(t)$ and the quotient lies in $\mathbb{Z}[t]$. In particular, $\prod_{i=1}^{r} \dim \mathfrak{B}(W_i)$ divides $\dim \mathfrak{B}(V)$.

An irreducible decomposition with $c_{ij} = c_{ji}^{-1}$, for all $i \neq j$ might be called a "quantum linear space of fat points". Observe the perfect analogy with Lemma 5.12. In general, one would consider first irreducible decompositions with $c_{ij} c_{ji} = c_{ii}^{a_{ij}}$, for all $i \neq j$ ... but before even daring of doing this, we should solve the problem for fat points, which is still far from our knowledge as we explain now.

Let Γ be a finite group. Let $V = M(g, \rho)$, for some $g \in \Gamma$ and $\rho : \Gamma^g \to \operatorname{End} W$ an irreducible representation of the isotropy subgroup Γ^g. Then $\rho(g)$ acts on W by a scalar, by Schur's lemma, say q. Consider $W \hookrightarrow V$, via $x \mapsto 1 \otimes x$. If $x, y \in W$, then
$$c(x \otimes y) = qy \otimes x.$$
A first immediate consequence is that $q \neq 1$ if $\mathfrak{B}(V)$ has finite dimension. But, arguing as in Proposition 5.24, Theorem 5.19 implies the following restrictions.

PROPOSITION 5.50. [**Gñ4**] *If* $\dim \mathfrak{B}(V)$ *is finite, then*
- *if* $\dim W \geq 3$ *then* $q = -1$;
- *if* $\dim W = 2$ *then* $q = -1$ *or it is a third root of unity.*

□

When $\dim W > 1$ and $q = -1$ or a a third root of unity, very little is known. For instance, if $\Gamma = \Gamma^g$, then $V = M(g, \rho)$ is not a fat point but a quantum linear space (if $q = -1$), or has type A_2 (if q is a third root of unity).

Let us now consider the case when $\dim W = 1$, i. e. when ρ is a character. We first describe, following [**Gñ4**], a relation between braided vector spaces of group type, whose indecomposable components can be realized as $M(g, \rho)$ with ρ a character, and set-theoretical solutions of the braid equation.

DEFINITION 5.51. A *crossed set* is a pair $(X, \triangleright)$, where X is a finite set and $\triangleright : X \times X \to X$ is a function, such that
- for each $i \in X$, the function $e^{\triangleright}(i) : X \to X$, $e^{\triangleright}(i)(j) = i \triangleright j$, is a bijection,
- $i \triangleright i = i$, for all $i \in X$,
- $j \triangleright i = i$ whenever $i \triangleright j = j$, and
- $i \triangleright (j \triangleright k) = (i \triangleright j) \triangleright (i \triangleright k) \; \forall i, j, k \in X$.

The archetypical example is a subset X of a finite group Γ stable under conjugation, with $i \triangleright j = iji^{-1}$. A crossed set provides a set-theoretical solution of the Braid Equation by

(5.49) $$c : X \times X \to X \times X, \quad c(i, j) = (i \triangleright j, i).$$

There is a large class of set-theoretical solutions of the Braid Equation which are equivalent, in a suitable sense, to solutions of this form [**So, LuYZ**].

Let A be an abelian group, denoted multiplicatively. Let $(C^\bullet(X, A), \delta)$ be the cochain complex defined by

- $C^n(X) = \{f : X^n \to A\}$, $n \geq 0$.

- $\delta^0 = 1$,

- $\delta^n(f)(x_0, \ldots, x_n) = \prod_{i=0}^{n-1} f(x_0, \ldots, x_{i-1}, x_{i+1}, \ldots, x_n)^{(-1)^i}$
  $\times f(x_0, \ldots, x_{i-1}, x_i \triangleright x_{i+1}, \ldots, x_i \triangleright x_n)^{(-1)^{i+1}}$.

The cohomology groups $H^n(X, A) = H^n(C^\bullet(X, A), \delta)$ have interesting interpretations for low n. First, $H^1(X, A) = A^{\pi_0(X)}$, where $\pi_0(X)$ is the set of equivalence classes of the relation generated by $j \sim i \triangleright j$.

Second, let $\mathbf{k}X$ denote the vector space with basis X. If $f \in C^2(X, \mathbf{k}^\times)$, we define a map $c^f : \mathbf{k}X \otimes \mathbf{k}X \to \mathbf{k}X \otimes \mathbf{k}X$ by

$$c^f(i \otimes j) = f(i,j) i \triangleright j \otimes i.$$

We then have

$$H^2(X, \mathbf{k}^\times) = \{f \in C^2(X, \mathbf{k}^\times) \mid c^f \text{ verifies the Braid Equation}\}/\sim;$$

where $\sim$ amounts for a change of basis of the form $i \mapsto \lambda_i i$, $\lambda_i \in \mathbf{k}^\times$ for all $i \in X$. We can then refer to a braiding c^f, for $f \in H^2(X, \mathbf{k}^\times)$. Similarly, we consider $H^2(X, \mathbb{G}_\infty)$, where $\mathbb{G}_\infty$ is the group of all roots of 1 in $\mathbf{k}$.

LEMMA 5.52. [Gñ4, Lemma 3.8]. *If $f \in H^2(X, \mathbb{G}_\infty)$ then $(\mathbf{k}X, c^f)$ is of finite group type. Conversely, a braided vector space of finite group type arising from a Yetter-Drinfeld module $\oplus_{1 \leq i \leq s} M(g_i, \rho_i)$ with ρ_i characters, is of the form $(\mathbf{k}X, c^f)$ for some $(X, \triangleright)$, $f \in H^2(X, \mathbb{G}_\infty)$.* □

In other words, the class of braided vector spaces $(\mathbf{k}X, c^f)$, $(X, \triangleright)$ a crossed set, $f \in H^2(X, \mathbb{G}_\infty)$, is one of the "suitable classes" we are looking for.

We are naturally led to the problems of, first, classifying all the crossed sets $(X, \triangleright)$, and second, computing $H^2(X, \mathbb{G}_\infty)$ for each of them. The answer is known only for crossed sets of cardinal 3 or 4 [Gñ4]. A first reduction is to determine the pairs $(\mathbf{k}X, c^f)$ which are fat points, or indecomposable. In this sense, it is natural to say that a crossed set $(X, \triangleright)$ is *indecomposable* if the following holds:

- For any $Y \subseteq X$ such that $Y \triangleright Y = Y$, $(X - Y) \triangleright (X - Y) = (X - Y)$, then necessarily $Y = X$ or $X - Y = X$.

If the braided vector space $(\mathbf{k}X, c^f)$ is indecomposable, then the crossed set $(X, \triangleright)$ is indecomposable. However, the converse is not true; see [Gñ4, Section 5.2].

QUESTION 5.53. *Given $(\mathbf{k}X, c^f)$, $(X, \triangleright)$ a crossed set, $f \in H^2(X, \mathbb{G}_\infty)$ compute the dimension of the associated $\mathfrak{B}(\mathbf{k}X)$. If finite, give a nice presentation of $\mathfrak{B}(\mathbf{k}X)$.*

Let us show now some explicit examples of braided vector spaces $(V,c) = (\mathbf{k}X, c^f)$. We assume that $\dim V \geq 3$, otherwise V is of diagonal type [**AnG**]. Notice that we always have a map $\mathbb{G}_\infty \to C^2(X, \mathbb{G}_\infty)$ by $q \mapsto f_q$, $f(i,j) = q$; we denote the corresponding cocycle by $c^q = c^{f_q}$.

EXAMPLE 5.54. [**MlS**]. Let (W,S) be a finite Coxeter group and let X be the union of the orbits of all elements of S; let $\triangleright$ be the restriction of the conjugation. The authors consider the braided vector space $(\mathbf{V}(W), c) = (\mathbf{k}X, c^q)$, $q = -1$. The dimension of $\mathfrak{B}(V)$ is known in the following cases:

- $\dim \mathfrak{B}(\mathbf{V}(\mathbb{S}_3)) = 12$.

- $\dim \mathfrak{B}(\mathbf{V}(\mathbb{S}_4)) = 24^2$.

- $\dim \mathfrak{B}(\mathbf{V}(\mathbb{D}_4)) = 48$.

- $\dim \mathfrak{B}(\mathbf{V}(\mathbb{S}_5)) < \infty$; this was done with a computer program [**FK**].

In the first three cases the explicit presentation by generators and relations is also given [**MlS**]. The computation of all the liftings of $\mathfrak{B}(\mathbf{V}(\mathbb{S}_3)) \# \mathbf{k}\mathbb{S}_3$ is not difficult and was done in [**AnG**].

EXAMPLE 5.55. [**Gñ4**, Section 5.2]. Let $X = \{1, 2, 3, 4\}$. Then there exists a structure of crossed set $\triangleright$ on X, such that the corresponding map $e^\triangleright$ is given by

$$e^\triangleright(1) = (2\,3\,4), \quad e^\triangleright(2) = (1\,4\,3), \quad e^\triangleright(3) = (1\,2\,4), \quad e^\triangleright(4) = (1\,3\,2).$$

Then there is a cocycle f such that $\mathfrak{B}(\mathbf{k}X)$ has dimension 72; a nice presentation by generators and relations is also given in [**Gñ4**].

These are the only examples were we know that the dimension is finite.

In the spirit of step (b), we also may ask whether, for a fixed finite group Γ, there exist "genuine" examples of finite dimensional pointed Hopf algebras H with $G(H) \simeq \Gamma$. The best way to state precisely what "genuine" means, is as follows.

DEFINITION 5.56. [**MlS**]. A Yetter-Drinfeld module $V \in {}_\Gamma^\Gamma \mathcal{YD}$ is *link-indecomposable* if Γ is generated by the elements g such that $V_g \neq 0$.

QUESTION 5.57. [**MlS**] Determine all finite groups Γ having a link-indecomposable Yetter-Drinfeld module V such that $\mathfrak{B}(V)$ is finite dimensional.

5.3.11. *Nichols algebras over semisimple Hopf algebras*. We know almost nothing about finite dimensional Hopf algebras whose coradical is a Hopf subalgebra, but are not pointed. In order to restrict our attention to "genuine" examples, we propose the following definition.

DEFINITION 5.58. Let H be a semisimple Hopf algebra and let $V \in {}_H^H \mathcal{YD}$. Let C_V be the subcoalgebra of H generated by the matrix coeficients of V. We say that V is *link-indecomposable* if H is generated as an algebra by $C_V + C_{V^*}$.

QUESTION 5.59. Determine all semisimple Hopf algebras H having a link-indecomposable Yetter-Drinfeld module V such that $\mathfrak{B}(V)$ is finite dimensional.

6. The general case

We assume in this Section that **k** is algebraically closed.

In this Section we are concerned with the following question: what is the classification of all Hopf algebras of a fixed dimension? Of course, this is a very difficult problem (it contains the classification of all groups of a fixed order). One should probably be happy to answer the following partial questions:

QUESTION 6.1. Classify all Hopf algebras of dimension d, where d is small; say, $d \leq 100$.

QUESTION 6.2. Classify all Hopf algebras of dimension d, where d factorizes in a simple way; say, $d = p^2, pq, pqr, pq^2, p^3, p^a q^b, \ldots$, where p, q, r are distinct prime numbers.

The purpose of addressing these Questions is, naturally, to gain insight into the structure of general Hopf algebras, before even daring to state general conjectures or questions about them.

About Question 6.2, the only general known result is Theorem 4.5, classifying Hopf algebras of prime dimension, when char **k** = 0. I am aware of the following two partial results.

THEOREM 6.3. [**AS1**]. *Assume that char* **k** $= 0$. *Let H be a Hopf algebra of dimension p^2. Assume that the order of S divides $2p$. Then H is either semisimple (hence, a group algebra) or pointed (hence a Taft algebra).* $\square$

Recall that by Radford's formula 3.5 and Nichols-Zöller Theorem 3.2, $S^{4p} = \text{id}$; so that the only remaining case is when the order of S is $4p$.

THEOREM 6.4. [**Na4**]. *Assume that char* **k** $= 0$. *Let H be a Hopf algebra of dimension p^2 or pq, where p and q are odd. Assume that H is quasitriangular. Then H is semisimple; hence, a group algebra.* $\square$

In general, Proposition 5.3 motivates the following Question:

QUESTION 6.5. Let N be a positive integer which is free of squares, and not divisible by the characteristic of **k**. Are all Hopf algebras of dimension N semisimple?

Here is what we know about Question 6.1.

THEOREM 6.6. *Assume that char* **k** $= 0$. *All Hopf algebras of dimension d are known for $d \leq 12$, or $d = 15, 21, 25, 35, 49$. They are either semisimple, or pointed, or dual to a pointed.* $\square$

For $d \leq 11$, this was proved in [**Wl**] and an alternative, more conceptual, proof was offered in [**St3**]; for $d = 12$, this was proved in [**Na3**]; the rest of the cases are in [**AN3**].

REMARK 6.7. It is natural to ask whether there exist Hopf algebras H which are neither semisimple, neither pointed, nor H^* is pointed. A first easy answer is to construct such Hopf algebras by extension; for instance, the tensor product of the Sweedler's 4-dimensional pointed Hopf algebra and the unique 8-dimensional nontrivial semisimple Hopf algebra is a Hopf algebra of the required sort. Infinite families of examples of such Hopf algebras are constructed in the very interesting paper [**Mul2**], as a byproduct of his study of the "finite quantum subgroups of

$GL(N)$", that is, finite dimensional Hopf algebra quotients of $\mathbf{k}_q[GL(N)]$. They can all be presented as extensions.

QUESTION 6.8. Is there any *simple* Hopf algebra which is neither semisimple, neither pointed, nor dual to a semisimple or pointed?

6.1. On the coradical filtration. The purpose of this Subsection is to state a description of the coradical filtration due to Nichols and some consequences, useful for problems in low dimension.

Let C be a coalgebra over $\mathbf{k}$. We denote by $\widehat{C}$ the set of isomorphism types of simple left C-comodules; and by V_τ (resp., V_τ^*) the simple left (resp. right) C-comodule corresponding to $\tau \in \widehat{C}$. We have $C_0 \simeq \oplus_{\tau \in \widehat{C}} C_\tau$, where C_τ is a simple subcoalgebra of dimension d_τ^2, $d_\tau \in \mathbb{Z}$. We also set $C_{0,d} := \oplus_{\tau \in \widehat{C}: d_\tau = d} C_\tau$.

A C_0-bicomodule is a vector space M with left and right C_0-coactions $\rho_L : M \to C_0 \otimes M$ and $\rho_R : M \to M \otimes C_0$ such that $(\rho_L \otimes \mathrm{id})\rho_R = (\mathrm{id} \otimes \rho_R)\rho_L$. Any C_0-bicomodule is a direct sum of simple C_0-sub-bicomodules and a simple C_0-bicomodule is of the form $V_\tau \otimes V_\mu^*$ and has dimension $d_\tau d_\mu$ for some $\tau, \mu \in \widehat{C}$. If M is a C_0-bicomodule, we set $M^{\tau,\mu}$ for the isotypic component of type $V_\tau \otimes V_\mu^*$.

There exists a coalgebra projection π of C onto C_0 [**Mo1**, 5.4.2]; let $I := \ker \pi$. Then C is a C_0-bicomodule via $\rho_L := (\pi \otimes \mathrm{id})\Delta : C \to C_0 \otimes C$, $\rho_R := (\mathrm{id} \otimes \pi)\Delta : C \to C \otimes C_0$; I and C_n, $n \geq 0$, are sub-bicomodules of C. Let $P_0 = 0$,

$$P_1 = \{x \in C : \Delta(x) = \rho_L(x) + \rho_R(x)\} = \Delta^{-1}(C_0 \otimes I + I \otimes C_0),$$
$$P_n = \{x \in C : \Delta(x) - \rho_L(x) - \rho_R(x) \in \sum_{1 \leq i \leq n-1} P_i \otimes P_{n-i}\}, \quad n \geq 2.$$

In particular the P_n's are C_0-sub-bicomodules of I, $n \geq 0$; but they are not intrinsic since they depend on the projection π. The following Lemma can be thought as a substitute of Theorem 5.2.

LEMMA 6.9. *(W. Nichols) (i).* $P_n = C_n \cap I$.
(ii). $C_1 = \sum_{\tau,\mu \in \widehat{C}} C_\tau \wedge C_\mu$.
(iii). $C_\tau \wedge C_\tau = C_\tau \oplus P_1^{\tau,\tau}$ and $C_\tau \wedge C_\mu = C_\tau \oplus C_\mu \oplus P_1^{\tau,\mu}$, if $\tau \neq \mu$. □

Assume in what follows that $C = H$ is a finite dimensional Hopf algebra. If $\tau \in \widehat{C}$ and $g \in G(H)$, we denote $C_{\tau^d} := \mathcal{S}(C_\tau)$, $g.C_\tau := C_{g.\tau}$, $C_\tau.g := C_{\tau.g}$; they are simple subcoalgebras of C.

COROLLARY 6.10. [**AN3**]. *(i). For any $g \in G(H)$,*

$$\dim P_1^{\tau,\mu} = \dim P_1^{\mu^d,\tau^d} = \dim P_1^{g.\tau,g.\mu} = \dim P_1^{\tau.g,\mu.g}.$$

(ii). If I is a direct sum of one-dimensional H_0-sub-bicomodules then $H_1 = H_0 + \sum_{g,h \in G(H)} \mathcal{P}_{g,h}(H)$. □

Consider the right action $\leftharpoonup: H^* \otimes H \to H^*$ given by $\alpha \leftharpoonup h = \langle \alpha_1, h \rangle \alpha_2$, $\forall h \in H$, $\alpha \in H^*$. Let $\int \in H^*$ be a non-zero left integral and let $g_0 \in G(H)$ be the distinguished group-like element, so that

$$\alpha \int = \langle \alpha, 1 \rangle \int \quad \text{and} \quad \int \alpha = \langle \alpha, g_0 \rangle \int, \qquad \forall \alpha \in H^*.$$

We shall assume in what follows that H is not cosemisimple, or equivalently, that $\langle \int, 1 \rangle = 0$; in particular $\int^2 = 0$ and if $g \in G(H)$, also $(\int \leftharpoonup g)^2 = \int^2 \leftharpoonup g = 0$. Observe that if $C \neq \mathbf{k}1$ is a simple subcoalgebra of H, and if $c \in C$, then

$$\langle \int, c \rangle 1 = \langle \int, c_2 \rangle c_1 \quad \in C \cap \mathbf{k}1,$$

whence $\int |_C = 0$, i. e., $\int$ belongs to the anihilator of H_0, $H_0^\perp = \operatorname{Jac} H^*$. Let $g \in G(H)$. Since the left (and right) multiplication by g is a coalgebra automorphism of H, it preserves H_0. This implies that also $\int \leftharpoonup g$ belongs to $\operatorname{Jac} H^*$. Also, for all $\alpha \in H^*$, we have

$$\alpha (\int \leftharpoonup g) = \langle \alpha, g^{-1} \rangle \int \leftharpoonup g, \quad \text{and} \quad (\int \leftharpoonup g) \alpha = \langle \alpha, g^{-1} g_0 \rangle \int \leftharpoonup g.$$

Hence $\mathbf{k}(\int \leftharpoonup g)$ is a two-sided ideal of H^* and $\mathbf{k}(\int \leftharpoonup g) \subseteq \operatorname{Jac} H^*$. Moreover, since distinct group-like elements are linearly independent and the map $H \to H^*$, $h \mapsto \int \leftharpoonup h$, is injective, the ideals $\mathbf{k}(\int \leftharpoonup g)$ and $\mathbf{k}(\int \leftharpoonup g')$ are distinct if $g \neq g'$.

LEMMA 6.11. [**AN3**] *Let H be a non-cosemisimple finite dimensional Hopf algebra. Let $L = (\int \leftharpoonup \mathbf{k}G(H))^\perp$, $s = |G(H)|$. Then $L \subseteq H$ is a subcoalgebra of H containing H_0 and there is an H_0-bicomodule decomposition*

$$H = L \oplus \bigoplus_{j=1}^{s} I_j,$$

where I_j are one-dimensional H_0-sub-bicomodules of I, $\forall j = 1, \ldots, s$. □

PROPOSITION 6.12. [**AN3**] *Let H be a non-cosemisimple finite dimensional Hopf algebra.*
(i). If $\dim H - \dim H_0 = |G(H)|$, then $\mathcal{P}_{g,h} \neq \mathbf{k}(g - h)$, for some $g, h \in G(H)$.
(ii). Suppose that $\mathcal{P}_{g,h} = \mathbf{k}(g - h)$, for all $g, h \in G(H)$. Then $\int \leftharpoonup \mathbf{k}G(H) \subseteq (\operatorname{Jac} H^*)^2$. In particular, $|G(H)| \leq \dim H - \dim H_1$.
(iii). If $H_1 = H$ then H has a non-trivial skew primitive element. In particular, if $\operatorname{char} \mathbf{k} = 0$, then $G(H)$ is non-trivial. □

6.2. Relation with quantum $s\ell(2)$. We briefly discuss a very interesting result from [**St3**] and some consequences. In this subsection, $\mathbf{k}$ is algebraically closed and $\operatorname{char} \mathbf{k} = 0$. If q is a root of 1, we denote by $\mathbf{k}_q[SL(2)]$ the "algebra of regular functions on the quantum group $SL(2)$", see for instance [**DCL**].

THEOREM 6.13. [**St3**, Th. 1.5] *Let A be a Hopf algebra containing a simple subcoalgebra C of dimension 4 stable under S. Assume that $1 < \operatorname{ord}(S^2|_C) = n < \infty$. Then there exists a root of unity ω, such that the order of ω^2 is n, and a surjective homomorphism of Hopf algebras $\mathbf{k}_{\sqrt{-\omega}}[SL(2)] \to B := \mathbf{k}<C>$.* □

Here, $\mathbf{k}<C>$ denotes the subalgebra of A generated by C, which is a Hopf subalgebra of A. Notice that the hypothesis on the order of the antipode always holds whenever $\mathbf{k}<C>$ is finite dimensional and non-semisimple.

Let $q := \sqrt{-\omega}$. Let N be the order of q. There is a central inclusion of Hopf algebras $\mathbf{k}[SL(2)] \to \mathbf{k}_q[SL(2)]$, $X_{ij} \mapsto x_{ij}^N$, where X_{ij} are the usual coordinates in $\mathbf{k}[SL(2)]$. The proof of the following Corollary is based on these ideas.

COROLLARY 6.14. [**Na4**] *Let A be a finite dimensional non-semisimple Hopf algebra. Suppose that A is generated by a simple subcoalgebra C of dimension 4 which is stable by the antipode. Then A fits into an extension $1 \to \mathbf{k}^G \to A \to H \to 1$, where G is a finite group and H^* is a pointed non-semisimple Hopf algebra.* □

It would be really interesting to generalize Theorem 6.13, and to understand the reasons behind it.

7. Forms

7.1. Forms of Hopf algebras.

The passage from the algebraically closed case to the general case follows in principle the general guidelines of Galois descent [**Ser2**]. Let $\mathbf{k}$ be a field with algebraic closure $\overline{\mathbf{k}}$ and suppose we have classified all Hopf algebras of certain type over $\overline{\mathbf{k}}$, say all the Hopf algebras of a fixed dimension d. For each Hopf algebra H over $\overline{\mathbf{k}}$ in the obtained list, one seeks to describe all Hopf algebras H_0 over $\mathbf{k}$ such that $H_0 \otimes_{\mathbf{k}} \overline{\mathbf{k}} \simeq H$; briefly, the $\mathbf{k}$-forms of H. [3] This problem splits into two parts: first one needs to show the existence of at least one form, second the set of all forms is described as a non-abelian H^1. The first part, which is trivial in the case of group algebras, is not all evident for general Hopf algebras. The literature on these questions is not very abundant: [**TO**] is concerned with commutative Hopf algebras of prime dimension, [**HaP**], [**P**] treat the case of finite group algebras. Recently, forms of Frobenius-Lusztig kernels were discussed in [**CDB**], over fields where the existence of a form is granted; so that the second question is answered in this case.

QUESTION 7.1. *Let A be a finite dimensional pointed Hopf algebra whose infinitesimal braiding is of Cartan type, say over $\mathbb{C}$. What are the number fields F such that there exist forms of A over F?*

Before discussing some explicit examples, we state some general remarks. Let us fix a subfield F of $\mathbb{C}$ and let H be a Hopf algebra over F. Then the coradical filtration of H is preserved by extension of scalars; namely

$$(H \otimes_F \mathbb{C})_j = H_j \otimes_F \mathbb{C}, \quad j \geq 0.$$

It is enough to show this for $j = 0$, *i. e.* for the coradical. The result for cosemisimple coalgebras is true more generally for separable extensions and follows easily from a well-known analogue for semisimple algebras, see *e. g.* [**CR**, Th. 69.4].

Let us assume that $(H \otimes_F \mathbb{C})_0$ is a Hopf subalgebra of $H \otimes_F \mathbb{C}$, for example that $H \otimes_F \mathbb{C}$ is pointed. Then H_0 is a Hopf subalgebra of H, and we conclude without effort that all the invariants of $H \otimes_F \mathbb{C}$ listed at the end of Subsection 5.3.2 (diagram, rank, infinitesimal braiding, ...) can be obtained from the corresponding invariants of H by extension of scalars.

On the other hand, recall that a semisimple algebra (resp. a cosemisimple coalgebra) over F is split if it is a direct product of matrix algebras over F (resp., a

---

[3]Sometimes the word "form" is applied in a different sense; given a Hopf algebras K and L over $\mathbf{k}$, it is said that K is a form of L if $K \otimes_{\mathbf{k}} \overline{\mathbf{k}} \simeq L \otimes_{\mathbf{k}} \overline{\mathbf{k}}$. For our purposes the present terminology is suitable.

direct sum of matrix coalgebras over F). Let A be a semisimple Hopf algebra over F. We shall use the following terminology:

- A is *a-split* if it is split as an algebra;

- A is *c-split* if it is split as a coalgebra;

- A is *split* if it is split as an algebra and as a coalgebra.

EXAMPLE 7.2. Let $b \in F$, $b \neq 0$ and let
$$H(b) = F[c, s]/\langle c^2 - bs^2 = 1, (c-1)(2c+1) = 0, s(2c+1) = 0 \rangle;$$
this is a semisimple Hopf algebra of dimension 3 with comultipication given by
$$\Delta(c) = c \otimes c + bs \otimes s, \quad \Delta(s) = c \otimes s + s \otimes c.$$
See [**HaP**]. It is a form of $\mathbb{C}(\mathbb{Z}/3)$ and all the forms can be presented in this way. Moreover, $H(b)$ is a-split if and only if b is a square in F. Now, it can be shown that $H(b)$ is c-split if and only if $-3b^{-1}$ is a square in F. Thus a necessary condition to have a split form of $\mathbb{C}(\mathbb{Z}/3)$ in F is that -3 is a square in F (this is also sufficient since -3 is the discriminant of the third cyclotomic polynomial).

EXAMPLE 7.3. Let $T_q(\mathbb{C})$ be a Taft algebra over $\mathbb{C}$ where $q \in \mathbb{C}$ is a primitive root of 1 of order N. Assume that F admits a form $\mathfrak{T}$ of T_q. Then $q \in F$ and $\mathfrak{T} \simeq T_q(F)$ as a Hopf algebra.

Indeed, by what we have said, the infinitesimal braiding (V, c) of $\mathfrak{T}$ has dimension 1. Therefore, there exist $g \in G(\mathfrak{T})$ and a character χ of $\mathfrak{T}$ (a one-dimensional representation) such that $V = V_g^\chi$. But then $q = \chi(g) \in F$; so that both χ and g have order N. The rest follows easily now.

EXAMPLE 7.4. Let $q \in \mathbb{C}$ be a primitive third root of 1. Let $\rho : \mathbb{Z}/3 \to \operatorname{End} V$ be the unique irreducible two-dimensional representation of the group $\mathbb{Z}/3$ over $\mathbb{Q}$. Let $g \in \mathbb{Z}/3$, $g \neq 1$, and define a Yetter-Drinfeld module structure on V by declaring $V = V_g$. Then the Hopf algebra $\mathfrak{B}(V) \# \mathbb{Q}(\mathbb{Z}/3)$ is a non-split form of the Hopf algebra
$$\mathbb{C}\langle g, x, y | g^3 = 1, gxg^{-1} = qx, gyg^{-1} = q^{-1}y, x^3 = 0, y^3 = 0, xy = q^{-1}y \rangle,$$
with comultiplication $\Delta(g) = g \otimes g$, $\Delta(x) = x \otimes 1 + g \otimes x$, $\Delta(y) = y \otimes 1 + g \otimes y$.

7.2. Finite compact quantum groups. There is another version of the notion of "form", with a more "quantum" flavor since it has origins in the theory of operator algebras. We shall work over $\mathbb{C}$, regarding $\mathbb{R}$-forms; the first general definitions and notions can be adapted *mutatis mutandis* to quadratic extensions. I do not know a formulation for more general field extensions, if there is any.

DEFINITION 7.5. A $*$-Hopf algebra is a pair $(H, *)$, where H is a Hopf algebra and $* : H \to H$ is a conjugate-linear map (that is, $(\lambda x)^* = \overline{\lambda} x^*$ for $\lambda \in \mathbb{C}$, $x \in H$) such that:

- $(xy)^* = y^* x^*$, i. e. $*$ is an anti-algebra map;

- $\Delta(x^*) = \sum_{(x)} x_{(1)}^* \otimes x_{(2)}^*$, i. e. $*$ is a coalgebra map.

It follows easily that $(\mathcal{S}*)^2 = \operatorname{id}$. Also, one could consider "$\circ$-Hopf algebra", where $\circ : H \to H$ is a conjugate-linear, algebra and anti-coalgebra, map. However, the two notions are equivalent by the relation $x^\circ = (\mathcal{S}(x))^*$.

For simplicity, assume now that $(H, *)$ is a finite dimensional $*$-Hopf algebra. The transpose of the $*$-operation is a $\circ$-operation in the dual Hopf algebra H^*; by the preceding H^* is a $*$-Hopf algebra in a natural way.

Let $\int : H \to \mathbb{C}$ be a right integral and let $(\,|\,) : H \times H \to \mathbb{C}$ be the sesquilinear form given by
$$(x|y) := \langle \int, y^*x \rangle.$$
It is easy to see that
$$(xy|z) = (y|x^*z), \qquad (p \rightharpoonup x|y) = (x|p^* \rightharpoonup y), \qquad x, y, z \in H, p \in H^*.$$
Here $\rightharpoonup$ is the action of H^* on H transpose to right multiplication. Assume further that H is semisimple; then we normalize $\int$ by $\langle \int, 1 \rangle = 1$. The corresponding $(\,|\,)$ is hermitian, since $\int$ is also a left integral.

DEFINITION 7.6. A *finite compact quantum group* is a $*$-Hopf algebra $(H, *)$ with H semisimple and $\int$ normalized by $\langle \int, 1 \rangle = 1$, such that $(\,|\,)$ is positive definite. We say that $*$ is a *compact involution*.

This definition is equivalent to the original definition of Woronowicz [**Wo1**], as shown in [**An2**, 2.4]. Finite compact quantum groups are equivalent to finite dimensional Kac algebras [**Ka1, Ka2, KaP**].

THEOREM 7.7. [**An2**, 2.4]. *Let $(H, *)$ be a finite compact quantum group and let $\# : H \to H$ be another structure of $*$-Hopf algebra. Then there exists a Hopf algebra automorphism $T : H \to H$ such that $\#$ and $T * T^{-1}$ commute.*

If $\#$ is another compact involution, then there exists a Hopf algebra automorphism $T : H \to H$ such that $\# = T * T^{-1}$. □

The proof of Theorem 7.7 is inspired by an analogous proof for semisimple Lie algebras due to Mostow; it actually applies also in the infinite dimensional case [**An2**]. In presence of this Theorem, the classification of finite compact quantum groups is equivalent to the following problem:

QUESTION 7.8. Given a semisimple Hopf algebra H, does it admit a compact involution?

8. Appendix. Questions

For convenience of the reader, we collect here the Questions discussed in the text.

QUESTION 2.1. Jordan-Hölder theorem for finite dimensional Hopf algebras.

QUESTION 2.3. Are all simple semisimple Hopf algebras twistings of group algebras of simple groups, or their duals?

QUESTION 2.4. Classify simple semisimple Hopf algebras of dimension 60.

QUESTION 2.6. Consider the subcategory of the category of all semisimple Hopf algebras which contains the group algebras and is closed by taking duals, by extensions and by twistings. Is this subcategory equal to the category of all semisimple Hopf algebras?

QUESTION 2.13. Find a semisimple Hopf algebra which is a bosonization but not an extension.

CONJECTURE 3.15. [**Kas1**]. If H is a semisimple and cosemisimple Hopf algebra over **k**, then $\exp H$ divides $\dim H$.

CONJECTURE 4.11. (Kaplansky). If H is a (semisimple) Hopf algebra over the algebraically closed field **k**, then the sizes of the matrices occuring in any full matrix constituent of H divide the dimension of H.

QUESTION 4.7. (Kaplansky). Classify all Hopf algebras of dimension p, in characteristic $q \leq p$.

QUESTION 4.17. (S. Montgomery). What is the analogue of Burnside's $p^a q^b$-Theorem for semisimple Hopf algebras?

QUESTION 4.19. (S. Montgomery). Classify all semisimple Hopf algebras H such that $\dim H < 60$; open for dimensions 24, 30, 32, 36, 40, 42, 48, 54, 56.

QUESTION 5.9. Given a braided vector space of diagonal type V, such that the entries of its matrix are roots of 1, compute the dimension of the associated Nichols algebra $\mathfrak{B}(V)$. If finite, give a nice presentation of $\mathfrak{B}(V)$.

QUESTION 5.28. Compute liftings of the quantum Serre relations for vertices in the same connected component, in the low order cases.

QUESTION 5.31. If $R = \oplus_{n \geq 0} R(n)$ is a graded braided Hopf algebra with $R(0) = \mathbf{k}1$ and $P(R) = R(1)$, is R generated in degree one, i. e. is R a Nichols algebra?

QUESTION 5.32. If $S = \oplus_{n \geq 0} S(n)$ is a graded braided Hopf algebra with $S(0) = \mathbf{k}1$ which is generated in degree one, is $P(S) = S(1)$, i. e. is S a Nichols algebra?

CONJECTURE 5.35. [**AS3**]. Any pointed finite dimensional Hopf algebra over an algebraically closed field **k** of characteristic 0 is generated by group-like and skew-primitive elements.

QUESTION 5.40. Same as 5.9 but in the particular case of rank 2.

QUESTION 5.53. Given $(\mathbf{k}X, c^f)$, for $(X, \triangleright)$ a crossed set, and $f \in H^2(X, \mathbb{G}_\infty)$ compute the dimension of the associated $\mathfrak{B}(\mathbf{k}X)$. If finite, give a nice presentation of $\mathfrak{B}(V)$.

QUESTION 5.57. [**MlS**]. Determine all finite groups Γ having a link-indecomposable Yetter-Drinfeld module V such that $\mathfrak{B}(V)$ is finite dimensional.

QUESTION 5.59. The same for semisimple Hopf algebras.

QUESTION 6.1. Classify all Hopf algebras of dimension $d \leq 100$.

QUESTION 6.2. Classify all Hopf algebras of dimension d, where d factorizes in a simple way.

QUESTION 6.5. Let N be a square-free positive integer, not divisible by char **k**. Are all Hopf algebras of dimension N, necessarily semisimple?

QUESTION 6.8. Is there any *simple* Hopf algebra H which is neither semisimple, neither cosemisimple, neither pointed, nor H^* is pointed?

QUESTION 7.1. Let A be a finite dimensional pointed Hopf algebra whose infinitesimal braiding is of Cartan type, say over $\mathbb{C}$. What are the minimal number fields F such that there exist forms of A over F?

QUESTION 7.8. Given a semisimple Hopf algebra H, does it admit a compact involution?

References

[AJS] H. H. Andersen, J. Jantzen and W. Sörgel, *Representations of quantum groups at a p-th root of unity and of semisimple groups in characteristic p: Independence of p*, Astérisque **220**, pp. 1994.
[An1] N. Andruskiewitsch, *Notes on extensions of Hopf algebras*, Canad. J. Math. **48** (1996), pp. 3–42.
[An2] _____, *Compact involutions on semisimple quantum groups*, Czechoslovak J. Phys. **44** (1994), pp. 963–972.
[AnDa] N. Andruskiewitsch and S. Dăscălescu, *On quantum groups at* -1, Algebr. Represent. Theory, to appear.
[AnDe] N. Andruskiewitsch and J. Devoto, *Extensions of Hopf algebras*, Algebra i Analiz **7** (1995), pp. 22–61; also in St. Petersburg Math. J. **7** (1996), pp. 17–52.
[AnG] N. Andruskiewitsch and M. Graña, *Braided Hopf algebras over non abelian groups*, Bol. Acad. Ciencias (Córdoba) **63** (1999), pp. 45-78.
[AN1] N. Andruskiewitsch and S. Natale, *Plancherel identity for semisimple Hopf algebras*, Comm. Algebra **25** (1997), pp. 3239-3254.
[AN2] _____, *Examples of self-dual Hopf algebras*, J. Math. Sci. Univ. Tokyo **6** (1999), pp. 181-215.
[AN3] _____, *Counting arguments for low dimensional Hopf algebras*, Tsukuba Math. J. **25**, pp. 187-201 (2001).
[AS0] N. Andruskiewitsch and H-J. Schneider, *Appendix to* [**An1**].
[AS1] _____, *Hopf algebras of order p^2 and braided Hopf algebras of order p*, J. Algebra, **199** (1998), pp. 430–454.
[AS2] _____, *Lifting of quantum linear spaces and pointed Hopf algebras of order p^3*, J. Algebra **209** (1998), pp. 659–691.
[AS3] _____ *Finite quantum groups and Cartan matrices*, Adv. Math. **154** (2000), 1–45.
[AS4] _____, *Lifting of Nichols algebras of type A_2 and Pointed Hopf Algebras of order p^4*, in "Hopf algebras and quantum groups", eds. S. Caeneppel and F. van Oystaeyen, M. Dekker, pp. 1–16.
[AS5] _____, *Finite quantum groups over abelian groups of prime exponent*, Ann. Sci. Ec. Norm. Super., to appear.
[AS6] _____, *Pointed Hopf Algebras*, in "New directions in Hopf algebras", MSRI series, Cambridge Univ. Press, to appear.
[AS7] _____, *A characterization of quantum groups*, in preparation.
[B] M. Beattie, *An isomorphism theorem for Ore extension Hopf algebras*, Comm. Algebra **28** (2000), pp. 569–584.
[BDG] M. Beattie, S. Dăscălescu, and L. Grünenfelder, *On the number of types of finite-dimensional Hopf algebras*, Inventiones Math. **136** (1999), pp. 1-7.
[BD] Yu. Bespalov and B. Drabant, *Hopf (bi-)modules and crossed modules in braided monoidal categories*, J. Pure Appl Alg. **123** (1998), pp. 105–129.
[Bi] J. Bichon, *Quelques nouvelles déformations du groupe symétrique*, C. R. Acad. Sci. Paris **330** (2000), pp. 761–764.
[BM] R. J. Blattner and S. Montgomery, *Crossed products and Galois extensions of Hopf algebras*, Pacific J. Math. **137** (1989), pp. 37–54.
[Bou] N. Bourbaki, *Groupes et algèbres de Lie*, Hermann, Paris (1968).
[By1] N. P. Byott, *Cleft extensions of Hopf algebras*, J. Algebra **157** (1993), pp. 405–429.
[By2] _____, *Cleft extensions of Hopf algebras, II*, Proc. London Math. Soc. (3) **67** (1993), pp. 277–304.
[CD1] S. Caenepeel and S. Dăscălescu, *Pointed Hopf algebras of dimension p^3*, J. Algebra **209** (1998), pp. 622–634.
[CD2] _____, *On pointed Hopf algebras of dimension 2^n*, Bull. London Math. Soc. **31** (1999), pp. 17–24.
[CDB] S. Caenepeel, S. Dăscălescu and L. Le Bruyn, *Forms of pointed Hopf algebras*, Manuscripta Math. **100** (1999), pp. 35–53.

[CDR] S. Caenepeel, S. Dăscălescu and S. Raianu, *Classifying Pointed Hopf algebras of dimension 16*, Comm. Algebra **28** (2000), pp. 541–568.

[Ca] P. Cartier, *Groupes algebriques et groupes formels* in "Colloq. Théorie des Groupes Algébriques (Bruxelles, 1962)", pp. 87–111, Librairie Universitaire, Louvain; Gauthier Villars, Paris.

[Ci] C. Cibils, *Tensor products of Hopf bimodules over a group algebra*, Proc. A.M.S. **125** (1997), pp. 1315–1321.

[Co] M. Cohen, *On characters of Hopf algebras*, in "New Trends in Hopf Algebra Theory"; N. Andruskiewitsch, W. R. Ferrer Santos and H.-J. Schneider (eds.), Contemp. Math. **267** (2000), pp. 55–66.

[CoZ] M. Cohen and S. Zhu, *Invariants of the adjoint action and Yetter-Drinfeld categories*, J. Pure and Applied Algebra 159 (2001), pp. 149-171.

[CR] C. Curtis and I. Reiner, *Representation theory of finite groups and associative algebras*, Wiley, New York (1962).

[D] S. Dăscălescu, *Pointed Hopf algebras with large coradical*, Comm. Algebra **27** (1999), pp. 4821–4826.

[DaNR] S. Dăscălescu, C. Năstăsescu and Ş. Raianu, *Hopf algebras: an introduction*, Marcel Dekker, 2000.

[DaNT] S. Dăscălescu, C. Nastasescu and B. Torrecillas, *Co-Frobenius Hopf algebras: Integrals, Doi-Koppinen Modules and Injective objects*, J. Algebra **220** (1999), pp. 542–560.

[DCL] C. De Concini, V. Lyubashenko, Quantum function algebra at roots of 1, Adv. Math. **108**, 205–262 (1994).

[DiPR] R. Dijkgraaf, V. Pasquier and P. Roche, *Quasi Hopf algebras, group cohomology and orbifold models*, Nuclear Phys. B Proc. Suppl. **18B** (1991), pp. 60–72.

[DoT] Y. Doi and M. Takeuchi, *Multiplication alteration by two-cocycles. The quantum version*, Comm. Algebra **22**, No.14, (1994), 5715-5732.

[Dr1] V. Drinfeld, *Quantum groups*, Proceedings of the International Congress of Mathematicians, (Berkeley, Calif., 1986), 798–820, Amer. Math. Soc., Providence, RI, 1987.

[Dr2] ———, *On almost cocommutative Hopf algebras*, Leningrad Math. J. **1** (1990), pp. 321–342.

[Dr3] ———, *Quasi-Hopf algebras*, Leningrad Math. J. **1** (1990), pp. 1419–1457.

[EG1] P. Etingof and S. Gelaki, *Some properties of finite-dimensional semisimple Hopf algebras*, Math. Res. Lett. **5** (1998), pp. 191–197.

[EG2] ———, *A method of construction of finite-dimensional triangular semisimple Hopf algebras*, Math. Res. Lett. **5** (1998), pp. 551–561.

[EG3] ———, *On finite-dimensional semisimple and cosemisimple Hopf algebras in positive characteristic*, Internat. Math. Res. Notices **1998**, no. 16, pp. 851–864.

[EG4] ———, *Semisimple Hopf algebras of dimension pq are trivial*, J. Algebra **210** (1998), pp. 664-669.

[EG5] ———, *On the exponent of finite-dimensional Hopf algebras*, Math. Res. Lett. **6** (1999), pp. 131–140.

[EG6] ———, *The representation theory of cotriangular semisimple Hopf algebras*, Internat. Math. Res. Notices **1999**, no. 7, pp. 387–394.

[EG7] ———, *The classification of triangular semisimple and cosemisimple Hopf algebras over an algebraically closed field*, Internat. Math. Res. Notices **2000**, no. 5, pp. 223–234.

[EV] M. Enock and L. Vainerman, *Deformation of a Kac algebra by an abelian subgroup*, Commun. Math. Phys. **178** (1996), pp. 571-596.

[FK] S. Fomin and K. N. Kirillov, *Quadratic algebras, Dunkl elements, and Schubert calculus*, Progr. Math. **172**, Birkhauser, (1999), pp. 146–182.

[F] N. Fukuda, *Semisimple Hopf algebras of dimension 12*, Tsukuba J. Math. **21** (1997), pp. 43–54.

[FV] R. Farnsteiner and D. Voigt, *Modules of solvable infinitesimal groups and the structure of representation-finite cocommutative Hopf algebras*, Math. Proc. Cambridge Philos. Soc. **127** (1999), pp. 441–459.

[Ga] P. Gabriel, *Étude infinitésimale des schémas en groupe et groupes formels* and *Groupes formels*,in Schémas en groupes. I: Propriétés générales des schémas en groupes. (French)

Séminaire de Géométrie Algébrique du Bois Marie 1962/64 (SGA 3). Dirigé par M. Demazure et A. Grothendieck. Lecture Notes in Mathematics **151** (1970), Springer-Verlag, Berlin-New York.

[Ge1] S. Gelaki, *Quantum Groups of Dimension pq^2*, Israel J. Math. **102** (1997), pp. 227-267.

[Ge2] _____, *On pointed Hopf algebras and Kaplansky's tenth conjecture*, J. Algebra **209** (1998), pp. 635-657.

[Ge3] _____, *Some properties and examples of triangular pointed Hopf algebras*, Math. Res. Lett. **6** (1999), pp. 563-572.

[GeW] S. Gelaki and S. Westreich, *On semisimple Hopf algebras of dimension pq*, Proc. Amer. Math. Soc. **128** (2000), pp. 39-47.

[Gñ1] M. Graña, *Pointed Hopf algebras of dimension 32*, Comm. Algebra **28** (2000), pp. 2935-2976.

[Gñ2] _____, *On Pointed Hopf algebras of dimension p^5*, Glasgow Math. J. **42** (2000), 405-419.

[Gñ3] _____, *A freeness theorem for Nichols algebras*, J. Algebra **231** (2000), pp. 235-257.

[Gñ4] _____, *On Nichols algebras of low dimension*, in "New Trends in Hopf Algebra Theory"; N. Andruskiewitsch, W. R. Ferrer Santos and H.-J. Schneider (eds.), Contemp. Math. **267** (2000), pp. 111-134.

[HaP] R. Haggenmüller and B. Pareigis, *Hopf algebra forms of the multiplicative group and other groups*, Manuscripta Math. **55** (1986), pp. 121-136.

[Hof] C. Hoffman, *On some examples of simple quantum groups*, Comm. Algebra **28** (2000), pp. 1867-1873.

[Ho] I. Hofstetter, *Extensions of Hopf algebras and their cohomological description*, J. Algebra **164** (1994), pp. 264-298.

[HS] J. H. Hong and W. Szymański, *Composition of subfactors and twisted bicrossed products*, J. Operator Theory **37** (1997), no. 2, 281-302.

[IK1] M. Izumi and H. Kosaki, *Finite-dimensional Kac algebras arising from certain group actions on a factor*, Internat. Math. Res. Notices **1996**, no. 8, 357-370.

[IK2] _____, *Kac algebras arising from compositions of subfactors: general theory and classification*, Mem. Amer. Math. Soc., to appear.

[J] M. Jimbo, *A q-difference analogue of $U(\mathfrak{g})$ and the Yang-Baxter equation*, Lett. Math. Phys. **10** (1985), pp. 63-69.

[K] V. Kac, *Infinite dimensional Lie algebras*, Cambridge Univ. Press (1995), Third edition.

[Ka1] G. I. Kac, *Extensions of groups to ring groups*, Math. USSR Sbornik **5** (1968), pp. 451-474.

[Ka2] _____, *Certain arithmetic properties of ring groups*, Funct. Anal. Appl. **6** (1972), pp. 158-160.

[KaP] G. I. Kac and V. G. Paljutkin, *Finite group rings*, Trans. Moscow. Math. Soc. **15** (1966), pp. 251-294.

[Kas1] Y. Kashina, *On the order of the antipode of Hopf algebras in $^H_H\mathcal{YD}$*, Comm. Algebra **27** (1999), pp. 1261-1273.

[Kas2] _____, *Classification of Semisimple Hopf Algebras of Dimension 16*, J. Algebra, **232** (2000), pp. 617-663.

[KMM] Y. Kashina, G. Mason and S. Montgomery, *Computing the Frobenius-Schur indicator for abelian extensions of Hopf algebras*, preprint (2001), `math 0108131`.

[KaSZ] Y. Kashina, Y. Sommerhäuser and Y. Zhu, *Self-dual modules of semisimple Hopf algebras*, preprint (2001), `math 0106254`.

[KM] T. Kobayashi and A. Masuoka, *A result extended from groups to Hopf algebras*, Tsukuba J. Math. **21** (1997), pp. 55-58.

[Ko] B. Kostant, *Graded manifolds, graded Lie theory, and prequantization*. Lect. Notes Math. **570** (1975), pp. 177-306.

[La] R. Larson, *Characters of Hopf algebras*, J. Algebra, **17** (1971), pp. 352-368.

[LaR1] R. Larson and D. Radford, *Finite dimensional cosemisimple Hopf algebras in characteristic 0 are semisimple*, J. Algebra, **117** (1988), pp. 267-289.

[LaR2] _____, *Semisimple Cosemisimple Hopf Algebras*, Amer. J. of Math. **110** (1988), pp. 187-195.

[LaS] R. Larson and M. Sweedler, *An associative orthogonal bilinear form for Hopf algebras*, Amer. J. of Math. **91** (1969), pp. 75-94.

[LiM]   V. Linchenko and S. Montgomery, *A Frobenius-Schur theorem for Hopf algebras*, Algebr. Represent. Theory **3** (2000), pp. 347–355.
[Lo1]   M. Lorenz, *On the class equation for Hopf algebras*, Proc. Amer. Math. Soc. **126** (1998), pp. 2841–2844.
[Lo2]   _____, *Representations of finite-dimensional Hopf algebras*, J. Algebra **188** (1997), pp. 476-505.
[LuYZ]   Jiang-Hua Lu, Min Yan & Yong-Chang Zhu, *On Set-theoretical Yang–Baxter equation*, Duke Math. J. **104** (2000), pp 1-18.
[Lu1]   G. Lusztig, *Finite dimensional Hopf algebras arising from quantized universal enveloping algebras*, J. of Amer. Math. Soc. **3**, pp. 257–296.
[Lu2]   _____, *Quantum groups at roots of 1*, Geom. Dedicata **35** (1990), pp. 89–114.
[Lu3]   _____, *Introduction to quantum groups*, Birkhäuser, 1993.
[Ma1]   S. Majid, *More examples of bicrossproduct and double cross product Hopf algebras*, Israel J. Math. **72** (1990), pp. 133–148.
[Ma2]   _____, *Crossed products by braided groups and bosonization*, J.Algebra **163** (1994), pp. 165–190.
[Ma3]   _____, *Foundations of Quantum Group Theory*. Cambridge Univ. Press, 1995.
[MaS]   S. Majid and Y. Soibelman, *Bicrossproduct structure of the quantum Weyl group*, J.Algebra **163** (1994), pp. 68–87.
[Mk1]   A. Masuoka, *Quotient theory of Hopf algebras*, in "Advances in Hopf algebras", Lect. Notes in Pure and Appl. Math. **158** (1994), ed. J. Bergen and S. Montgomery, pp. 107–133.
[Mk2]   _____, *Self dual Hopf algebras of dimension p^3 obtained by extension*, J. Algebra **178** (1995), pp. 791–806.
[Mk3]   _____, *The p^n-th Theorem for Hopf algebras*, Proc. Amer. Math. Soc. **124** (1996), pp. 187–195.
[Mk4]   _____, *Semisimple Hopf algebras of dimension 2p*, Comm. Algebra **23** (1995), pp. 1931–1940.
[Mk5]   _____, *Some Further Classification Results on Semisimple Hopf algebras*, Comm. Algebra **21** (1996), pp. 307–329.
[Mk6]   _____, *Calculations of some groups of Hopf algebra extensions*, J. Algebra **191** (1997), pp. 568–588; *Corrigendum*, J. Algebra **197** (1997), pp. 656.
[Mk7]   _____, *Extensions of Hopf algebras and Lie bialgebras*, Trans. Amer. Math. Soc. **352** (2000), no. 8, pp. 3837–3879.
[Mk8]   _____, *Extensions of Hopf algebras*, (1999), Trab. Mat. 41/99 (FaMAF).
[Mk9]   _____, *Defending the negated Kaplansky conjecture*, Proc. Amer. Math. Soc. **129** (2001), 3185–3192.
[Mk10]   _____, *Cocycle deformations and Galois objects for some cosemisimple Hopf algebras of finite dimension*, in "New Trends in Hopf Algebra Theory"; N. Andruskiewitsch, W. R. Ferrer Santos and H.-J. Schneider (eds.), Contemp. Math. **267** (2000), pp. 195–214.
[MlS]   A. Milinski and H.-J. Schneider, *Pointed Indecomposable Hopf Algebras over Coxeter Groups*, in "New Trends in Hopf Algebra Theory"; N. Andruskiewitsch, W. R. Ferrer Santos and H.-J. Schneider (eds.), Contemp. Math. **267** (2000), pp. 215–236.
[MM]   J. Milnor and J.C. Moore, *On the structure of Hopf algebras*, Ann. of Math. **81** (1965), pp. 211–264.
[Mo1]   S. Montgomery, *Hopf Algebras and their Actions on Rings*, AMS (1993), CMBS **82**.
[Mo2]   _____, *Classifying finite dimensional semisimple Hopf algebras*, Contemp. Math **229** (1998), AMS, pp. 265–279.
[MonW]   S. Montgomery and S. Witherspoon, *Irreducible representations of Crossed Products*, J. Pure Appl. Alg. **129** (1998), pp. 315–326.
[Mov]   M. Movshev, *Twisting in group algebras of finite groups*, Funct. Anal. Appl. **27** (1994), pp. 240–244.
[Mul1]   E. Müller, *Some topics on Frobenius-Lusztig kernels, I*, J. Algebra **206** (1998), pp. 624–658.
[Mul2]   _____, *Finite subgroups of the quantum general linear group*, Proc. London Math. Soc. (3) **81** (2000), no. 1, pp. 190–210.
[Mus]   I. Musson, *Finite Quantum Groups and Pointed Hopf Algebras*, preprint (1999).

[Na1] S. Natale, *On semisimple Hopf algebras of dimension pq^2*, J. Algebra **221** (1999), pp. 242–278.
[Na2] _____, *On semisimple Hopf algebras of dimension pq^2, II*, Algebr. Represent. Theory **5**, 277–291 (2001).
[Na3] _____, *Hopf algebras of dimension 12*, Algebr. Represent. Theory, to appear.
[Na4] _____, *Quasitriangular Hopf algebras of dimension pq*, Bull. London Math. Soc., to appear.
[Na5] _____, *On semisimple Hopf algebras of dimension pq^r*, Algebr. Represent. Theory, to appear.
[Ni] W.D. Nichols, *Bialgebras of type one*, Commun. Alg. **6** (1978), pp. 1521–1552.
[NiR] W.D. Nichols and M. B. Richmond, *The Grothendieck group of a Hopf algebra*, J. Pure Appl. Alg. **106** (1996), pp. 297–306.
[NiZ] W.D. Nichols and M. B. Zöller, *A Hopf algebra freeness Theorem*, Amer. J. Math. **111** (1989), pp. 381–385.
[Nk] D. Nikshych, *K_0-rings and twisting of finite dimensional semisimple Hopf algebras*, Commun. Alg.**26** (1998), pp. 321–342; *Corrigendum* Commun. Alg. **26** (1998), p. 2019.
[OS1] U. Oberst and H.-J. Schneider, *Über Untergruppen endlicher algebraischer Gruppen*, Manuscripta Math. **8** (1973), pp. 217–241.
[OS2] _____, *Untergruppen formeller Gruppen von endlichen Index*, J. Algebra **31** (1974), pp. 10–44.
[P] B. Pareigis, *Forms of Hopf algebras and Galois theory*, in Topics in algebra, Part 1 (Warsaw, 1988), Banach Center Publ. **26** (1990), pp. 75–93, Part 1, PWN, Warsaw, 1990.
[R1] D. Radford, *The order of the antipode of a finite dimensional Hopf algebra is finite*, Amer. J. of Math. **98** (1976), pp. 333–355.
[R2] _____, *Hopf algebras with projection*, J. Algebra **92** (1985), pp. 322–347.
[R3] _____, *On Kauffman's knot invariants arising from finite dimensional Hopf algebras*, in "Advances in Hopf algebras", Lect. Notes in Pure and Appl. Math. **158** (1994), ed. J. Bergen and S. Montgomery, pp. 205–266.
[R4] _____, *Minimal Quasitriangular Hopf algebras*, J. Algebra **157** (1993), pp. 285–315.
[Re] N. Reshetikhin, *Multiparameter quantum groups and twisted quasitriangular Hopf algebras*, Lett. Math. Phys. **20**, (1990), pp. 331–335.
[ReS] N. Reshetikhin and M. Semenov-Tian-Shansky, *Quantum R-matrices and factorization problems*, J. Geom. Phys. **5**, (1988), pp. 533–550.
[Ri] C. Ringel, *PBW-bases of quantum groups*, J. Reine Angew. Math. **470** (1996), pp. 51–88.
[Ro1] M. Rosso, *Groupes quantiques et algebres de battage quantiques*, C.R.A.S. (Paris) **320** (1995), pp. 145–148
[Ro2] _____, *Quantum groups and quantum shuffles*, Inventiones Math. **133** (1998), pp. 399–416.
[Rz] J. Rozanski, *Braided antisymmetrizer as bialgebras homomorphism*, Reports on Math. Phys. **38** (1996), pp. 273–277.
[Scf] B. Scharfschwerdt, *The Nichols-Zöller theorem for braided Hopf algebras*, Comm. Algebra, to appear.
[Scg1] P. Schauenburg, *Hopf bimodules and Yetter-Drinfeld modules*, J. Algebra **169** (1994), pp. 874–890.
[Scg2] _____, *A characterization of the Borel-like subalgebras of quantum enveloping algebras*, Comm. Algebra **24** (1996), pp. 2811–2823.
[Scg3] _____, *Faithful flatness over Hopf subalgebras - counterexamples*, in "Interactions between ring theory and representations of algebras", ed. F. van Oystaeyen and M. Saorin, Lect. Notes Pure Appl. Math., vol. 210, Dekker, New York, 2000.
[Sch1] H.-J. Schneider, *Normal basis and transitivity for crossed products of Hopf algebras*, J. Algebra **151** (1992), pp. 289–312.
[Sch2] _____, *Some remarks on exact sequences of quantum groups*, Comm. Algebra **21** (1993), pp. 3337–3358.
[Sch3] _____, *Lectures on Hopf algebras*, (1995), Trab. Mat. 31/95 (FaMAF).
[Sch4] _____, *Finiteness results for semisimple Hopf algebras*, Lectures at the Universidad de Córdoba (1996).
[Sch5] _____, *Some properties of factorizable Hopf algebras*, Proc. Amer. Math. Soc. **129** (2001), 1891-1898.

[Sek]   Y. Sekine, *An example of finite-dimensional Kac algebras of Kac-Paljutkin type*, Proc. Amer. Math. Soc. **124** (1996), no. 4, 1139–1147.

[Ser1] J.-P. Serre, *Représentations linéaires des groupes finis* (1971), Hermann, Paris, Deuxiémé édition, refondue.

[Ser2] _____, *Cohomologie Galoisienne* (1994), Springer, Berlin, Cinquième édition, révisée et complétée, Lecture Notes in Maths. **5**.

[Ser3] _____, *Gèbres*, Enseign. Math. (2) **39** (1993), pp. 33–85.

[Si]   W. Singer, *Extension theory for connected Hopf algebras*, J. Algebra **21** (1972), pp. 1–16.

[So]   A. Soloviev, *Non-unitary set-theoretical solutions to the quantum Yang–Baxter equation*, q-alg 0003194.

[Sr1]   Y. Sommerhäuser, *Yetter-Drinfeld Hopf algebras over groups of prime order*, Preprint gk-mp-9905/59, München (1999). Lecture Notes in Math., Springer-Verlag, to appear.

[Sr2]   _____, *On Kaplansky's conjectures*, in "Interactions between ring theory and representations of algebras", ed. F. van Oystaeyen and M. Saorin, Lect. Notes Pure Appl. Math., vol. 210, Dekker, New York, 2000, 393-412.

[Sr3]   _____, *Ribbon transformations, Integrals, and Triangular Decompositions*, preprint gk-mp-9707/52, J. Algebra, to appear.

[St1]   D. Ştefan, *Hopf subalgebras of pointed Hopf algebras and applications*, Proc. Amer. Math. Soc. **125** (1997), pp. 3191–3193.

[St2]   _____, *The set of types of n-dimensional semisimple and cosemisimple Hopf algebras is finite*, J. Algebra **193** (1997), pp. 571–580.

[St3]   _____, *Hopf algebras of low dimension*, J. Algebra **211** (1999), pp. 343–361.

[StvO] D. Ştefan and F. van Oystaeyen, *Hochschild cohomology and coradical filtration of pointed Hopf algebras*, J. Algebra **210** (1998), 535–556.

[Sul]   J. Sullivan, *The uniqueness of integrals for Hopf algebras and some existence theorems of integrals for commutative Hopf algebra*, J. Algebra **19** (1971), pp. 426-440.

[Sw]   M.E. Sweedler, *Hopf algebras*, Benjamin, New York, 1969.

[TaW]   E. Taft and R. L. Wilson, *On antipodes in pointed Hopf algebras*, J. Algebra **29** (1974), pp. 27–32.

[Tk1]   M. Takeuchi, *The Nichols-Zöller theorem for braided Hopf algebras*, preprint (1999).

[Tk2]   _____, *Survey on braided Hopf algebras*, in "New Trends in Hopf Algebra Theory"; N. Andruskiewitsch, W. R. Ferrer Santos and H.-J. Schneider (eds.), Contemp. Math. **267** (2000), pp. 301–324.

[TO]   J. Tate and F. Oort, *Group schemes of prime order*, Ann. Sci. École Norm. Sup. (4) **3** (1970), pp. 1–21.

[TZ]   Y. Tsang and Y. Zhu, *On the Drinfeld double of a Hopf algebra*, Preprint (1998).

[V]   L. Vainerman, *D2-cocycles and twisting of Kac algebras*, Commun. Math. Phys. **191** (1998), pp. 697–721.

[Wl]   R. Williams, *Finite dimensional Hopf algebras*, Ph. D. Thesis, Florida State University (1988).

[Wi1]   S. J. Witherspoon, *The representation ring of the quantum double of a finite group*, J. Algebra **179** (1996), pp. 305–329.

[Wi2]   _____, *The representation ring and the centre of a Hopf algebra*, Canad. J. Math. **51** (1999), pp. 881–896.

[Wo1]   S. L. Woronowicz, *Compact matrix pseudogroups*, Comm. Math. Phys. **111** (1987), pp. 613–665.

[Wo2]   _____, *Differential calculus on compact matrix pseudogroups (quantum groups)*, Comm. Math. Phys. **122** (1989), pp. 125–170.

[ZS]   S. Zhu, *On finite dimensional Hopf algebras*, Comm. Algebra **21** (1993), pp. 3871–3885.

[Z1]   Y. Zhu, *Hopf algebras of prime dimension*, Internat. Math. Res. Notes **1** (1994), pp. 53–59.

[Z2]   _____, *A commuting pair in Hopf algebras*, Proc. Amer. Math. Soc. **125** (1997), pp. 2847-2851.

FaMAF. Medina Allende y Haya de la Torre. Universidad Nacional de Córdoba. (5000) Ciudad Universitaria. Córdoba. Argentina

E-mail address: andrus@mate.uncor.edu

URL: http://www.mate.uncor.edu/andrus

Lectures on differentials, generalized differentials and on some examples related to theoretical physics

Michel Dubois-Violette

ABSTRACT. These notes contain a survey of some aspects of the theory of differential modules and complexes as well as of their generalization, that is, the theory of N-differential modules and N-complexes. Several applications and examples coming from physics are discussed. The commun feature of these physical applications is that they deal with the theory of constrained or gauge systems. In particular different aspects of the BRS methods are explained and a detailed account of the N-complexes arising in the theory of higher spin gauge fields is given.

1. Introduction

Differential algebraic and (co)homological methods have rapidly sprung up in theoretical physics in connection with the development of gauge theories. Their interventions occur at two levels, firstly at a classical level under a more systematic use of the calculus of differential forms, secondly under the emergence of the BRS methods in connection with the quantization of gauge theories. In fact the BRS technique provides an explicitely local and relativistic invariant way to develop perturbation theory for quantum gauge theories [2], [3]. It is worth noticing here that one cannot overestimate the role of the locality principle in perturbative renormalization [24]. Independently of these perturbative developments, methods for quantizing constrained systems on phase space have been developed using the path integral [28] which were obviously related. In both cases enter "ghosts" [25] and the occurrence of a differential, i.e. an endomorphism of square zero. It turns out that the latter construction essentially reduces to a "homological" description of classical constrained systems [37] in which the ghosts and the differential have a natural interpretation in terms of standard mathematical concepts [48], [52], [56], [12].

Here we shall not give a systematic exposition of the above topics but, instead, we shall follow a sort of transversal way. These notes give a survey of appropriate concepts and results in homology which will be illustrated at each level with examples of application in theoretical physics. Furthermore recent developments in a generalization of homology will be reviewed as well as some physical applications.

The plan is the following. In Section 2 we give the basic definitions and results on homology of differential modules. In Section 3 we introduce graduation, that is we discuss complexes and give several examples; in this section we explain the constructions connected with simplicial modules and we describe the tensor product of complexes. Section 4 is a physical illustration of the fact that there is no natural tensor product of differential modules whereas there is one for complexes; we show there that the introduction of ghosts at the one-particle level in the free field theory is worthwhile to render the theory natural over the physical space. In Section 5 we introduce N-differentials and discuss the generalization of homology associated with N-differential modules; we give there several examples of constructions some of which are related to physics (e.g. parafermions). Section 6 is devoted to the corresponding graded situation i.e. to N-complexes; we recall there the constructions of N-complexes associated to simplicial modules and the result which expresses in these cases the generalized homology in terms of the ordinary one (Theorem 2) [**14**]. In Section 7 which summarizes results of [**17**], [**18**], we introduce N-complexes of tensor fields on $\mathbb{R}^D$ generalizing the complex of differential forms and we state the corresponding generalization of the Poincaré lemma (Theorem 3); we also explain why these N-complexes naturally enter the theory of higher spin gauge fields. In Section 8 we discuss graded differential algebras and their "N-generalization" and give a universal N-construction generalizing the usual universal differential calculus over a unital associative algebra [**20**], [**14**]. Section 9 describes the homological approach to "subquotients" and applies it to constrained systems (BRS-method). The main result, Theorem 4, is slightly more general than the results of [**12**] (more general context), so we give a sketch of proof of Lemma 10 on which it relies. Finally in Section 10 we generalize constructions of the previous section to N-differential modules in connection with a quantum gauge group problem arising for the zero modes in the Wess-Zumino-Novikov-Witten model; this section is a summary of [**23**] (see also [**22**]) .

These notes contain almost no proof, many results are classical or easy. There are two notable exceptions, namely Theorem 2 and Theorem 3 the proof of which are absolutely non trivial although their meaning is transparent.

Let us say some words on our conventions. For sake of completeness we have given the formulation in terms of modules over a commutative ring **k**; the tensor product symbol $\otimes$ without other specification means the tensor product over **k** (of **k**-modules), i.e. $\otimes = \otimes_{\mathbf{k}}$. In the physical examples **k** is either the field $\mathbb{R}$ of the field $\mathbb{C}$, so the reader may well understand **k** like that and then the **k**-modules are vector spaces over $\mathbb{R}$ or $\mathbb{C}$. We use throughout the Einstein convention of summation of repeated up-down indices. A diagram of mappings between sets is said to be a *commutative diagram* if given two path of mappings between (two vertex) two sets of the diagram, the corresponding compositions of mappings coincide. A *Young diagram* is a finite collection of boxes, or cells, arranged in left-justified rows, with a weakly decreasing number of cells in each row. Given a Young diagram of n cells Y, one associates to it a projector **Y**, the Young symmetrizer, on the space of covariant tensors of degree n on $\mathbb{R}^D$ by the following procedure. Let $T_{\mu_1 \cdots \mu_n}$ be the components of T, then the components $\mathbf{Y}(T)_{\mu_1 \cdots \mu_n}$ of $\mathbf{Y}(T)$ are obtained by filling successively the cells of the rows of Y with $\mu_1, \cdots, \mu_n$, then by symmetrizing the μ's which belong to the same rows and then by antisymmetrizing the μ's which belong to the same columns. For Young diagrams etc., we use the notations of [**30**]. We also mention that many subjects of these lectures are also treated in [**16**] so, although the aims of [**16**] are different, it is a complement for the present notes.

2. Differential modules

Throughout these notes, **k** is a commutative ring with a unit and by a module without other specification, we always mean a **k**-module; the same convention is adopted for homomorphisms, endomorphisms, etc.. A module E equipped with an endomorphism d satisfying $d^2 = 0$ will be referred to as a *differential module* and the endomorphism d as its *differential*. Given two differential modules (E, d) and (E', d'), a *homomorphism of differential modules* of E into E' is a homomorphism (of **k**-modules) $\varphi : E \to E'$ satisfying $\varphi \circ d = d' \circ \varphi$.

A sequence of homomorphisms of modules (resp. of differential modules)
$$\cdots \longrightarrow E_i \xrightarrow{\varphi_i} E_{i+1} \xrightarrow{\varphi_{i+1}} E_{i+2} \longrightarrow \cdots$$
is said to be *exact* if $\mathrm{Im}(\varphi_i) = \mathrm{Ker}(\varphi_{i+1})$. In particular the sequence $0 \to E \xrightarrow{\varphi} F$ is exact if and only if φ is injective and the sequence $E \xrightarrow{\varphi} F \to 0$ is exact if and only if φ is surjective.

Let E be a differential module with differential d, then by definition one has $\mathrm{Im}(d) \subset \mathrm{Ker}(d)$ so the non exactness of the sequence $E \xrightarrow{d} E \xrightarrow{d} E$ is measured by the module $H(E) = \mathrm{Ker}(d)/\mathrm{Im}(d)$ which is referred to as the *homology* of the differential module E. Let $\varphi : E \to F$ be a homomorphism of differential modules, then one has $\varphi(\mathrm{Im}(d)) \subset \mathrm{Im}(d)$ and $\varphi(\mathrm{Ker}(d)) \subset \mathrm{Ker}(d)$ (with an obvious abuse of notations) so φ induces a homomorphism $\varphi_* : H(E) \to H(F)$ in homology. An important result for the computations of homology is given by the following proposition.

PROPOSITION 1. *Let $0 \to E \xrightarrow{\varphi} F \xrightarrow{\psi} G \to 0$ be an exact sequence of differential modules; then there is a homomorphism $\partial : H(G) \to H(E)$ such that the triangle of homomorphisms*

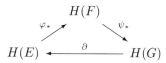

is exact.

The exactness at $H(F)$ is easy and we only sketch the construction of ∂. Let $z \in G$ be such that $dz = 0$ and let us denote by $[z] \in H(G)$ the class of z. Since ψ is surjective there is a $y \in F$ such that $\psi(y) = z$; one has $\psi(dy) = d\psi(y) = dz = 0$ so $dy \in \mathrm{Ker}(\psi)$. By exactness at F, there is an $x \in E$ such that $\varphi(x) = dy$ and one has $\varphi(dx) = d\varphi(x) = d^2 y = 0$. Since φ is injective it follows that $dx = 0$ and we denote by $[x] \in H(E)$ the class of x. It turns out (and this is not difficult to verify) that $[x] \in H(E)$ does only depend on $[z] \in H(G)$ and that the mapping $[z] \mapsto \partial[z] = [x]$ is a homomorphism $\partial : H(G) \to H(E)$ which satisfies the statement of the proposition.

Quite generally, a five terms exact sequence of the form

$$0 \longrightarrow E \xrightarrow{\varphi} F \xrightarrow{\psi} G \longrightarrow 0$$

is called a *short exact sequence* and given a short exact sequence of differential modules as in Proposition 1, the homomorphism $\partial : H(G) \to H(E)$ is called the *connecting homomorphism* of the short exact sequence of differential modules. The connecting homomorphism is natural (i.e. functorial) in the following sense: For any commutative diagram of differential modules

$$\begin{array}{ccccccccc} 0 & \longrightarrow & E & \xrightarrow{\varphi} & F & \xrightarrow{\psi} & G & \longrightarrow & 0 \\ & & \lambda \downarrow & & \mu \downarrow & & \nu \downarrow & & \\ 0 & \longrightarrow & E' & \xrightarrow{\varphi'} & F' & \xrightarrow{\psi'} & G' & \longrightarrow & 0 \end{array}$$

with exact rows, the diagram

$$\begin{array}{ccc} H(G) & \xrightarrow{\partial} & H(E) \\ \nu_* \downarrow & & \lambda_* \downarrow \\ H(G') & \xrightarrow{\partial} & H(E') \end{array}$$

is commutative.

It is worth noticing here that although direct sums of differential modules are well defined, there is no natural tensor product of differential modules. A natural tensor product will be only obtained in the graded case, that is for complexes (see below).

In the case where **k** is a field, a differential module will be called a *differential vector space* or simply a *differential space*. In the examples connected with physics, **k** will always be either the field $\mathbb{R}$ or the field $\mathbb{C}$.

3. Complexes

By a *complex*, without other specification, we always mean a differential module E which is $\mathbb{Z}$-graded, $E = \underset{n \in \mathbb{Z}}{\oplus} E^n$, with a differential d which is of degree 1 or -1. When d is of degree -1, E is referred to as a *chain complex* and when d is of degree 1, E is referred to as a *cochain complex*. One passes from the chain complexes to the cochain ones by changing the signs of the degrees ($n \mapsto -n$). In the following we shall only consider the cochain case. The homology of a cochain complex E is usually referred to as the *cohomology* of E. Since d is homogeneous, the homology of a complex E is $\mathbb{Z}$-graded : $H(E) = \underset{n \in \mathbb{Z}}{\oplus} H^n(E)$ with $H^n(E) = \text{Ker}(d) \cap E^n / \text{Im}(d) \cap E^n$. Many notions for complexes do only depend on the underlying $\mathbb{Z}_2$ graduation ($\mathbb{Z}_2 = \mathbb{Z}/2\mathbb{Z}$) so let us define a $\mathbb{Z}_2$-*complex* to be a differential module E which is $\mathbb{Z}_2$-graded, $E = E^0 \oplus E^1$, with a differential d which is of degree 1 ($=-1$). Again, the homology $H(E)$ of a $\mathbb{Z}_2$-complex is $\mathbb{Z}_2$-graded, that is $H(E) = H^0(E) \oplus H^1(E)$. A *homomorphism of complexes* or *of $\mathbb{Z}_2$-complexes* is a homomorphism of differential modules which is homogeneous of degree 0.

Let $0 \longrightarrow E \xrightarrow{\varphi} F \xrightarrow{\psi} G \longrightarrow 0$ be a short exact sequence of cochain complexes; it follows from the definition of the connecting homomorphism ∂ that the exact triangle of Proposition 1 gives rise to the long exact sequence of homomorphisms

$$\cdots \xrightarrow{\partial} H^n(E) \xrightarrow{\varphi_*} H^n(F) \xrightarrow{\psi_*} H^n(G) \xrightarrow{\partial} H^{n+1}(E) \xrightarrow{\varphi_*} \cdots$$

in cohomology. Similarly if $0 \longrightarrow E \xrightarrow{\varphi} F \xrightarrow{\psi} G \longrightarrow 0$ is a short exact sequence of $\mathbb{Z}_2$-complexes, the exact triangle of Proposition 1 gives rise to the exact hexagon of homomorphisms

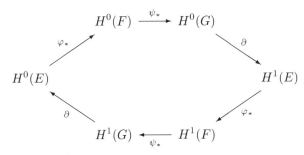

for the corresponding homologies.

Let E and F be two cochain complexes, (resp. $\mathbb{Z}_2$-complexes), *their tensor product* $E \otimes F$ is the graded module $E \otimes F = \oplus_n (E \otimes F)^n$ with $(E \otimes F)^n = \bigoplus_{r+s=n} E^r \otimes F^s$ equipped with the differential d defined by

$$d(e \otimes f) = de \otimes f + (-1)^n e \otimes df,$$

for any $e \in E^n$ and $f \in F$. One verifies that so defined on $E \otimes F$, d is homogeneous of degree 1 and satisfies $d^2 = 0$ so that $E \otimes F$ is again a cochain complex, (resp. a $\mathbb{Z}_2$-complex). The virtue of this definition is the Künneth formula which we describe only for complexes of vector spaces in the following proposition, [**36**], [**66**].

PROPOSITION 2. *Assume that the ring* **k** *is a field then one has* $H(E \otimes F) = H(E) \otimes H(F)$.

The above tensor product being the tensor product of graded vector spaces (over **k**) i.e. $H^n(E \otimes F) = \bigoplus_{r+s=n} H^r(E) \otimes H^s(F)$. This formula applies as well to the (co)chain complexes case and to the $\mathbb{Z}_2$-complexes case (whenever **k** is a field).

In the next section we shall describe a physical application of Proposition 2 combined with the remark that there is no such tensor product for differential spaces. We now achieve this section by the description of some classical constructions which will be used later.

Let $\mathfrak{g}$ be a Lie algebra, let R be a representation space of $\mathfrak{g}$ and denote by $X \mapsto \pi(X) \in \mathrm{End}(R)$ the action of $\mathfrak{g}$ on R. An R-valued (Lie algebra) n-cochain of $\mathfrak{g}$ is a linear mapping $X_1 \wedge \cdots \wedge X_n \mapsto \omega(X_1, \ldots, X_n)$ of $\bigwedge^n \mathfrak{g}$ into R. The vector space of these n-cochains will be denoted by $C^n_\wedge(\mathfrak{g}, R)$. One defines a homogeneous endomorphism d of degree 1 of the $\mathbb{N}$-graded vector space $C_\wedge(\mathfrak{g}, R) = \oplus_n C^n_\wedge(\mathfrak{g}, R)$ of all R-valued cochains of $\mathfrak{g}$ by setting

$$d(\omega)(X_0, \ldots, X_n) = \sum_{k=0}^n (-1)^k \pi(X_k) \omega(X_0, \overset{k}{.\check{.}.}, X_n)$$
$$+ \sum_{0 \leq r < s \leq n} (-1)^{r+s} \omega([X_r, X_s], X_0 \overset{r}{.\check{.}} \overset{s}{.\check{.}} X_n)$$

for $\omega \in C^n_\wedge(\mathfrak{g}, R)$ and $X_i \in \mathfrak{g}$. It follows from the Jacobi identity and from $\pi(X)\pi(Y) - \pi(Y)\pi(X) = \pi([X, Y])$ that $d^2 = 0$. Thus equipped with d, $C_\wedge(\mathfrak{g}, R)$ is a cochain complex and its cohomology, denoted by $H(\mathfrak{g}, R)$, is called the *R-valued cohomology of $\mathfrak{g}$*. The complexes $C_\wedge(\mathfrak{g}, R)$ are also called *Chevalley-Eilenberg complexes* and the differential d is the *Chevalley-Eilenberg differential*.

There is a standard way to produce positive complexes (i.e. complexes $E = \oplus E^n$ with $E^n = 0$ for $n < 0$) starting from (co)simplicial modules, (see e.g. [47], [66]). A *pre-cosimplicial module* (or *semi-cosimplicial* in the terminology of [66]) is a sequence of modules $(E^n)_{n \in \mathbb{N}}$ together with *coface homomorphisms* $\mathfrak{f}_i : E^n \to E^{n+1}$, $i \in \{0, 1, \ldots, n+1\}$, satisfying

($\mathfrak{F}$) $\qquad \mathfrak{f}_j \mathfrak{f}_i = \mathfrak{f}_i \mathfrak{f}_{j-1} \quad$ if $i < j$.

Given a pre-cosimplicial module (E^n), one associates to it a positive complex (E, d) by setting $E = \oplus_{n \in \mathbb{N}} E^n$ and $d = \sum_{i=0}^{n+1} (-1)^i \mathfrak{f}_i : E^n \to E^{n+1}$. One verifies that $d^2 = 0$ is implied by the coface relations ($\mathfrak{F}$). The differential d will be referred to as the *simplicial differential of (E^n)*. The cohomology $H(E) = \oplus H^n(E)$ with $H^n(E) = \mathrm{Ker}(d : E^n \to E^{n+1})/dE^{n-1}$ of (E, d) will be referred to as *the cohomology of the pre-cosimplicial module (E^n)*. A *cosimplicial module* is a pre-cosimplicial module (E^n) with coface homomorphisms $\mathfrak{f}_i$ as before together with *codegeneracy homomorphisms* $\mathfrak{s}_i : E^{n+1} \to E^n$, $i \in \{0, \ldots, n\}$, satisfying

($\mathfrak{S}$) $\qquad \mathfrak{s}_j \mathfrak{s}_i = \mathfrak{s}_i \mathfrak{s}_{j+1} \quad$ if $i \leq j$

and

($\mathfrak{S}\mathfrak{F}$) $\qquad \mathfrak{s}_j \mathfrak{f}_i = \begin{cases} \mathfrak{f}_i \mathfrak{s}_{j-1} & \text{if } i < j \\ I & \text{if } i = j \text{ or } i = j+1 \\ \mathfrak{f}_{i-1} \mathfrak{s}_j & \text{if } i > j+1 \end{cases}$

Given a cosimplicial module (E^n) the elements ω of E^n such that $\mathfrak{s}_i(\omega) = 0$ for $i \in \{0, \cdots, n\}$ are called *normalized cochains of degree n* and the graded module $N(E) = \oplus_n N^n(E)$ of all normalized cochains is a subcomplex of E which has the same cohomology as the one of E, i.e. $H(E)$. The correspondence $(E^n) \mapsto N(E)$ defines an equivalence between the category of cosimplicial modules and the category of positive cochain complexes [66] which is referred to as *the Dold-Kan correspondence* (for the category of **k**-modules).

Let $\mathcal{A}$ be an associative unital **k**-algebra and let $\mathcal{M}$ be an $(\mathcal{A}, \mathcal{A})$-bimodule. A *$\mathcal{M}$-valued Hochschild cochain of degree n* or *Hochschild n-cochain of $\mathcal{A}$* is a linear mapping $x_1 \otimes \cdots \otimes x_n \mapsto \omega(x_1, \cdots, x_n)$ of $\otimes^n \mathcal{A}$ into $\mathcal{M}$. The **k**-module of all $\mathcal{M}$-valued Hochschild n-cochains is denoted by $C^n(\mathcal{A}, \mathcal{M})$. The sequence $(C^n(\mathcal{A}, \mathcal{M}))$ is a cosimplicial module with cofaces $\mathfrak{f}_i$ and codegeneracies $\mathfrak{s}_i$ defined by [47], [66]
$\mathfrak{f}_0(\omega)(x_0, \ldots, x_n) = x_0 \omega(x_1, \ldots, x_n)$
$\mathfrak{f}_i(\omega)(x_0, \ldots, x_n) = \omega(x_0, \ldots, x_{i-1} x_i, \ldots, x_n) \qquad$ for $i \in \{1, \ldots, n\}$
$\mathfrak{f}_{n+1}(\omega)(x_0, \ldots, x_n) = \omega(x_0, \ldots, x_{n-1}) x_n$
and
$\mathfrak{s}_i(\omega)(x_1, \ldots, x_{n-1}) = \omega(x_1, \ldots, x_i, \mathbf{1}, x_{i+1}, \ldots, x_{n-1}) \qquad$ for $i \in \{0, \ldots, n-1\}$

for $\omega \in C^n(\mathcal{A}, \mathcal{M})$ and $x_i \in \mathcal{A}$. The cohomology $H(\mathcal{A}, \mathcal{M})$ of this cosimplicial module is the $\mathcal{M}$-valued *Hochschild cohomology* of $\mathcal{A}$. In his case the simplicial differential is called the *Hochschild differential*.

There is a relation between the cohomology of a Lie algebra $\mathfrak{g}$ and the Hochschild cohomology of its universal enveloping algebra $U(\mathfrak{g})$ which we now describe again in the case where **k** is a field. Given a bimodule $\mathcal{M}$ over $U(\mathfrak{g})$ (that is a $(U(\mathfrak{g}), U(\mathfrak{g}))$-bimodule), let us define the representation $X \mapsto \mathrm{ad}(X)$ of $\mathfrak{g}$ in the vector space $\mathcal{M}$ by $\mathrm{ad}(X)m = Xm - mX$ for $X \in \mathfrak{g}$ and $m \in \mathcal{M}$. Let $H(\mathfrak{g}, \mathcal{M}^{\mathrm{ad}})$ denote the Lie algebra cohomology of $\mathfrak{g}$ with values in $\mathcal{M}$ for the ad representation; its relation with the $\mathcal{M}$-valued Hochschild cohomology of $U(\mathfrak{g})$, $H(U(\mathfrak{g}), \mathcal{M})$ is given by the following theorem [7], [47].

THEOREM 1. *Assume that* **k** *is a field, let* $\mathfrak{g}$ *be a Lie algebra over* **k** *and let* $\mathcal{M}$ *be a bimodule over* $U(\mathfrak{g})$. *Then there is a canonical isomorphism* $H(\mathfrak{g}, \mathcal{M}^{\mathrm{ad}}) \simeq H(U(\mathfrak{g}), \mathcal{M})$.

If R is a representation space of $\mathfrak{g}$ with action $X \mapsto \pi(X)$, then by the very definition of $U(\mathfrak{g})$, π extends as a representation of $U(\mathfrak{g})$ so R is canonically a left $U(\mathfrak{g})$-module. One converts R into a $(U(\mathfrak{g}), U(\mathfrak{g}))$-bimodule $\mathcal{R}$ by acting on the right with the trivial action given by the counit of $U(\mathfrak{g})$ (recall that $U(\mathfrak{g})$ is a Hopf algebra); one then has $R = \mathcal{R}^{\mathrm{ad}}$.

4. A physical example: Naturality of ghosts

The Wigner one-particle space for mass zero and spin one is the direct hilbertian integral $\int_{C_+} d\mu_0(p) \mathcal{H}(p)$ of 2-dimensional Hilbert spaces $\mathcal{H}(p)$ over the future light cone
$$C_+ = \{p | g^{\mu\nu} p_\mu p_\nu = p_0^2 - \vec{p}^2 = 0, \ p_0 > 0\}$$
with respect to the invariant measure $d\mu_0(p) = \frac{1}{(2\pi)^3} \frac{d^3\vec{p}}{2p^0}$, where $\mathcal{H}(p)$ is the quotient of the subspace $\mathcal{Z}(p) = \{A_\mu \in \mathbb{C}^4 | p^\mu A_\mu = 0\}$ of $\mathcal{C}(p) = \mathbb{C}^4$ by the subspace $\mathcal{B}(p) = \{p_\mu \varphi | \varphi \in \mathbb{C}\}$ spanned by p, the scalar product of $\mathcal{H}(p)$ being induced by the indefinite scalar product of $\mathcal{C}(p)$ defined by $\langle A | A' \rangle = -g^{\mu\nu} \bar{A}_\mu A'_\nu$. The scalar product of $\mathcal{C}(p)$ is positive semi-definite on $\mathcal{Z}(p)$ and $\mathcal{B}(p)$ is its isotropic subspace whereas $\mathcal{Z}(p)$ is the orthogonal of $\mathcal{B}(p)$ in $\mathcal{C}(p)$. Notice that the indefinite metric space $\mathcal{C}(p)$ does not depend on p; we keep the reference to p in order to remember that it carries a representation of *the little group at* p. The little group at p here means the subgroup $\mathcal{L}_p$ of the Lorentz group which consists of the Lorentz tranformations Λ preserving the (quadri) vector p, that is
$$\mathcal{L}_p = \{\Lambda \in GL(4, \mathbb{R}) \ | \ \Lambda^\mu_\lambda \Lambda^\nu_\rho g^{\lambda\rho} = g^{\mu\nu} \text{ and } \Lambda^\mu_\nu p^\nu = p^\mu\}.$$
The occurrence of such a triplet $(\mathcal{C}(p), \mathcal{Z}(p), \mathcal{B}(p))$ where $\mathcal{C}(p)$ has an indefinite scalar product with $\mathcal{B}(p)$ isotropic having $\mathcal{Z}(p)$ as orthogonal, etc. is familiar in connection with indecomposable representations of groups (here the little group) [51], [1] and the indefinite metric is furthermore required to get a local covariant description of the electromagnetic gauge potential [61], [62], see also [46] in this context.

Let $Q(p) = Q$ be the linear endomorphism of $\mathcal{C}(p)$ defined by $Q(A)_\mu = p_\mu p^\nu A_\nu$. Then Q is hermitian, i.e. $\langle A | QA' \rangle = \langle QA | A' \rangle$, and one has $Q^2 = 0$ in view of $p_\mu p^\mu = 0$. Furthermore the image of Q is $\mathcal{B}(p)$ and its kernel is $\mathcal{Z}(p)$. In other words $(\mathcal{C}(p), Q(p))$ is a differential space and $\mathcal{H}(p)$ is its homology, i.e. one has $\mathcal{H}(p) = \text{Ker}(Q)/\text{Im}(Q)$. Thus, apart from questions of domain and function spaces, everything is perfect at the "one-particle" level: Namely one has an indefinite metric space $\mathcal{C}$ which consists of functions $p \mapsto A_\mu(p) \in \mathcal{C}(p)$ on the light cone C_+ and which is equipped with a differential Q (i.e. an endomorphism satisfying $Q^2 = 0$) such that the physical one-particle space, (i.e. the Wigner space), is the homology $\text{Ker}(Q)/\text{Im}(Q)$ of $\mathcal{C}$.

As is well known, the role of $\mathcal{C}$ is to provide, via the Fock space constructions, an indefinite metric space on which the local covariant gauge potential (free) field operator acts; the corresponding space of physical states being of course the Fock space constructed over the one-particle Wigner space. However it turns out that the above one-particle (homological) picture does not generalize naively at the n-particle level for $n \geq 2$. To show what is involved here, let us analyze the situation at the two-particle level. In order to avoid complications connected with the problem of the choice of the function space and with the problem of symmetrization, let us work at fixed momenta p_1 and p_2 on the light cone C_+ with $p_1 \neq p_2$. The indefinite metric space is then the 16-dimensional space $\mathcal{C}(p_1) \otimes \mathcal{C}(p_2)$ whereas the space of physical states is the 4-dimensional Hilbert space $\mathcal{H}(p_1) \otimes \mathcal{H}(p_2)$. The point now is that there is no canonical way to construct $\mathcal{H}(p_1) \otimes \mathcal{H}(p_2)$ from $\mathcal{C}(p_1) \otimes \mathcal{C}(p_2)$. More precisely there is no canonical way to build a differential on $\mathcal{C}(p_1) \otimes \mathcal{C}(p_2)$ out of the differentials $Q(p_1)$ and $Q(p_2)$ of $\mathcal{C}(p_1)$ and $\mathcal{C}(p_2)$ in such a way that its homology is $\mathcal{H}(p_1) \otimes \mathcal{H}(p_2)$. In fact the most natural candidate would be $Q_{12} = Q(p_1) \otimes \text{Id}_2 + \text{Id}_1 \otimes Q(p_2)$ but this is not of square zero, only its third power vanishes, (for the "n-particle" case it would be the $(n + 1)$-th power). Notice that with Q_{12} satisfying $(Q_{12})^3 = 0$ one can associate the generalized homologies (see below) $H_{(1)}(Q_{12}) = \text{Ker}(Q_{12})/\text{Im}((Q_{12})^2)$ and $H_{(2)}(Q_{12}) = \text{Ker}((Q_{12})^2)/\text{Im}(Q_{12})$ however it is easy to show that one canonically has $H_{(1)}(Q_{12}) = \mathcal{Z}(p_1) \otimes \mathcal{Z}(p_2)$ and that $H_{(2)}(Q_{12})$ is isomorphic to $H_{(1)}(Q_{12})$. Thus $H_{(1)}(Q_{12})$ is a subspace of $\mathcal{C}(p_1) \otimes \mathcal{C}(p_2)$ on which the metric is positive semi-definite but it is still not the physical space $\mathcal{H}(p_1) \otimes \mathcal{H}(p_2)$.

Notice that we do not claim that there is no differential on $\mathcal{C}(p_1) \otimes \mathcal{C}(p_2)$ such that the corresponding homology is $\mathcal{H}(p_1) \otimes \mathcal{H}(p_2)$ but that we claim that there is no canonical one, that is no reasonable expression for such a differential in terms of the differentials $Q(p_1)$ and $Q(p_2)$. We refer to Appendix A for the precise statement.

As pointed out above, the origin of the difficulty is the non-existence of a good tensor product between differential spaces, i.e. between vector spaces equipped with endomorphisms of square zero. If instead of differential spaces one has complexes (of vector spaces), then the situation is much better; namely one has a canonical tensor product of complexes which is such that the homology of the tensor product is the tensor product of the homologies, (see last section). Furthermore one can show that the symmetrization-antisymmetrization involved in the Fock space construction does not spoil this picture.

Fortunately there is a canonical way (related to Theorem 4) to construct a complex $C(p) = C^{-1}(p) \oplus C^0(p) \oplus C^1(p)$ with a differential of degree 1 such that $C^0(p) = \mathcal{C}(p)$ and such that its (co)homology is again $\mathcal{H}(p)$. We now describe this construction. Let ε^μ be the (real) canonical base of $\mathcal{C}(p) = C^0(p) = \mathbb{C}^4$ and let $\omega^{(+)}$ and $\omega^{(-)}$ be the basis of the one dimensional spaces $C^1(p)$ and $C^{-1}(p)$ ($\cong \mathbb{C}$). Define the homogeneous linear endomorphism $\delta(p) = \delta$ of degree 1 of $C(p)$ by $\delta\omega^{(+)} = 0$, $\delta\varepsilon^\mu = \alpha p^\mu \omega^{(+)}$ and $\delta\omega^{(-)} = p_\mu \varepsilon^\mu$, ($\alpha$ being a non-vanishing constant). It is clear that $\delta^2 = 0$ and it is straightforward to verify that the (co)homology $H(C(p)) = \text{Ker}(\delta)/\text{Im}(\delta)$ of $C(p)$ is given by $H(C(p)) = H^0(C(p)) = \mathcal{H}(p)$. Notice that if $c\omega^{(-)} + A_\mu \varepsilon^\mu + \tilde{c}\omega^{(+)}$ is an arbitrary element of $C(p)$, δ reads in components $\delta A_\mu = p_\mu c$, $\delta c = 0$ and $\delta \tilde{c} = \alpha p^\lambda A_\lambda$. One defines an indefinite hermitian scalar product on $C(p)$ extending the one of $C^0(p) = \mathcal{C}(p)$ for which δ is hermitian by setting $\langle \varepsilon^\mu | \varepsilon^\nu \rangle = -g^{\mu\nu}$, $\langle \omega^{(+)} | \varepsilon^\mu \rangle = 0$, $\langle \omega^{(+)} | \omega^{(+)} \rangle = 0$, $\langle \omega^{(-)} | \varepsilon^\mu \rangle = 0$, $\langle \omega^{(-)} | \omega^{(-)} \rangle = 0$ and $\langle \omega^{(-)} | \omega^{(+)} \rangle = -\alpha^{-1}$. One can now construct the generalized Fock space $\mathfrak{F}(C)$ over the graded space C of "functions" $p \mapsto (\tilde{c}(p), A_\mu(p), c(p)) \in C(p)$ on the future light cone. The space $\mathfrak{F}(C)$ is the graded-commutative algebra (freely) generated by the graded vector space C and one extends δ as an antiderivation of $\mathfrak{F}(C)$, again denoted by δ, which still satisfies $\delta^2 = 0$. The scalar product of C extends canonically into an indefinite scalar product of $\mathfrak{F}(C)$ for which δ is hermitian and the cohomology $H^0(\delta)$ is (a dense subspace of) the physical space (i.e. the Fock space over the Wigner one-particle space). One then constructs as usual the local gauge potential field operator corresponding to the above one-particle A_μ as well as the fermionic ghost and antighost field operators corresponding to the above one-particle c and $\tilde{c}$. In order that the ghost and the antighost fields be relatively local, it is necessary to take α purely imaginary, i.e. $\alpha = i\lambda$ with $\lambda \in \mathbb{R}_*$, otherwise one would obtain a factor $D^{(1)}$ in their anticommutators. With this choice ($\alpha = i\lambda, \lambda \in \mathbb{R}_*$) the gauge potential, the ghost and the antighost field operators are local and relatively local, (see e.g. in [50]). Moreover these fields are hermitian by their very definition.

Let us say a few words on the case of spin two (and zero mass). In this case, the Wigner one-particle space is again the direct hilbertian integral $\int_{C_+} d\mu_0(p)\mathcal{H}(p)$ of two-dimensional Hilbert spaces $\mathcal{H}(p)$ over the future light cone with respect to $d\mu_0$ with $\mathcal{H}(p) = \mathcal{Z}(p)/\mathcal{B}(p)$ and $\mathcal{Z}(p) \subset \mathcal{C}(p)$ as above but now, $\mathcal{C}(p)$ is the 10-dimensional space of symmetric tensors $h_{\mu\nu} = h_{\nu\mu}$,

$$\mathcal{Z}(p) = \{h_{\mu\nu} \in \mathcal{C}(p) | p^\mu(h_{\mu\nu} - \tfrac{1}{2}g_{\mu\nu}g^{\alpha\beta}h_{\alpha\beta}) = 0\},$$

$$\mathcal{B}(p) = \{p_\mu \varphi_\nu + p_\nu \varphi_\mu | \varphi_\lambda \in \mathbb{C}^4\}$$

and the scalar product of $\mathcal{H}(p)$ is induced by the indefinite scalar product of $\mathcal{C}(p)$ defined by $\langle h | h' \rangle = g^{\mu\nu}g^{\lambda\rho}\bar{h}_{\mu\lambda}h'_{\nu\rho} - \tfrac{1}{2}g^{\alpha\beta}\bar{h}_{\alpha\beta}g^{\gamma\delta}h'_{\gamma\delta}$. Again $\mathcal{B}(p)$ is a completely isotropic (4-dimensional) subspace of $\mathcal{C}(p)$ whereas the 6-dimensional space $\mathcal{Z}(p)$ is its orthogonal in $\mathcal{C}(p)$, ($\mathcal{Z}(p) = \mathcal{B}(p)^\perp$). It is worth noticing here that, apart from a multiplicative constant, the scalar product $\langle h | h' \rangle$ is the unique non-trivial covariant scalar product on $\mathcal{C}(p)$ for which $\mathcal{B}(p)$ is isotropic; equivalently, the condition $p^\mu(h_{\mu\nu} - \tfrac{1}{2}g_{\mu\nu}g^{\alpha\beta}h_{\alpha\beta}) = 0$ is the unique covariant linear (gauge) condition preserved by the translations of $\mathcal{B}(p)$. In view of the connection between the classical linearized gravity theory and the massless spin two particle, it is natural to interpret $h_{\mu\nu} \in \mathcal{C}(p)$ as the positive frequency part of the Fourier transform at p of a perturbation $\underline{g}_{\mu\nu}(x) = g_{\mu\nu} + \varepsilon h_{\mu\nu}(x)$ of the Minkowskian metric $g_{\mu\nu}$. Translations by $\mathcal{B}(p)$ then read $h_{\mu\nu}(x) \mapsto h_{\mu\nu}(x) + \partial_\mu \varphi_\nu(x) + \partial_\nu \varphi_\mu(x)$ which corresponds to the first order in ε (i.e. the linearization) of the action of infinitesimal diffeomorphisms (i.e. vector fields) whereas the condition to be in $\mathcal{Z}(p)$ translates into $\partial^\mu(h_{\mu\nu}(x) - \frac{1}{2}g_{\mu\nu}g^{\alpha\beta}h_{\alpha\beta}(x)) = 0$ which is the first order in ε of the de Donder harmonic coordinates condition $\frac{1}{\sqrt{|g|}}\partial_\mu\left(\sqrt{|g|}g^{\mu\nu}\right) = \Delta_g(x^\nu) = 0$. It may well be that this observation (i.e. connection between Poincaré covariant Wigner analysis and de Donder harmonic coordinates condition) is a little more than a curiosity. In any case, we can now proceed as for the spin one case. One defines the graded vector space $C(p) = C^{-1}(p) \oplus C^0(p) \oplus C^1(p)$ by $C^0(p) = \mathcal{C}(p)$ and $C^{-1}(p) \simeq \mathbb{C}^4 \simeq C^1(p)$ and we let $\omega^{(-)\mu}$ and $\omega^{(+)\mu}$ be the basis of $C^{-1}(p)$ and $C^{+1}(p)$ corresponding to the canonical base ε^μ of $\mathbb{C}^4$ and $\varepsilon^{\mu\nu} = \frac{1}{2}(\varepsilon^\mu \otimes \varepsilon^\nu + \varepsilon^\nu \otimes \varepsilon^\mu)$ be the corresponding basis of $C^0(p) = \mathcal{C}(p)$. One defines then a differential δ of degree 1 of $C(p)$ by setting $\delta\omega^{(+)\mu} = 0$, $\delta\varepsilon^{\mu\nu} = \alpha(p^\mu\omega^{(+)\nu} + p^\nu\omega^{(+)\mu})$ and $\delta\omega^{(-)\mu} = p_\nu(\varepsilon^{\mu\nu} - \frac{1}{2}g^{\mu\nu}g_{\alpha\beta}\varepsilon^{\alpha\beta})$, $\alpha \in \mathbb{C}_*$. Again one verifies that the cohomology $H(C(p)) = \mathrm{Ker}(\delta)/\mathrm{Im}(\delta)$ of $C(p)$ is given by $H(C(p)) = H^0(C(p)) = \mathcal{H}(p)$. If we let $c_\rho\omega^{(-)\rho} + h_{\mu\nu}\varepsilon^{\mu\nu} + \tilde{c}_\lambda\omega^{(+)\lambda}$ be an arbitrary element of $C(p)$, δ reads in components $\delta h_{\mu\nu} = p_\mu c_\nu + p_\nu c_\mu$, $\delta c_\mu = 0$ and $\delta\tilde{c}_\mu = \alpha p^\nu(h_{\mu\nu} - \frac{1}{2}g_{\mu\nu}g^{\alpha\beta}h_{\alpha\beta})$. Finally, one defines an indefinite hermitian scalar product on $C(p)$ extending the one of $C^0(p) = \mathcal{C}(p)$ for which δ is hermitian by setting $\langle\varepsilon^{\lambda\rho}|\varepsilon^{\mu\nu}\rangle = \frac{1}{2}(g^{\lambda\mu}g^{\rho\nu} + g^{\lambda\nu}g^{\rho\mu} - g^{\lambda\rho}g^{\mu\nu})$, $\langle\omega^{(+)\lambda}|\varepsilon^{\mu\nu}\rangle = 0$, $\langle\omega^{(+)\mu}|\omega^{(+)\nu}\rangle = 0$, $\langle\omega^{(-)\lambda}|\varepsilon^{\mu\nu}\rangle = 0$, $\langle\omega^{(-)\mu}|\omega^{(-)\nu}\rangle = 0$ and $\langle\omega^{(-)\mu}|\omega^{(+)\nu}\rangle = \frac{1}{2\alpha}g^{\mu\nu}$. Thus, apart from numbers of components, everything works as in the case of spin one, in particular one must again take $\alpha = i\lambda$ with $\lambda \in \mathbb{R}_*$ in order to have locality and relative locality between the hermitian free fields corresponding to $h_{\mu\nu}, c_\lambda$ and $\tilde{c}_\rho$.

The main message of this section is "the natural necessity" of ghosts (i.e. of graduations) in order to have a canonical local construction over the physical space and the fact that, in the previous examples (and others), there is a canonical way to introduce their counterpart at the one-particle level. This rewriting of the free field theory for zero mass and spin ≥ 1 is certainly needed in order to start to introduce consistently interactions between abelian gauge fields. In particular this reformulation can be considered as the zero-step for the perturbative construction of quantum operatorial Yang-Mills theory.

5. N-differential modules

In the following, N is a positive integer with $N \geq 2$. A module E equipped with an endomorphism d satisfying $d^N = 0$ will be referred to as an N-*differential module* and the endomorphism d as its N-*differential*. With this terminology, a 2-differential module is just a differential module. For each integer m with $1 \leq m \leq N-1$, one defines the sub-modules $Z_{(m)}(E)$ and $B_{(m)}(E)$ by setting $Z_{(m)}(E) = \mathrm{Ker}(d^m)$ and $B_{(m)}(E) = \mathrm{Im}(d^{N-m})$. It follows from the equation $d^N = 0$ that $B_{(m)}(E)$ is a submodule of $Z_{(m)}(E)$ and the quotient modules $H_{(m)}(E) = Z_{(m)}(E)/B_{(m)}(E)$, $m \in \{1, \ldots, N-1\}$, will be referred to as *the (generalized) homology* of the N-differential module E.

Let m be an integer with $1 \leq m \leq N-2$ and let E be an N-differential module. One has the inclusions $Z_{(m)}(E) \subset Z_{(m+1)}(E)$ and $B_{(m)}(E) \subset B_{(m+1)}(E)$ which induces a homomorphism $[i] : H_{(m)}(E) \to H_{(m+1)}(E)$. One has also the inclusions $dZ_{(m+1)}(E) \subset Z_{(m)}(E)$ and $dB_{(m+1)}(E) \subset B_{(m)}(E)$ which induces a homomorphism $[d] : H_{(m+1)}(E) \to H_{(m)}(E)$. The following basic result show that the $H_{(m)}(E)$ are not independent [20], [14].

LEMMA 1. *Let ℓ and m be integers with $\ell \geq 1$, $m \geq 1$ and $\ell + m \leq N - 1$. Then the following hexagon $(\mathcal{H}^{\ell,m})$ of homomorphisms*

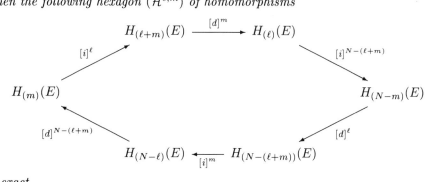

is exact.

One has obvious notions of homomorphism of N-differential modules, of N-differential submodule of an N-differential module, etc.. Let $\varphi : E \to E'$ be a homomorphism of N-differential modules. Then one has $\varphi(Z_{(m)}(E)) \subset Z_{(m)}(E')$ and $\varphi(B_{(m)}(E)) \subset B_{(m)}(E')$ which implies that φ induces a homomorphism $\varphi_* : H_{(m)}(E) \to H_{(m)}(E')$, $\forall m \in \{1, \ldots, N-1\}$. Moreover φ_* satisfies $\varphi_* \circ [i] = [i] \circ \varphi_*$ and $\varphi_* \circ [d] = [d] \circ \varphi_*$. Proposition 1 has the following generalization for N-differential modules.

PROPOSITION 3. *Let $0 \to E \xrightarrow{\varphi} F \xrightarrow{\psi} G \to 0$ be a short exact sequence of N-differential modules. Then there are homomorphisms $\partial : H_{(m)}(G) \to H_{(N-m)}(E)$ for $m \in \{1, \ldots, N-1\}$ such that the following hexagons $(\mathcal{H}_n)$ of homomorphisms*

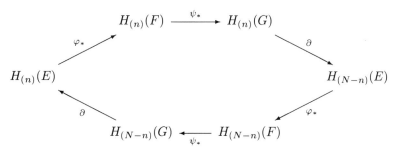

are exact, for $n \in \{1, \ldots, N-1\}$.

For a proof, we refer to [43], [14], [15]. In fact, there is a way to interpret $(\mathcal{H}_n)$ as the exact hexagon corresponding to a short exact sequence of $\mathbb{Z}_2$-complexes $0 \to C_{(n)}(E) \to C_{(n)}(F) \to C_{(n)}(G) \to 0$ associated with the N-complexes, [15].

Let us now give some criteria ensuring the vanishing of the $H_{(n)}(E)$. The first criterion is extracted from [40].

LEMMA 2. *Let E be an N-differential module such that $H_{(k)}(E) = 0$ for some integer k with $1 \leq k \leq N - 1$. Then one has $H_{(n)}(E) = 0$ for any integer n with $1 \leq n \leq N - 1$.*

A short proof of this lemma using Lemma 1 is given in [**14**]. The next criterion which is in [**40**] is connected with an appropriate generalization of homotopy, see in [**43**] and in [**14**]. It is given by the following lemma the proof of which is easy.

LEMMA 3. *Let E be an N-differential module such that there are endomorphisms of modules $h_k : E \to E$ for $k = 0, 1, \ldots, N-1$ satisfying $\sum_{k=0}^{N-1} d^{N-1-k} h_k d^k = Id_E$; then one has $H_{(n)}(E) = 0$ for each integer n with $1 \leq n \leq N-1$.*

In order to formulate the last criterion, we recall the definition of q-numbers. With $q \in \mathbf{k}$, one associates a mapping $[.]_q : \mathbb{N} \to \mathbf{k}$, $n \mapsto [n]_q$, which is defined by setting $[0]_q = 0$ and $[n]_q = 1 + \cdots + q^{n-1} = \sum_{k=0}^{n-1} q^k$ for $n \geq 1$, ($q^0 = 1$). For $n \in \mathbb{N}$ with $n \geq 1$, one defines the q-factorial $[n]_q! \in \mathbf{k}$ by $[n]_q \ldots 1 = \prod_{k=1}^{n} [k]_q$. For integers n and m with $n \geq 1$ and $0 \leq m \leq n$, one defines inductively the q-binomial coefficients $\begin{bmatrix} n \\ m \end{bmatrix}_q \in \mathbf{k}$ by setting $\begin{bmatrix} n \\ 0 \end{bmatrix}_q = \begin{bmatrix} n \\ n \end{bmatrix}_q = 1$ and $\begin{bmatrix} n \\ m \end{bmatrix}_q + q^{m+1} \begin{bmatrix} n \\ m+1 \end{bmatrix}_q = \begin{bmatrix} n+1 \\ m+1 \end{bmatrix}_q$ for $0 \leq m \leq n-1$. As in [**43**] let us introduce the following assumptions (A_0) and (A_1) on the ring $\mathbf{k}$ and the element q of $\mathbf{k}$:

(A_0) $\quad [N]_q = 0$
(A_1) $\quad [N]_q = 0$ and $[n]_q$ is invertible for $1 \leq n \leq N-1$, ($n \in \mathbb{N}$).

Notice that $[N]_q = 0$ implies that $q^N = 1$ and therefore that q is invertible. Furthermore if q is invertible one has $[n]_{q^{-1}} = q^{-n+1} [n]_q$, $\forall n \in \mathbb{N}$. Therefore Assumption (A_0), (resp. (A_1)), for $\mathbf{k}$ and $q \in \mathbf{k}$ is equivalent to Assumption (A_0), (resp. (A_1)), for $\mathbf{k}$ and $q^{-1} \in \mathbf{k}$. Let us give two typical examples:

1. $\mathbf{k} = \mathbb{C}$, $q \in \mathbb{C}$. Then Assumption (A_0) means that q is an N-th root of unity distinct of 1 and Assumption (A_1) means that q is a primitive N-th root of unity.
2. $\mathbf{k} = \mathbb{Z}_N = \mathbb{Z}/N\mathbb{Z}$, then $1 \in \mathbf{k}$ satisfies Assumption (A_0) and Assumption (A_1) is satisfied if and only if N is a prime number.

A useful result is that if $\mathbf{k}$ and $q \in \mathbf{k}$ satisfy Assumption (A_1) then one has $\begin{bmatrix} N \\ m \end{bmatrix}_q = 0$ for $m \in \{1, \ldots, N-1\}$; notice that Assumption (A_0) is not sufficient in order to have this result.

We are now ready to state the last criterion [**13**].

LEMMA 4. *Suppose that $\mathbf{k}$ and $q \in \mathbf{k}$ satisfy (A_1) and let E be an N-differential module. Assume that there is a module-endomorphism h of E such that $hd - qdh = Id_E$. Then one has $H_{(n)}(E) = 0$ for each integer n with $1 \leq n \leq N-1$.*

In order to proof this lemma, one shows that in the unital $\mathbf{k}$-algebra generated by h and d with the relation $hd - qdh = \mathbf{1}$ one has $\sum_{k=0}^{N-1} d^{N-1-k} h^{N-1} d^k = [N-1]_q! \mathbf{1}$, which implies the result in view of Lemma 3 since $[N-1]_q!$ is invertible in $\mathbf{k}$ (see in [**43**] and in [**14**]).

It is obvious that Lemma 4 above is closely related to the theory of q-oscillators, (e.g. d corresponds to the creation operator whereas h corresponds to the annihilation operator), and this is the essence of the proof of [**14**]. As well known in physics, there is another natural way to produce creation operators with vanishing N-th powers which consists in considering parafermions of order $N-1$; this has the generalization we now describe.

As already pointed out (in Section 2 and Section 4) there is no natural tensor product between differential modules. The same is true for N-differential modules with N fixed. However, if (E', d') is an N'-differential module and if (E'', d'') is an N''-differential module $(N', N'' \geq 2)$ then one defines an $(N' + N'' - 1)$-differential d on $E' \otimes E''$ by setting

$$d = d' \otimes I'' + I' \otimes d''$$

where I' (resp. I'') denotes the identity mapping $Id_{E'}$ (resp. $Id_{E''}$) of E' (resp. of E''). Therefore, a natural construction of an N-differential module consists in starting with $(N-1)$ ordinary differential modules (E_i, d_i) and equipping their tensor product $E = E_1 \otimes \cdots \otimes E_{N-1}$ with the N-differential

$$d = d_1 \otimes I_2 \otimes \cdots \otimes I_{N-1} + \cdots + I_1 \otimes \cdots \otimes I_{N-2} \otimes d_{N-1}.$$

If all the (E_i, d_i) are identical, with d_i being a fermionic creation operator, the above formula is the Green ansatz [**34**] for the parafermionic creation operator of order $N-1$.

In the case where **k** is a field, an N-differential module will be referred to as an *N-differential vector space*. Assume that E is a finite-dimensional N-differential vector space. Then one has $E \simeq \text{Ker}(d^n) \oplus \text{Im}(d^n) = Z_{(n)}(E) \oplus B_{(N-n)}(E)$ and $E \simeq \text{Ker}(d^{N-n}) \oplus \text{Im}(d^{N-n}) = Z_{(N-n)}(E) \oplus B_{(n)}(E)$ which together with $Z_{(n)}(E) \simeq B_{(n)}(E) \oplus H_{(n)}(E)$ and $Z_{(N-n)}(E) \simeq B_{(N-n)}(E) \oplus H_{(N-n)}(E)$ implies (since $\dim(E) < \infty$) that $H_{(n)}(E)$ and $H_{(N-n)}(E)$ are isomorphic. In the case where E is a finite-dimensional N-differential vector space over $\mathbf{k} = \mathbb{R}$ or $\mathbb{C}$, one can show (see e.g. in [**35**]) by decomposing E into indecomposable factor for the action of the N-differential d that one has an isomorphism $E \simeq \oplus_{n=1}^N \mathbf{k}^n \otimes \mathbf{k}^{m_n}$, $d \simeq \oplus_{n=2}^N D_n \otimes Id_{\mathbf{k}^{m_n}}$ with

$$D_n = \begin{pmatrix} 0 & 1 & 0 & \cdots & & 0 \\ & \cdot & \cdot & \cdot & & \cdot \\ \cdot & & \cdot & \cdot & \cdot & \cdot \\ \cdot & & & \cdot & \cdot & \cdot \\ \cdot & & & & \cdot & 0 \\ \cdot & & & & & 1 \\ 0 & \cdot & \cdot & \cdot & \cdot & 0 \end{pmatrix} \in M_n(\mathbf{k})$$

where the *multiplicities* m_n, $n \in \{1, \ldots, N\}$, are invariants of (E, d) with $\sum_{n=1}^N n m_n = \dim(E)$. Notice that one has $m_N \geq 1$ whenever $d^{N-1} \neq 0$. Notice also that the above decomposition of d is its *Jordan normal-form*. In terms of the multiplicities, one can easily compute the dimensions of the vector spaces $H_{(k)}(E)$. The result is given by the following proposition.

PROPOSITION 4. *Let E be a finite dimensional N-differential vector space over $\mathbb{R}$ or $\mathbb{C}$ with multiplicities m_n, $n \in \{1, 2, \ldots, N\}$, then one has for each integer k with $1 \leq k \leq N/2$*

$$\dim H_{(k)}(E) = \dim H_{(N-k)}(E) = \sum_{j=1}^{k} \sum_{i=j}^{N-j} m_i.$$

Although easy, that kind of results is useful for applications (see below).

6. N-complexes

An N-complex of modules [40] or simply an N-complex is an N-differential module E which is $\mathbb{Z}$-graded, i.e. $E = \oplus_{n \in \mathbb{Z}} E^n$, with a homogeneous N-differential d of degree 1 or -1. When d is of degree 1 then E is referred to as a *cochain N-complex* and when d is of degree -1 then E is referred to as a *chain N-complex*. Here we adopt the cochain language and therefore in the following an N-complex, without other specification, always means a cochain N-complex of modules. If E is an N-complex then the $H_{(m)}(E)$ are $\mathbb{Z}$-graded modules; $H_{(m)}(E) = \oplus_{n \in \mathbb{Z}} H^n_{(m)}(E)$ with $H^n_{(m)}(E) = \mathrm{Ker}(d^m : E^n \to E^{n+m})/d^{N-m}(E^{n+m-N})$. In this case the hexagon $(\mathcal{H}^{\ell,m})$ of Lemma 1 splits into long exact sequences $(\mathcal{S}^{\ell,m}_p)$, $p \in \mathbb{Z}$

$$\cdots \to H^{Nr+p}_{(m)}(E) \xrightarrow{[i]^\ell} H^{Nr+p}_{(\ell+m)}(E) \xrightarrow{[d]^m} H^{Nr+p+m}_{(\ell)}(E)$$

$$(\mathcal{S}^{\ell,m}_p) \qquad \xrightarrow{[i]^{N-(\ell+m)}} H^{Nr+p+m}_{(N-m)}(E) \xrightarrow{[d]^\ell} H^{Nr+p+\ell+m}_{(N-(\ell+m))}(E)$$

$$\xrightarrow{[i]^m} H^{Nr+p+\ell+m}_{(N-\ell)}(E) \xrightarrow{[d]^{N-(\ell+m)}} H^{N(r+1)+p}_{(m)}(E) \xrightarrow{[i]^\ell} \cdots$$

One has $(\mathcal{S}^{\ell,m}_p) = (\mathcal{S}^{\ell,m}_{p+N})$.

Let E and E' be N-complexes, a *homomorphism of N-complexes* of E into E' is a homomorphism of N-differential modules $\varphi : E \to E'$ which is homogeneous of degree 0, (i.e. $\varphi(E^n) \subset E'^n$). Such a homomorphism of N-complexes induces module-homomorphisms $\varphi_* : H^n_{(m)}(E) \to H^n_{(m)}(E')$ for $n \in \mathbb{Z}$ and $1 \leq m \leq N-1$. Let $0 \to E \xrightarrow{\varphi} F \xrightarrow{\psi} G \to 0$ be a short exact sequence of N-complexes, then the hexagon $(\mathcal{H}_n)$ of Lemma 2 splits into long exact sequences $(\mathcal{S}_{n,p})$, $p \in \mathbb{Z}$

$$\cdots \to H^{Nr+p}_{(n)}(E) \xrightarrow{\varphi_*} H^{Nr+p}_{(n)}(F) \xrightarrow{\psi_*} H^{Nr+p}_{(n)}(G)$$

$$(\mathcal{S}_{n,p}) \qquad \xrightarrow{\partial} H^{Nr+p+n}_{(N-n)}(E) \xrightarrow{\varphi_*} H^{Nr+p+n}_{(N-n)}(F) \xrightarrow{\psi_*} H^{Nr+p+n}_{(N-n)}(G)$$

$$\xrightarrow{\partial} H^{N(r+1)+p}_{(n)}(E) \xrightarrow{\varphi_*} \cdots$$

One has again $(\mathcal{S}_{n,p}) = (\mathcal{S}_{n,p+N})$.

In the following of this section, $(E^n)_{n\in\mathbb{N}}$ is a pre-cosimplicial module (see in Section 3), E denotes the (positively) graded module $\oplus_n E^n$ and $q \in \mathbf{k}$ is such that $[N]_q = 0$, i.e. such that $\mathbf{k}$ and $q \in \mathbf{k}$ satisfy the assumption (A_0) of Section 5. One can construct a sequence $(d_n)_{n\in\mathbb{N}}$ of N-differentials of degree 1 on E by using $q \in \mathbf{k}$ as above [**14**]. Here we shall only consider the first two d_0 and d_1 which are the most natural ones. They are defined by setting for $n \in \mathbb{N}$

$$d_0 = \sum_{i=0}^{n+1} q^i \mathfrak{f}_i : E^n \to E^{n+1}$$

and

$$d_1 = \sum_{i=0}^{n} q^i \mathfrak{f}_i - q^n \mathfrak{f}_{n+1} : E^n \to E^{n+1}.$$

LEMMA 5. *One has $d_0^N = 0$ and $d_1^N = 0$.*

This is a consequence of $[N]_q = 0$ and of the relations ($\mathfrak{F}$); for a proof we refer to [**14**].
Thus (E, d_0) and (E, d_1) are N-complexes and, as shown in [**14**], there are natural homomorphisms of the cohomology $H(E)$ of the pre-cosimplicial module (E^n) into the generalized cohomologies of these N-complexes. In order to compute completely these generalized cohomologies we shall need some more assumptions. We shall need Assumption (A_1) for $\mathbf{k}$ and $q \in \mathbf{k}$ and we shall restrict attention to cosimplicial modules. The generalized cohomologies of (E, d_0) and (E, d_1) are then given by the following theorem [**14**].

THEOREM 2. *Let $\mathbf{k}$ and $q \in \mathbf{k}$ satisfy Assumption (A_1) and let (E^n) be a cosimplicial module. Then one has:*

(0) $H_{(m)}^{Nr-1}(E, d_0) = H^{2r-1}(E)$, $H_{(m)}^{N(r+1)-m-1}(E, d_0) = H^{2r}(E)$ *and* $H_{(m)}^n(E, d_0) = 0$ *otherwise,*

(1) $H_{(m)}^{Nr}(E, d_1) = H^{2r}(E)$, $H_{(m)}^{N(r+1)-m}(E, d_1) = H^{2r+1}(E)$ *and* $H_{(m)}^n(E, d_1) = 0$ *otherwise,*

for $r \in \mathbb{N}$ and $m \in \{1, \ldots, N-1\}$.

There is of course a dual statement for simplicial modules and the analogs of d_0 and d_1 which are then of degree -1, see in [**14**]. The above theorem and its dual version cover all the (co)simplicial cases investigated so far that I know ([**49**], [**20**], [**13**], [**43**] and [**14**]). In [**14**] the generalized cohomology of E for every d_n ($n \in \mathbb{N}$) was also computed in the case of a cosimplicial module (E^n) as well as the generalized homologies of their chain analogs in the case of a simplicial module (E_n) under assumption (A_1) for $\mathbf{k}$ and $q \in \mathbf{k}$. As a rule, we found there that (in the (co)simplicial case) these generalized (co)homologies do only depend on the ordinary (co)homology of the (co)simplicial module. In fact one of the ingredients in the proof of the above theorem is to use the whole sequence of N-differentials $(d_n)_{n\in\mathbb{N}}$ because, for any $p \in \mathbb{N}$ there is a $n_p \in \mathbb{N}$ such that d_n coincides with the simplicial differential in degree r (i.e. on E^r) whenever $n \geq n_p$ and $r \leq p$; the proof is nevertheless highly non trivial, (see in [**14**]).

Many notions for N-complexes do only depend on the underlying $\mathbb{Z}_N$-graduation ($\mathbb{Z}_N = \mathbb{Z}/N\mathbb{Z}$) so let us define a $\mathbb{Z}_N$-*complex* to be an N-differential module which is $\mathbb{Z}_N$-graded with an N-differential which is homogeneous of degree 1. We have avoided the terminology $\mathbb{Z}_N$-N-complex since we shall not consider differential modules equipped with $\mathbb{Z}_N$ graduations with $N \geq 3$ and since a $\mathbb{Z}_2$-complex with the above definition is a $\mathbb{Z}_2$-complex according to the definition of Section 3. We now give an example of $\mathbb{Z}_N$-complex.

Let $\mathbf{k}$ and $q \in \mathbf{k}$ satisfy Assumption (A_1) and let us introduce the standard basis E_ℓ^k, ($k, \ell \in \{1, \ldots, N\}$), of the algebra $M_N(\mathbf{k})$ of $N \times N$ matrices defined by $(E_\ell^k)_j^i = \delta_j^k \delta_\ell^i$. One has $E_\ell^k E_s^r = \delta_s^k E_\ell^r$ and $\sum_{n=1}^N E_n^n = \mathbb{1}$. It follows that one can equip $M_N(\mathbf{k})$ with a structure of $\mathbb{Z}_N$-graded algebra, $M_N(\mathbf{k}) = \oplus_{a \in \mathbb{Z}_N} M_N(\mathbf{k})^a$, by giving to E_ℓ^k the degree $k - \ell \bmod(N)$. Let $e = \lambda_1 E_1^2 + \cdots + \lambda_{N-1} E_{N-1}^N + \lambda_N E_N^1$ be an element of degree 1 of $M_N(\mathbf{k})$ and define the endomorphism d by $d(A) = eA - q^a Ae$ for $A \in M_N(\mathbf{k})^a$. One has $d^N = 0$ so $(M_N(\mathbf{k}), d)$ is a $\mathbb{Z}_N$-complex. One verifies that $e^N = \lambda_1 \ldots \lambda_N \mathbb{1}$ and that $e^{N-1} d(A) - q d(e^{N-1} A) = (1-q) \lambda_1 \ldots \lambda_N A$. Therefore if $1-q$ and the λ_i are invertible in $\mathbf{k}$, Lemma 4 implies that $H_{(n)}(M_N(\mathbf{k}), d) = 0$ for $n \in \{1, \ldots, N-1\}$. It is worth noticing that the above N-differential satisfies the *graded q-Leibniz rule* $d(AB) = d(A)B + q^a A d(B)$, $\forall A \in M_N(\mathbf{k})^a$, $\forall B \in M_N(\mathbf{k})$.

It is clear that for any N-complex one has an underlying $\mathbb{Z}_N$-complex which is obtained by retaining only the degree modulo N. On the other hand starting from a $\mathbb{Z}_N$-complex $E = \oplus_{n \in \mathbb{Z}_N} E^n$ like $(M_N(\mathbf{k}), d)$ above, one can construct an N-complex $\tilde{E} = \oplus_{n \in \mathbb{Z}} \tilde{E}^n$ by setting $\tilde{E}^n = E^{\pi(n)}$ where π is the canonical projection of $\mathbb{Z}$ onto $\mathbb{Z}_N$, the definition of the N-differential on $\tilde{E}$ being obvious in terms of the one of E.

The content of Section 5 and Section 6 is based on [14] (see also in [20] and in [13]). Particular N-complexes were introduced and analysed in [49] for $\mathbf{k} = \mathbb{Z}_N$, ($N$ prime). Several mathematicians wrote on N-complexes at the end of the 40's, beginning of the 50's. The subject was reconsidered in [40] and developed more recently in [20], [13], [43] and [14]. In [43] an approach in the line of modern homological algebra [7] was developed with the introduction of generalizations of the functors Ext and Tor.

7. N-complexes of tensor fields

In this section we shall describe N-complexes of tensor fields on $\mathbb{R}^D$ which generalize the complex $\Omega(\mathbb{R}^D)$ of differential forms [17], [18]. Therefore here the ring $\mathbf{k}$ is the field $\mathbb{R}$ (or eventually $\mathbb{C}$ if one considers complex tensors). Furthermore, in such an N-complex, for each degree the tensor fields will be smooth mapping $x \mapsto T(x)$ of $\mathbb{R}^D$ into the vector space of covariant tensors of a given Young symmetry. Let us recall that this implies that the representation of GL_D in the corresponding space of tensors is irreducible. For Young diagrams, etc. we refer to [30] and for more details and developments we refer to [17], [18].

Throughout the following $(x^\mu) = (x^1, \ldots, x^D)$ denotes the canonical coordinates of $\mathbb{R}^D$ and ∂_μ are the corresponding partial derivatives which we identify with the corresponding covariant derivatives associated to the canonical flat linear connection of $\mathbb{R}^D$. Thus, for instance, if T is a covariant tensor field of degree p on $\mathbb{R}^D$ with components $T_{\mu_1\ldots\mu_p}(x)$, then ∂T denotes the covariant tensor field of degree $p+1$ with components $\partial_{\mu_1} T_{\mu_2\ldots\mu_{p+1}}(x)$. The operator ∂ is a first-order differential operator which increases by one the tensorial degree.

In this context, the space $\Omega(\mathbb{R}^D)$ of differential forms on $\mathbb{R}^D$ is the graded vector space of (covariant) antisymmetric tensor fields on $\mathbb{R}^D$ with graduation induced by the tensorial degree whereas the exterior differential d is the composition of the above ∂ with antisymmetrisation, i.e.

$$d = \mathbf{A}_{p+1} \circ \partial : \Omega^p(\mathbb{R}^D) \to \Omega^{p+1}(\mathbb{R}^D)$$

where $\mathbf{A}_p$ denotes the antisymmetrizer on tensors of degree p. One has $d^2 = 0$ and the Poincaré lemma asserts that the cohomology of the complex $(\Omega(\mathbb{R}^D), d)$ is trivial, i.e. that one has $H^p(\Omega(\mathbb{R}^D)) = 0$, $\forall p \geq 1$ and $H^0(\Omega(\mathbb{R}^D)) = \mathbb{R}$.

From the point of view of Young symmetry, antisymmetric tensors correspond to Young diagrams (partitions) described by one column of cells, i.e. the space of values of p-forms corresponds to one column of p cells, (1^p), whereas $\mathbf{A}_p$ is the associated Young symmetrizer, (see e.g. in [**30**]).

There is a relatively easy way to generalize the pair $(\Omega(\mathbb{R}^D), d)$ which we now describe. Let $Y = (Y_p)_{p\in\mathbb{N}}$ be a sequence of Young diagrams such that the number of cells of Y_p is p, $\forall p \in \mathbb{N}$ (i.e. such that Y_p is a partition of the integer p for any p). We define $\Omega_Y^p(\mathbb{R}^D)$ to be the vector space of smooth covariant tensor fields of degree p on $\mathbb{R}^D$ which have the Young symmetry type Y_p and we let $\Omega_Y(\mathbb{R}^D)$ be the graded vector space $\oplus_p \Omega_Y^p(\mathbb{R}^D)$. We then generalize the exterior differential by setting $d = \mathbf{Y} \circ \partial$, i.e.

$$d = \mathbf{Y}_{p+1} \circ \partial : \Omega_Y^p(\mathbb{R}^D) \to \Omega_Y^{p+1}(\mathbb{R}^D)$$

where $\mathbf{Y}_p$ is now the Young symmetrizer on tensor of degree p associated to the Young symmetry Y_p. This d is again a first order differential operator which is of degree one, (i.e. it increases the tensorial degree by one), but now, $d^2 \neq 0$ in general. Instead, one has the following result.

LEMMA 6. *Let N be an integer with $N \geq 2$ and assume that Y is such that the number of columns of the Young diagram Y_p is strictly smaller than N (i.e. $\leq N-1$) for any $p \in \mathbb{N}$. Then one has $d^N = 0$.*

In fact the indices in one column are antisymmetrized and $d^N \omega$ involves necessarily at least two partial derivatives ∂ in one of the columns since there are N partial derivatives involved and at most $N - 1$ columns.

Thus if Y satisfies the condition of Lemma 6, $(\Omega_Y(\mathbb{R}^D), d)$ is an N-complex. Notice that $\Omega_Y^p(\mathbb{R}^D) = 0$ if the first column of Y_p contains more than D cells and that therefore, if Y satisfies the condition of Lemma 6, then $\Omega_Y^p(\mathbb{R}^D) = 0$ for $p > (N-1)D$.

One can also define a graded bilinear product on $\Omega_Y(\mathbb{R}^D)$ by setting

$$(\alpha\beta)(x) = \mathbf{Y}_{a+b}(\alpha(x) \otimes \beta(x))$$

for $\alpha \in \Omega_Y^a(\mathbb{R}^D)$, $\beta \in \Omega_Y^b(\mathbb{R}^D)$ and $x \in \mathbb{R}^D$. This product is by construction bilinear with respect to the $C^\infty(\mathbb{R}^D)$-module structure of $\Omega_Y(\mathbb{R}^D)$, $(\Omega_Y^0(\mathbb{R}^D) = C^\infty(\mathbb{R}^D))$. However it is generically non associative.

In the following we shall not stay at this level of generality but, for each $N \geq 2$ we shall choose a particular Y, denoted by $Y^N = (Y_p^N)_{p\in\mathbb{N}}$, satisfying the condition of Lemma 6 which is maximal in the sense that all the rows are of maximal length $N-1$ except the last one (eventually). In other words the Young diagram with p cells Y_p^N is defined in the following manner: write the division of p by $N-1$, i.e. write $p = (N-1)n_p + r_p$ where n_p and r_p are (the unique) integers with $0 \leq n_p$ and $0 \leq r_p \leq N-2$ (n_p is the quotient whereas r_p is the remainder), and let Y_p^N be the Young diagram with n_p rows of $N-1$ cells and the last row with r_p cells (if $r_p \neq 0$). One has $Y_p^N = ((N-1)^{n_p}, r_p)$, that is we fill the rows maximally.

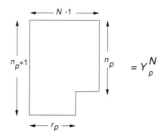

We shall denote $\Omega_{Y^N}(\mathbb{R}^D)$ and $\Omega_{Y^N}^p(\mathbb{R}^D)$ by $\Omega_N(\mathbb{R}^D)$ and $\Omega_N^p(\mathbb{R}^D)$. It is clear that $(\Omega_2(\mathbb{R}^D), d)$ is the usual complex $(\Omega(\mathbb{R}^D), d)$ of differential forms on $\mathbb{R}^D$. The N-complex $(\Omega_N(\mathbb{R}^D), d)$ will be simply denoted by $\Omega_N(\mathbb{R}^D)$. The Poincaré lemma admits the following generalization [17], [18].

THEOREM 3. *One has $H_{(k)}^{(N-1)n}(\Omega_N(\mathbb{R}^D)) = 0$, $\forall n \geq 1$ and $H_{(k)}^0(\Omega_N(\mathbb{R}^D))$ is the space of real polynomial functions on $\mathbb{R}^D$ of degree strictly less than k (i.e. $\leq k-1$) for $k \in \{1, \ldots, N-1\}$.*

This statement reduces to the Poincaré lemma for $N = 2$ but it is a nontrivial generalization for $N \geq 3$ in the sense that, the spaces $H_{(k)}^p(\Omega_N(\mathbb{R}^D))$ are nontrivial for $p \neq (N-1)n$ and in fact generically infinite dimensional for $D \geq 3$, $p \geq N$.

The connection between the complex of differential forms on $\mathbb{R}^D$ and the theory of classical gauge field of spin 1 is well known. Namely the subcomplex

(1) $$\Omega^0(\mathbb{R}^D) \stackrel{d}{\to} \Omega^1(\mathbb{R}^D) \stackrel{d}{\to} \Omega^2(\mathbb{R}^D) \stackrel{d}{\to} \Omega^3(\mathbb{R}^D)$$

has the following interpretation in terms of spin 1 gauge field theory. The space $\Omega^0(\mathbb{R}^D)(= C^\infty(\mathbb{R}^D))$ is the space of infinitesimal gauge transformations, the space $\Omega^1(\mathbb{R}^D)$ is the space of gauge potentials (which are the appropriate description of spin 1 gauge fields to introduce local interactions). The subspace $d\Omega^0(\mathbb{R}^D)$ of $\Omega^1(\mathbb{R}^D)$ is the space of pure gauge configurations (which are physically irrelevant), $d\Omega^1(\mathbb{R}^D)$ is the space of field strengths or curvatures of gauge potentials. The identity $d^2 = 0$ ensures that the curvatures do not see the irrelevant pure gauge potentials whereas, at this level, the Poincaré lemma ensures that it is only these irrelevant configurations which are forgotten when one passes from gauge potentials to curvatures (by applying d). Finally $d^2 = 0$ also ensures that curvatures of gauge potentials satisfy the Bianchi identity, i.e. are in $\mathrm{Ker}(d : \Omega^2(\mathbb{R}^D) \to \Omega^3(\mathbb{R}^D))$, whereas at this level the Poincaré lemma implies that conversely the Bianchi identity characterizes the elements of $\Omega^2(\mathbb{R}^D)$ which are curvatures of gauge potentials.

Classical spin 2 gauge field theory is the linearization of Einstein geometric theory. In this case, and more generally in the linearization of (pseudo)riemannian geometry, the analog of (1) is a complex $\mathcal{E}^1 \overset{d_1}{\to} \mathcal{E}^2 \overset{d_2}{\to} \mathcal{E}^3 \overset{d_3}{\to} \mathcal{E}^4$ where $\mathcal{E}^1$ is the space of covariant vector field $(x \mapsto X_\mu(x))$ on $\mathbb{R}^D$, $\mathcal{E}^2$ is the space of covariant symmetric tensor fields of degree 2 $(x \mapsto h_{\mu\nu}(x))$ on $\mathbb{R}^D$, $\mathcal{E}^3$ is the space of covariant tensor fields of degree 4 $(x \mapsto R_{\lambda\mu,\rho\nu}(x))$ on $\mathbb{R}^D$ having the symmetries of the Riemann curvature tensor and where $\mathcal{E}^4$ is the space of covariant tensor fields of degree 5 on $\mathbb{R}^D$ having the symmetries of the left-hand side of the Bianchi identity. The arrows d_1, d_2, d_3 are given by

$$(d_1 X)_{\mu\nu}(x) = \partial_\mu X_\nu(x) + \partial_\nu X_\mu(x)$$
$$(d_2 h)_{\lambda\mu,\rho\nu}(x) = \partial_\lambda \partial_\rho h_{\mu\nu}(x) + \partial_\mu \partial_\nu h_{\lambda\rho}(x) - \partial_\mu \partial_\rho h_{\lambda\nu}(x) - \partial_\lambda \partial_\nu h_{\mu\rho}(x)$$
$$(d_3 R)_{\lambda\mu\nu,\alpha\beta}(x) = \partial_\lambda R_{\mu\nu,\alpha\beta}(x) + \partial_\mu R_{\nu\lambda,\alpha\beta}(x) + \partial_\nu R_{\lambda\mu,\alpha\beta}(x).$$

The symmetry of $x \mapsto R_{\lambda\mu,\rho\nu}(x)$, $\left(\begin{array}{|c|c|}\hline \lambda & \rho \\ \hline \mu & \nu \\ \hline \end{array}\right)$, shows that $\mathcal{E}^3 = \Omega_3^4(\mathbb{R}^D)$ and that $\mathcal{E}^4 = \Omega_3^5(\mathbb{R}^D)$; furthermore one canonically has $\mathcal{E}^1 = \Omega_3^1(\mathbb{R}^D)$ and $\mathcal{E}^2 = \Omega_3^2(\mathbb{R}^D)$. One also sees that d_1 and d_3 are proportional to the 3-differential d of $\Omega_3(\mathbb{R}^D)$, i.e. $d_1 \sim d : \Omega_3^1(\mathbb{R}^D) \to \Omega_3^2(\mathbb{R}^D)$ and $d_3 \sim d : \Omega_3^4(\mathbb{R}^D) \to \Omega_3^5(\mathbb{R}^D)$. The structure of d_2 looks different, it is of second order and increases by 2 the tensorial degree. However it is easy to see that it is proportional to $d^2 : \Omega_3^2(\mathbb{R}^D) \to \Omega_3^4(\mathbb{R}^D)$. Thus the analog of (1) is (for spin 2 gauge field theory)

$$(2) \qquad \Omega_3^1(\mathbb{R}^D) \overset{d}{\to} \Omega_3^2(\mathbb{R}^D) \overset{d^2}{\to} \Omega_3^4(\mathbb{R}^D) \overset{d}{\to} \Omega_3^5(\mathbb{R}^D)$$

and the fact that it is a complex follows from $d^3 = 0$ whereas the generalized Poincaré lemma (Theorem 3) implies that it is in fact an exact sequence. Exactness at $\Omega_3^2(\mathbb{R}^D)$ is $H_{(2)}^2(\Omega_3(\mathbb{R}^D)) = 0$ and exactness at $\Omega_3^4(\mathbb{R}^D)$ is $H_{(1)}^4(\Omega_3(\mathbb{R}^D)) = 0$, (the exactness at $\Omega_3^4(\mathbb{R}^D)$ is the main statement of [33]).

Thus what plays the role of the complex of differential forms for the spin 1 (i.e. $\Omega_2(\mathbb{R}^D)$) is the 3-complex $\Omega_3(\mathbb{R}^D)$ for the spin 2. More generally, for the spin $S \in \mathbb{N}$, this role is played by the $(S+1)$-complex $\Omega_{S+1}(\mathbb{R}^D)$. In particular, the analog of Sequence (1) for spin 1 is the complex

$$(3) \qquad \Omega_{S+1}^{S-1}(\mathbb{R}^D) \overset{d}{\to} \Omega_{S+1}^{S}(\mathbb{R}^D) \overset{d^S}{\to} \Omega_{S+1}^{2S}(\mathbb{R}^D) \overset{d}{\to} \Omega_{S+1}^{2S+1}(\mathbb{R}^D)$$

for the spin S. The fact that (3) is a complex was known, [10], here it follows from $d^{S+1} = 0$. One easily recognizes that $d^S : \Omega_{S+1}^S(\mathbb{R}^D) \to \Omega_{S+1}^{2S}(\mathbb{R}^D)$ is the generalized (linearized) curvature of [10]. Theorem 3 implies that sequence (3) is exact: exactness at $\Omega_{S+1}^S(\mathbb{R}^D)$ is $H_{(S)}^S(\Omega_{S+1}(\mathbb{R}^D)) = 0$ whereas exactness at $\Omega_{S+1}^{2S}(\mathbb{R}^D)$ is $H_{(1)}^{2S}(\Omega_{S+1}(\mathbb{R}^D)) = 0$, (exactness at $\Omega_{S+1}^S(\mathbb{R}^D)$ was directly proved in [9] for the case $S = 3$).

Finally, there is a generalization of Hodge duality for $\Omega_N(\mathbb{R}^D)$, which is obtained by contractions of the columns with the Kroneker tensor $\varepsilon^{\mu_1 \cdots \mu_D}$ of $\mathbb{R}^D$ [17],

[18]. When combined with Theorem 3, this duality leads to another kind of results. A typical result of this kind is the following one. Let $T^{\mu\nu}$ be a symmetric contravariant tensor field of degree 2 on $\mathbb{R}^D$ satisfying $\partial_\mu T^{\mu\nu} = 0$, (like e.g. the stress energy tensor), then there is a contravariant tensor field $R^{\lambda\mu\rho\nu}$ of degree 4 with the symmetry $\begin{array}{|c|c|}\hline \lambda & \rho \\ \hline \mu & \nu \\ \hline \end{array}$, (i.e. the symmetry of Riemann curvature tensor), such that

$$T^{\mu\nu} = \partial_\lambda \partial_\rho R^{\lambda\mu\rho\nu}$$

In order to connect this result with Theorem 3, define $\tau_{\mu_1\ldots\mu_{D-1}\nu_1\ldots\nu_{D-1}} = T^{\mu\nu}\varepsilon_{\mu\mu_1\ldots\mu_{D-1}}\varepsilon_{\nu\nu_1\ldots\nu_{D-1}}$. Then one has $\tau \in \Omega_3^{2(D-1)}(\mathbb{R}^D)$ and conversely, any $\tau \in \Omega_3^{2(D-1)}(\mathbb{R}^D)$ can be expressed in this form in terms of a symmetric contravariant 2-tensor. It is easy to verify that $d\tau = 0$ (in $\Omega_3(\mathbb{R}^D)$) is equivalent to $\partial_\mu T^{\mu\nu} = 0$. On the other hand, Theorem 3 implies that $H_{(1)}^{2(D-1)}(\Omega_3(\mathbb{R}^D)) = 0$ and therefore $\partial_\mu T^{\mu\nu} = 0$ implies that there is a $\rho \in \Omega_3^{2(D-2)}(\mathbb{R}^D)$ such that $\tau = d^2\rho$. The latter is equivalent to the above equation with $R^{\mu_1\mu_2\ \nu_1\nu_2}$ proportional to $\varepsilon^{\mu_1\mu_2\ldots\mu_D}\varepsilon^{\nu_1\nu_2\ldots\nu_D}\rho_{\mu_3\ldots\mu_D\nu_3\ldots\nu_D}$ and one verifies that, so defined, R has the correct symmetry. This result has been used in [65] in the investigation of the consistent deformations of the free spin two gauge field action.

8. Graded differential algebras and generalizations

A *graded differential algebra* is a (cochain) complex $\mathfrak{A} = \oplus_{n\in\mathbb{Z}}\mathfrak{A}^n$ with differential d such that $\mathfrak{A}$ is a $\mathbb{Z}$-graded associative unital **k**-algebra and such that d is an antiderivation i.e. satisfies the graded Leibniz rule

$$d(\alpha\beta) = d(\alpha)\beta + (-1)^a \alpha d(\beta)$$

for any $\alpha \in \mathfrak{A}^a$, $\beta \in \mathfrak{A}$ and where $(\alpha,\beta) \mapsto \alpha\beta$ denotes the product of $\mathfrak{A}$. If $\mathfrak{A}$ is such a graded differential algebra with differential d, $\mathrm{Ker}(d)$ is a graded unital subalgebra of $\mathfrak{A}$ whereas $\mathrm{Im}(d)$ is a graded two-sided ideal of $\mathrm{Ker}(d)$ so the cohomology $H(\mathfrak{A})$ is a (unital associative) graded algebra. If $\mathfrak{A}$ and $\mathfrak{B}$ are two graded differential algebras, the tensor product $\mathfrak{A} \otimes \mathfrak{B}$ of the underlying complexes (as defined in Section 3) is again a graded differential algebra with product defined by

$$(\alpha \otimes \beta)(\alpha' \otimes \beta') = (-1)^{ba'}\alpha\alpha' \otimes \beta\beta'$$

for $\alpha \in \mathfrak{A}$, $\beta \in \mathfrak{B}^b$, $\alpha' \in \mathfrak{A}^{a'}$ and $\beta' \in \mathfrak{B}$. In the following, the product of a tensor product of graded algebras will be always the above one. With this convention, if **k** is a field one has $H(\mathfrak{A} \otimes \mathfrak{B}) = H(\mathfrak{A}) \otimes H(\mathfrak{B})$ for the corresponding cohomology algebras (which is the refined counterpart of Proposition 2 for graded differential algebras).

Let $(\mathfrak{A}^n)_{n\in\mathbb{N}}$ be a pre-cosimplicial module (see in Section 3) such that $\mathfrak{A} = \oplus \mathfrak{A}^n$ is a (positively) graded algebra and assume that the cofaces homomorphisms $\mathfrak{f}_i$ satisfy the following assumptions $(\mathfrak{MF})$:

$(\mathfrak{MF}_1)$ $\qquad \mathfrak{f}_i(\alpha\beta) = \begin{cases} \mathfrak{f}_i(\alpha)\beta & \text{if } i \leq a \\ \alpha\mathfrak{f}_{i-a}(\beta) & \text{if } i > a \end{cases}, i \in \{0,\ldots,a+b+1\}$

and

$(\mathfrak{MF}_2)$ $\qquad \mathfrak{f}_{a+1}(\alpha)\beta = \alpha\mathfrak{f}_0(\beta)$

for $\alpha \in \mathfrak{A}^a$ and $\beta \in \mathfrak{A}^b$ (where $(\alpha,\beta) \mapsto \alpha\beta$ denote the product of $\mathfrak{A}$). Then the corresponding complex $(\mathfrak{A}, d)$ is a graded differential algebra. If furthermore $(\mathfrak{A}^n)$ is a cosimplicial module with codegeneracy homomorphisms $\mathfrak{s}_i$ satisfying the following assumption ($\mathfrak{MS}$)

$$(\mathfrak{MS}) \qquad \mathfrak{s}_i(\alpha\beta) = \begin{cases} \mathfrak{s}_i(\alpha)\beta & \text{if } i < a \\ \alpha\mathfrak{s}_{i-a}(\beta) & \text{if } i \geq a \end{cases}$$

$i \in \{0, \ldots, a + b - 1\}$, then the subcomplex $N(\mathfrak{A})$ of normalized cochains of $\mathfrak{A}$ is a graded differential subalgebra of $\mathfrak{A}$. In [14], a pre-cosimplicial module $(\mathfrak{A}^n)$ as above with cofaces satisfying ($\mathfrak{MF}$) (which was denoted there by $(\mathfrak{AF})$) was called a *pre-cosimplicial algebra* and in the case where $(\mathfrak{A}^n)$ is furthermore a cosimplicial module with codegeneracies satisfying ($\mathfrak{MS}$) (which was denoted there $(\mathfrak{AS})$) it was called a *cosimplicial algebra*, however it has been remarked by Max Karoubi that this terminology is misleading so we shall speak in the following of a $\mathfrak{M}$-*pre-cosimplicial module* in the first case and of a $\mathfrak{M}$-*cosimplicial module* in the second case. In fact $\mathfrak{M}$-cosimplicial modules is what corresponds to graded differential algebras in an appropriate specific version of the Dold-Kan correspondence.

Let $\mathcal{A}$ be an associative unital **k**-algebra and $\mathcal{M}$ be a $(\mathcal{A}, \mathcal{A})$-bimodule. As pointed out in Section 3, the $\mathcal{M}$-valued Hochschild cochains give rise to a cosimplicial module $(C^n(\mathcal{A}, \mathcal{M}))_{n \in \mathbb{N}}$. In the case $\mathcal{M} = \mathcal{A}$, $C(\mathcal{A}, \mathcal{A})$ has a natural structure of $\mathbb{N}$-graded associative unital **k**-algebra with product $(\alpha, \beta) \mapsto \alpha\beta$ given by

$$\alpha\beta(x_1, \ldots, x_{a+b}) = \alpha(x_1, \ldots, x_a)\beta(x_{a+1}, \ldots, x_{a+b}),$$

for $\alpha \in C^a(\mathcal{A}, \mathcal{A})$, $\beta \in C^b(\mathcal{A}, \mathcal{A}), x_i \in \mathcal{A}$.
It is easily verified that the assumptions ($\mathfrak{MF}$) and ($\mathfrak{MS}$) are satisfied so that $(C^n(\mathcal{A}, \mathcal{A}))$ is a $\mathfrak{M}$-cosimplicial module. Thus $C(\mathcal{A}, \mathcal{A})$ equipped with the simplicial (Hochschild) differential (as in Section 3) is a graded differential algebra and the submodule of normalized cochains is a graded differential subalgebra of $C(\mathcal{A}, \mathcal{A})$.

Let again $\mathcal{A}$ be an associative unital **k**-algebra and let us denote by $\mathfrak{T}(\mathcal{A}) = \oplus_{n \in \mathbb{N}} \mathfrak{T}^n(\mathcal{A})$ the tensor algebra over $\mathcal{A}$ of the $(\mathcal{A}, \mathcal{A})$-bimodule $\mathcal{A} \otimes \mathcal{A}$. This is a (positively) graded associative unital **k**-algebra with $\mathfrak{T}^n(\mathcal{A}) = \otimes^{n+1}\mathcal{A}$ and product $(x_0 \otimes \cdots \otimes x_n)(y_0 \otimes \cdots \otimes y_m) = x_0 \otimes \cdots \otimes x_{n-1} \otimes x_n y_0 \otimes y_1 \otimes \cdots \otimes y_m$ for $x_i, y_j \in \mathcal{A}$. One verifies that one defines a structure of $\mathfrak{M}$-cosimplicial module for $(\mathfrak{T}^n(\mathcal{A}))$ by setting
$\mathfrak{f}_0(x_0 \otimes \cdots \otimes x_n) = \mathbf{1} \otimes x_0 \otimes \cdots \otimes x_n$
$\mathfrak{f}_i(x_0 \otimes \cdots \otimes x_n) = x_0 \otimes \cdots \otimes x_{i-1} \otimes \mathbf{1} \otimes x_i \otimes \cdots \otimes x_n$ for $1 \leq i \leq n$
$\mathfrak{f}_{n+1}(x_0 \otimes \cdots \otimes x_n) = x_0 \otimes \cdots \otimes x_n \otimes \mathbf{1}$
and
$\mathfrak{s}_i(x_0 \otimes \cdots \otimes x_n) = x_0 \otimes \cdots \otimes x_i x_{i+1} \otimes \cdots \otimes x_n$ for $0 \leq i \leq n - 1$.
It follows that, equipped with the corresponding simplicial differential, $\mathfrak{T}(\mathcal{A})$ is a graded differential algebra and that the submodule of normalized cochains is a graded differential subalgebra of $\mathfrak{T}(\mathcal{A})$. This latter graded differential algebra will be denoted by $\Omega(\mathcal{A})$ and referred to as the *universal graded differential envelope* of $\mathcal{A}$ or simply the *universal differential envelope* of $\mathcal{A}$; it is characterized by the following universal property [41], [42] (see also e.g. in [8] and [16]).

PROPOSITION 5. *Any homomorphism φ of unital algebras of $\mathcal{A}$ into the subalgebra Ω^0 of elements of degree 0 of a graded differential algebra Ω has a unique extension $\tilde{\varphi} : \Omega(\mathcal{A}) \to \Omega$ as a homomorphism of graded differential algebras.*

The graded differential algebra $\Omega(\mathcal{A})$ is usually constructed in a different manner; the fact that it identifies with the graded differential algebra of normalized cochains of $\mathfrak{T}(\mathcal{A})$ is well known. It is worth noticing here that $\Omega(\mathcal{A})$ is also the graded differential subalgebra of $\mathfrak{T}(\mathcal{A})$ generated by $\mathcal{A}$ (i.e. the smallest graded differential subalgebra which contains $\mathcal{A}$).

We now come to an N-complex version of graded differential algebra ($N \geq 2$). For that we shall need $q \in \mathbf{k}$ such that Assumption (A_1) of Section 5 is satisfied i.e. $[N]_q = 0$ and $[n]_q$ invertible in $\mathbf{k}$ for $n \in \{1, \cdots, N-1\}$. Throughout the following of this section, N and $q \in \mathbf{k}$ are fixed and such that (A_1) is satisfied. The following lemma is basic for the generalization, [14]. In this lemma, (and in the following) d_1 is the N-differential defined in Section 6 for any pre-cosimplicial module.

LEMMA 7. *Suppose that $\mathbf{k}$ and $q \in \mathbf{k}$ satisfy Assumption (A_1) and let ($\mathfrak{A}^n$) be a $\mathfrak{M}$-pre-cosimplicial module. Then the N-differential d_1 satisfies the graded q-Leibniz rule, that is*

$$d_1(\alpha\beta) = d_1(\alpha)\beta + q^a \alpha d_1(\beta)$$

for $\alpha \in \mathfrak{A}^a$ and $\beta \in \mathfrak{A} = \oplus_n \mathfrak{A}^n$.

A unital associative graded algebra equipped with an N-differential satisfying the (above) *graded q-Leibniz rule* will be referred to as a *graded q-differential algebra* [20], [14]. The content of the above lemma is that if ($\mathfrak{A}^n$) is a $\mathfrak{M}$-pre-cosimplicial module then ($\mathfrak{A}, d_1$) is a graded q-differential algebra which is positively graded. If, furthermore ($\mathfrak{A}^n$) is a $\mathfrak{M}$-cosimplicial module then the generalized cohomology of ($\mathfrak{A}, d_1$) is given in terms of the ordinary cohomology of ($\mathfrak{A}^n$) by Theorem 2.

Let $\mathcal{A}$ be as above an associative unital algebra. It follows from Lemma 7 that $\mathfrak{T}(\mathcal{A})$ equipped with the N-differential d_1 is a graded q-differential algebra (which is $\mathbb{N}$-graded). Let $\Omega_q(\mathcal{A})$ be the graded q-differential subalgebra of $\mathfrak{T}(\mathcal{A})$ generated by $\mathcal{A}$, i.e. the smallest subalgebra of $\mathfrak{T}(\mathcal{A})$ which contains $\mathcal{A}$ and which is stable by the N-differential d_1. As graded q-differential algebra, $\Omega_q(\mathcal{A})$ is characterized uniquely up to an isomorphism by the following universal property [20], [14].

PROPOSITION 6. *Any homomorphism φ of unital algebras of $\mathcal{A}$ into the subalgebra Ω^0 of elements of degree 0 of a graded q-differential algebra Ω has a unique extension $\tilde{\varphi} : \Omega_q(\mathcal{A}) \to \Omega$ as a homomorphism of graded q-differential algebras.*

This is the q-analog of Proposition 5, (a homomorphism of graded q-differential algebra being a homomorphism of graded algebras permuting the N-differentials). For $N = 2$, $\Omega_q(\mathcal{A})$ reduces to $\Omega(\mathcal{A})$. The graded q-differential algebra $\Omega_q(\mathcal{A})$ is referred to as the *universal q-differential envelope* of $\mathcal{A}$ [20], [14]. The generalized cohomologies of ($\mathfrak{T}(\mathcal{A}), d_1$) and of $\Omega_q(\mathcal{A})$ are generically trivial; one has the following result, [14].

PROPOSITION 7. *Assume that $\mathcal{A}$ admits a linear form $\omega \in \mathcal{A}^*$ such that $\omega(\mathbf{1}) = 1$. Then the generalized cohomologies of ($\mathfrak{T}(\mathcal{A}), d_1$) and of $\Omega_q(\mathcal{A})$ are given* by:

$$H_{(k)}^n(\mathfrak{T}(\mathcal{A}), d_1) = H_{(k)}^n(\Omega_q(\mathcal{A})) = 0 \quad \text{for } n \geq 1 \text{ and}$$
$$H_{(k)}^0(\mathfrak{T}(\mathcal{A}), d_1) = H_{(k)}^0(\Omega_q(\mathcal{A})) = \mathbf{k}, \quad \forall k \in \{1, \cdots, N-1\}.$$

Notice that the assumption of this proposition is satisfied if $\mathbf{k}$ is a field and that the case $N = 2$ means, under the same assumption, the triviality of the cohomologies of $\mathfrak{T}(\mathcal{A})$ and $\Omega(\mathcal{A})$, (a well known fact, [42]).

The above discussion shows the naturality of the notion of graded q-differential algebra as "N-generalization" or q-analog of the notion of graded differential algebra. This notion has a slight drawback which is the non existence of natural tensor products [55]; let us discuss this point. It was shown in [40] that if $q \in \mathbf{k}$ is such that Assumption (A_1) is satisfied then one can construct a tensor product for N-complexes in the following manner. Let (E', d') and (E'', d'') be two N-complexes and let us define d on $E' \otimes E''$ by setting

$$d(\alpha' \otimes \alpha'') = d'(\alpha') \otimes \alpha'' + q^{a'} \alpha' \otimes d''(\alpha''), \quad \forall \alpha' \in E'^{a'}, \forall \alpha'' \in E'',$$

one has by induction on $n \in \mathbb{N}$

$$d^n(\alpha' \otimes \alpha'') = \sum_{m=0}^{n} q^{a'(n-m)} \begin{bmatrix} n \\ m \end{bmatrix}_q d'^m(\alpha') \otimes d''^{n-m}(\alpha''),$$

therefore Assumption (A_1) implies $d^N(\alpha' \otimes \alpha'') = d'^N(\alpha') \otimes \alpha'' + \alpha' \otimes d''^N(\alpha'') = 0$. Unfortunately, as pointed out in [55], when (E', d') and (E'', d'') are furthermore two graded q-differential algebras, d fails to be a q-differential in that it does not satisfy the graded q-Leibniz rule except for $q = 1$ or $q = -1$.

As for N-complexes, many notions for graded q-differential algebras do only depend on the underlying $\mathbb{Z}_N$-graduation so it is natural to consider the following $\mathbb{Z}_N$-graded version. A $\mathbb{Z}_N$-graded q-differential algebra is a $\mathbb{Z}_N$-graded algebra equipped with a homogeneous endomorphism d of degree 1 which is an N-differential, i.e. $d^N = 0$, and which satisfies the graded q-Leibniz rule $d(\alpha\beta) = d(\alpha)\beta + q^a \alpha d(\beta)$ for α homogeneous of degree $a \in \mathbb{Z}_N$, (let us remind that N and $q \in \mathbf{k}$ are connected by Assumption (A_1)). We have already met such a $\mathbb{Z}_N$-graded q-differential algebra at the end of Section 6, (namely $M_N(\mathbf{k})$).

The notion of graded q-differential algebra was introduced in [20] for $\mathbf{k} = \mathbb{C}$ as well as the construction of the universal q-differential envelopes. Here, we have followed the presentation of [14].

9. Subquotients and constraints

Let E be a module and let $u \in E^*$ be a linear form on E. To these data, one associates a graded differential algebra $K(u)$ which is constructed in the following manner. As an algebra, $K(u)$ is the exterior algebra (over $\mathbf{k}$) $\wedge E$ of E but it is equipped with the opposite graduation, i.e. $K(u) = \oplus_n K^n(u)$ with $K^n(u) = \wedge^{-n} E$ if $n \leq 0$ and $K^n(u) = 0$ if $n > 0$. The differential d_u of $K(u)$ is then defined to be the unique homogeneous $\mathbf{k}$-linear endomorphism of degree 1 of $K(u)$ satisfying the graded Leibniz rule and such that $d_u(e) = u(e) \in \mathbf{k} = K^0(u)$ for any $e \in E = K^{-1}(u)$. One has $d_u^2 = 0$ so $K(u)$ is a graded differential algebra; the underlying complex is a *Koszul complex* and will be referred to as the Koszul complex associated to the pair (E, u).

Let M be a smooth (finite-dimensional, connected, paracompact) manifold and let V be a closed submanifold of M. The $\mathbb{R}$-algebra $C^\infty(M)$ of smooth functions on M will play the role of $\mathbf{k}$ and $I(V)$ will denote the ideal of a smooth function on M vanishing on V. We introduce the following regularity assumption (R_0) for the data (M, V):

(R_0) $I(V)$ is generated by m functions $u_\alpha \in C^\infty(M)$, $\alpha \in \{1, \cdots, m\}$, which are independent on V in the sense $du_1(x) \wedge \cdots \wedge du_m(x) \neq 0$, $\forall x \in V$.

Let $E = \mathbb{R}^m \otimes_\mathbb{R} C^\infty(M)$ be the free $C^\infty(M)$-module of rank m with canonical basis denoted by $\pi_\alpha, \alpha \in \{1, \cdots, m\}$, and let $u \in E^*$ be the $(C^\infty(M)$-$)$linear form on E defined by $u(\pi_\alpha) = u_\alpha \in C^\infty(M)$, for $\alpha \in \{1, \cdots, m\}$. The Koszul complex $K(u)$ associated to the pair (E, u) identifies with the free graded-commutative unital $C^\infty(M)$-algebra generated by the π_α in degree -1 equipped with the differential d_u above; This is a graded differential $C^\infty(M)$-algebra. Under Assumption (R_0) one has with these notations the following result [12].

LEMMA 8. *The cohomology $H(K(u))$ of $K(u)$ is given by $H^n(K(u)) = 0$ if $n \neq 0$ and $H^0(K(u))$ identifies canonically with the algebra $C^\infty(V)$ of smooth functions on V.*

In fact, $d_u(K^{-1}(u)) = I(V)$ so one has $H^0(K(u)) = C^\infty(M)/I(V)$ which is canonically $C^\infty(V)$; notice that $C^\infty(V)$ is a $C^\infty(M)$-algebra. Notice also that, since the $\mathbb{R}$-algebra $C^\infty(M)$ is unital, $K(u)$ is also a graded differential algebra over $\mathbb{R}$.

Lemma 8 gives a homological description of the algebra of functions on a submanifold (under assumption (R_0)). Our aim is now to give a homological description of the algebra of functions on a quotient manifold and finally to mix both descriptions to obtain a homological description of the algebra of functions on a quotient of a submanifold (subquotient).

Let V be a smooth manifold. Recall that a *foliation* of V is a vector subbundle F of the tangent bundle $T(V)$ of V which is such that the $C^\infty(V)$-module $\mathcal{F}$ of sections of F is also a Lie subalgebra of the Lie algebra of vector fields on V. In the following, we shall identify the foliation with $\mathcal{F}$. The ideal $(\wedge \mathcal{F})^\perp$ of the algebra $\Omega(V)$ of differential forms on V of forms vanishing on $\mathcal{F}$ is a differential ideal so the quotient $\Omega(V)/(\wedge \mathcal{F})^\perp$ is a graded differential algebra over $\mathbb{R}$ which will be denoted by $\Omega(V, \mathcal{F})$ and referred to as the graded differential algebra of *longitudinal forms* of $\mathcal{F}$; its differential will be denoted by $d_\mathcal{F}$. In fact $\Omega(V, \mathcal{F})$ is a subcomplex of the Chevalley-Eilenberg complex $C_\wedge(\mathcal{F}, C^\infty(V))$, (see Appendix B), where $\mathcal{F}$ acts by derivations on $C^\infty(V)$. Let $H(V, \mathcal{F})$ be the cohomology of longitudinal forms; in degree 0, $H^0(V, \mathcal{F})$ identifies with the $\mathbb{R}$-algebra of smooth functions on V invariant by the action of the vector fields belonging to $\mathcal{F}$. In the case where the quotient $V/\mathcal{F}$ exists as smooth manifold and is such that the canonical projection $p : V \to V/\mathcal{F}$ is a submersion, $H^0(V, \mathcal{F})$ identifies with the algebra $C^\infty(V/\mathcal{F})$ of smooth functions on the quotient. Let us introduce the following regularity assumption (R_1) for $(V, \mathcal{F})$:

(R_1) $\mathcal{F}$ is a free $C^\infty(V)$-module of rank m', or equivalently F is a trivial vector bundle of rank m'.

If (R_1) is satisfied $\Omega(V,\mathcal{F})$ identifies with the graded-commutative algebra $C^\infty(V)\otimes_\mathbb{R}\wedge\mathbb{R}^{m'}$; in fact, $\Omega(V,\mathcal{F})$ is then the free graded-commutative unital $C^\infty(V)$-algebra generated by the $\chi^{\alpha'}$ in degree 1, $\alpha'\in\{1,\cdots,m'\}$, where $(\chi^{\alpha'})$ is the dual basis of the basis $(\xi_{\alpha'})$ of $\mathcal{F}$. With these conventions, $d_\mathcal{F}$ is given on the generators by

$$\text{(4)}\quad \begin{cases} d_\mathcal{F} f &= \xi_{\alpha'}(f)\chi^{\alpha'},\ \forall f\in C^\infty(V) \\ d_\mathcal{F}\chi^{\alpha'} &= -\tfrac{1}{2}C^{\alpha'}_{\beta'\gamma'}\chi^{\beta'}\chi^{\gamma'} \end{cases}$$

the $C^{\alpha'}_{\beta'\gamma'}\in C^\infty(V)$ being given by $[\xi_{\beta'},\xi_{\gamma'}]=C^{\alpha'}_{\beta'\gamma'}\xi_{\alpha'}$ i.e. $C^{\alpha'}_{\beta'\gamma'}=\chi^{\alpha'}([\xi_{\beta'},\xi_{\gamma'}])$. One must be aware of the fact that $\Omega(V,\mathcal{F})$ is a graded differential algebra over $\mathbb{R}$ (and not over $C^\infty(V)$).

It is worth noticing here that an infinite dimensional analog of the above appears in gauge theory; there, V is replaced by the affine space of gauge potentials (connections), $\mathcal{F}$ is replaced by the Lie algebra of the group of gauge transformations acting on gauge potentials whereas the analog of the $\chi^{\alpha'}$ are components of the ghost field $\chi(x)$. In this context the BRS differential [2] corresponds to the longitudinal differential $d_\mathcal{F}$, [64], [59].

Let now M be a smooth manifold, V be a closed submanifold of M and assume that V is equipped with a foliation $\mathcal{F}$. We want to combine the above constructions to produce a homological description of $C^\infty(V/\mathcal{F})$. More precisely, our aim is to produce a graded differential algebra which contains $C^\infty(M)$ and which has the longitudinal cohomology $H(V,\mathcal{F})$ as cohomology. We assume in the following that the assumption (R_0) is satisfied by (M,V) and that the assumption (R_1) is satisfied by $(V,\mathcal{F})$. With (M,V) satisfying (R_0) is associated as above the Koszul complex $K(u)$ with differential d_u. Let $\mathcal{K}=\oplus_{i,j}\mathcal{K}^{i,j}$ be the bigraded algebra $K(u)\otimes_\mathbb{R}\wedge\mathbb{R}^{m'}$ with $\mathcal{K}^{i,j}=K^i(u)\otimes_\mathbb{R}\wedge^j\mathbb{R}^{m'}$ i.e. $\mathcal{K}^{i,j}=\wedge^{-i}\mathbb{R}^m\otimes_\mathbb{R} C^\infty(M)\otimes_\mathbb{R}\wedge^j\mathbb{R}^{m'}$ if $i\leq 0\leq j$ and $\mathcal{K}^{i,j}=0$ otherwise. One can also consider that $\mathcal{K}$ is a $\mathbb{Z}$-graded algebra, $\mathcal{K}=\oplus_n\mathcal{K}^n$, for the total degree $\mathcal{K}^n=\oplus_{i+j=n}\mathcal{K}^{i,j}$. We shall again denote by $\pi_\alpha, \alpha\in\{1,\cdots,m\}$, and $\chi^{\alpha'}, \alpha'\in\{1,\cdots,m'\}$ the elements of $\mathcal{K}$ corresponding to the canonical basis of $\mathbb{R}^m$ and of $\mathbb{R}^{m'}$. As graded $C^\infty(M)$-algebra, $\mathcal{K}$ is the free graded-commutative unital $C^\infty(M)$-algebra generated by the π_α in degree -1 and the $\chi^{\alpha'}$ in degree 1. One recovers the bidegree by giving the bidegree $(-1,0)$ to the π_α and the bidegree $(0,1)$ to the $\chi^{\alpha'}$. Let us extend the differential d_u of $K(u)$ as the unique antiderivation δ_0 of $\mathcal{K}$ such that $\delta_0\chi^{\alpha'}=0$, $\delta_0 f=0$ for $f\in C^\infty(M)$ and $\delta_0\pi_\alpha=u_\alpha$; one still has $\delta_0^2=0$ so $\mathcal{K}$ equipped with δ_0 is a graded differential algebra. Furthermore since δ_0 is homogeneous for the bidegree (of bidegree $(1,0)$) the cohomology $H(\delta_0)$ of $(\mathcal{K},\delta_0)$ is bigraded, $H(\delta_0)=\oplus_{i,j}H^{i,j}(\delta_0)$, and Lemma 8 implies that $H^{i,j}(\delta_0)=0$ if $i\neq 0$ and that $H^{0,j}(\delta_0)=C^\infty(V)\otimes_\mathbb{R}\wedge^j\mathbb{R}^{m'}$ in other words one has the following lemma.

LEMMA 9. *As a graded $\mathbb{R}$-algebra, the cohomology $H(\delta_0)$ of $(\mathcal{K},\delta_0)$ identifies with the graded algebra $\Omega(V,\mathcal{F})$ of longitudinal forms.*

The following lemma states that there is an antiderivation of $\mathcal{K}$ which induces the longitudinal differential $d_\mathcal{F}$ on $H(\delta_0)$.

LEMMA 10. *There is an antiderivation of δ_1 of degree 1 of $\mathcal{K}$ which is homogeneous for the bidegree of bidegree $(0,1)$, which satisfies $\delta_0\delta_1 + \delta_1\delta_0 = 0$ and which induces the longitudinal differential $d_\mathcal{F}$ on $H(\delta_0) = \Omega(V, \mathcal{F})$.*

Proof. The longitudinal differential is given by (4) on $C(V) \otimes_\mathbb{R} \wedge \mathbb{R}^{m'}$. It follows from our assumptions that there are vector fields $\tilde{\xi}_{\alpha'}$ on M such that their restrictions to V are tangent to V and coincide with the $\xi_{\alpha'}$, $\tilde{\xi}_{\alpha'} \upharpoonright V = \xi_{\alpha'}$ for $\alpha' \in \{1, \cdots, m'\}$. Similarly there are $\tilde{C}^{\alpha'}_{\beta'\gamma'} \in C^\infty(M)$ such that $\tilde{C}^{\alpha'}_{\beta'\gamma'} \upharpoonright V = C^{\alpha'}_{\beta'\gamma'} \in C^\infty(V)$. Define then δ_1 on $C^\infty(M) \otimes_\mathbb{R} \wedge \mathbb{R}^{m'}$ by

$$(5) \quad \begin{cases} \delta_1 f &= \tilde{\xi}_{\alpha'}(f)\chi^{\alpha'}, \quad \forall f \in C^\infty(M) \\ \delta_1 \chi^{\alpha'} &= -\frac{1}{2}\tilde{C}^{\alpha'}_{\beta'\gamma'}\chi^{\beta'}\chi^{\gamma'} \end{cases}$$

One has $(\delta_0\delta_1 + \delta_1\delta_0)f = \delta_0\delta_1 f = 0$ and $(\delta_0\delta_1 + \delta_1\delta_0)\chi^{\alpha'} = \delta_0\delta_1\chi^{\alpha'} = 0$ for $f \in C^\infty(M)$ and $\alpha' \in \{1, \cdots, m'\}$. On the other hand, one has $\delta_1\delta_0\pi_\alpha = \delta_1 u_\alpha = \tilde{\xi}_{\alpha'}(u_\alpha)\chi^{\alpha'}$ and, by construction $\tilde{\xi}_{\alpha'}(u_\alpha)$ vanishes on V so $\tilde{\xi}_{\alpha'}(u_\alpha) = A^\beta_{\alpha'\alpha}u_\beta$ for some $A^\beta_{\alpha'\alpha} \in C^\infty(M)$. By setting $\delta_1\pi_\alpha = -A^\beta_{\alpha'\alpha}\pi_\beta\chi^{\alpha'}$ and by extending δ_1 to $\mathcal{K}$ by the antiderivation property, one has $\delta_0\delta_1 + \delta_1\delta_0 = 0$ so δ_1 induces an antiderivation of degree 1 of $H(\delta_0) = C(V) \otimes_\mathbb{R} \wedge \mathbb{R}^{m'}$ which coincides with $d_\mathcal{F}$ in view of (4). $\square$
As it is apparent in Formula (5), δ_1 is an antiderivation of $\mathcal{K}$ considered as a graded algebra over $\mathbb{R}$ (and not over $C^\infty(M)$ in contrast with δ_0).

LEMMA 11. *There are antiderivations δ_r of degree 1 of the graded $\mathbb{R}$-algebra $\mathcal{K}$ with δ_r homogeneous for the bidegree of bidegree $(1-r, r)$ for $r \geq 2$, such that one has with δ_0 and δ_1 as above $\sum_{r+s=n} \delta_r\delta_s = 0$ for any $n \in \mathbb{N}$.*

For the proof we refer to the proof of Theorem 3.7 of **[12]**. This is a proof by induction on n using $H^{1-r,r+1}(\delta_0) = 0$ and $H^{1-r,r+2}(\delta_0) = 0$ for $r \geq 2$. Notice that $\delta_r = 0$ if $r > m'$ or $r > m + 1$.

THEOREM 4. *Let δ_r ($r \geq 0$) be as above then $\delta = \sum_{r \geq 0} \delta_r$ is a differential of the graded $\mathbb{R}$-algebra $\mathcal{K}$ and the cohomology $H(\delta)$ of the graded differential algebra $(\mathcal{K}, \delta)$ identifies with the longitudinal cohomology $H(V, \mathcal{F})$.*

Again we refer to **[12]** (the proof of Theorem 3.8 there); the first part of the statement is obvious, the identification of $H(\delta)$ with $H(V, \mathcal{F})$ follows essentially from an elementary spectral sequence argument.

Let (M, ω) be a *symplectic manifold* (i.e. a smooth manifold M equipped with a closed nondegenerate 2-form ω) and let V be a closed submanifold of M. We denote by ω_V the closed 2-form $i^*(\omega)$ on V induced by the inclusion $i : V \to M$. In general ω_V is degenerate; its characteristic distribution F is the set of tangent vectors X of V such that $i_X\omega = 0$. It follows from the equation $d\omega_V = 0$ that the $C^\infty(V)$-module $\mathcal{F}$ of vector fields on V which are valued in F is a Lie subalgebra of the Lie algebra of vector fields. Therefore if ω_V is of constant rank, which will be assumed in the sequel, $\mathcal{F}$ is a foliation of V. In fact we shall assume not only that ω_V is of constant rank but also that the quotient $V/\mathcal{F} = M_0$ is a smooth manifold and that the canonical projection $p : V \to M_0$ is a submersion. With these regularity assumptions, ω_V has a projection ω_0 on M_0 which is, by construction, a closed nondegenerate 2-form. Thus (M_0, ω_0) is a symplectic manifold which is referred to as *the reduced phase space* and which is the natural phase space for a hamiltonian system on M which is constrained to move on V. One has $i^*(\omega) = \omega_V = p^*(\omega_0)$. The algebra of observables of such a constrained system is $C^\infty(M_0)$ which identifies with the longitudinal cohomology of degree 0, $C^\infty(M_0) = H^0(V, \mathcal{F})$. Thus if (M, ω) and V are such that (R_0) is satisfied for (M, V) and (R_1) is satisfied for $(V, \mathcal{F})$, one can use Theorem 4 to compute $C^\infty(M_0)$ and more generally $H(V, \mathcal{F})$. The graded differential algebra $\mathcal{K}$ is the ghost complex appropriate to the situation and δ is the corresponding BRS differential.

The specificity of the above situation is that $\mathcal{F}$ does only depend on the submanifold V of the symplectic manifold (M, ω); in particular if assumptions (R_0) and (R_1) are satisfied, one can show easily that $m \geq m'$ and, on the other hand $m + m' = \dim(M) - \dim(M_0)$ is necessarily even since M and M_0 are both symplectic (and finite-dimensional). The case where the ideal $I(V)$ of smooth functions on M which vanish on V is stable by the Poisson bracket (associated to ω) is referred to as the case of *first class constraints* or the *coisotropic case*. In such a case, one has $m = m'$ and Assumption (R_0) implies (R_1); indeed in this case with Assumption (R_0) the hamiltonian vector fields $\text{Ham}(u_\alpha)$ of the u_α have restrictions to V which are tangent to V and form a basis of the $C^\infty(V)$-module $\mathcal{F}$, (see e.g. in [**12**]). This case has the further property that one can extend the Poisson bracket in a superbracket on $\mathcal{K}$ by setting $\{\pi_\alpha, \chi^\beta\} = \delta_\alpha^\beta$, $\{\chi^\alpha, \chi^\beta\} = \{\pi_\alpha, \pi_\beta\} = 0$, $\{\pi_\alpha, f\} = \{\chi^\alpha, f\} = 0$ for $f \in C^\infty(M)$ and that $\mathcal{K}$ can then be interpreted as the algebra of "functions" on a "super phase space". Moreover in this case the BRS differential δ can be realized as superhamiltonian, i.e. $\delta\varphi = \{Q, \varphi\}, \forall \varphi \in \mathcal{K}$, for some $Q \in \mathcal{K}$ of total degree 1, [**37**]. In this case it has been shown in [**37**] that the arbitrariness of the whole construction is a canonical transformation of the super phase space.

In gauge theory, the usual ghost complex without antighosts was understood early as a Lie algebra cochain complex (see e.g. in [**4**], [**58**], [**21**], [**11**]) or as a complex of longitudinal forms (see in [**64**], [**59**]). This led through the Koszul formula [**45**] to the interpretation of the corresponding ghosts as components of the Maurer-Cartan form of the gauge group (see also in [**63**]). The key of the understanding of the antighost or conjugate ghost part in terms of usual mathematical concepts appears in [**48**] where it was shown that they provide Koszul resolutions. This led to the homological approach to constrained systems developed e.g. in [**52**], [**56**], [**12**] and [**44**] in terms of standard mathematical objects which is partly described above.

Assumption (R_0) means regular submanifold V. One can generalize the above constructions in several directions without such a regularity. In the case of the first class constrained hamiltonian systems this has been investigated in [**56**] and in [**27**] where BRS cohomology with ghosts of ghosts has been applied. Finally, it is worth noticing here that an "infinite dimensional" form of Theorem 4 applies directly to the antifield formalism [**26**], [**38**] and is also implicit behind the ghost lagrangian formalism of gauge theory [**25**], [**2**].

10. N-complex versions of BRS methods

The canonical approach to the quantum Wess-Zumino-Novikov-Witten (WZNW) model gives rise to a finite-dimensional quantum group gauge problem for the zero modes. This has been studied in a convenient form for us in [**31**], [**32**]. The result is a finite-dimensional gauge model in which the physical state space appears as a quotient $\mathcal{H}'/\mathcal{H}''$ where $\mathcal{H}'$ is a subspace of the original finite-dimensional indefinite metric space whereas $\mathcal{H}''$ is the subspace of "null vectors" (isotropic subspace) of $\mathcal{H}'$. Using the results of [**31**], [**32**], it was shown in [**22**] that, in the case of the $SU(2)$ WZNW model, the physical state space can be realized as a direct sum $\oplus_{n=1}^{N-1} H_{(n)}(\mathcal{H}_I, A)$ where $(H_{(n)}(\mathcal{H}_I, A))$ is the generalized homology of an N-differential vector space $\mathcal{H}_I$ with N-differential A. In fact, for the level k representation of the $\widehat{\mathfrak{su}}(2)$ Kac-Moody algebra, A satisfies $A^N = 0$ with $N = k+2$. The N^4-dimensional space $\mathcal{F} \otimes \bar{\mathcal{F}} = \mathcal{H}$ of chiral zero modes carries a representation of the quantum group $U_q(\mathfrak{sl}_2) \otimes U_q(\mathfrak{sl}_2)$ where $q = e^{i(\pi/N)}$; it is a representation of the usual finite-dimensional quotient $\mathcal{U}_q$ of $U_q(\mathfrak{sl}_2) \otimes U_q(\mathfrak{sl}_2)$ at the primitive root of unity q ($q^{2N} = 1$). The N-differential A of $\mathcal{H}$ commutes with the action of the Hopf algebra $\mathcal{U}_q$ so the $(2N-1)$-dimensional subspace $\mathcal{H}_I$ of $\mathcal{U}_q$-invariant vectors is stable by A and it is the generalized homology of the N-differential vector space $(\mathcal{H}_I, A)$ which is of interest. In [**23**] we produced an N-differential vector space which contains $\mathcal{H}$ and has the same generalized homology as $(\mathcal{H}_I, A)$. It is this construction which will be explained in a very general setting in what follows.

In short, one has a vector space $\mathcal{H}$ on which act a Hopf algebra $\mathcal{U}_q$ and a nilpotent endomorphism A satisfying $A^N = 0$. The action of the algebra $\mathcal{U}_q$ commutes with A, i.e. one has on $\mathcal{H}$: $[A, X] = 0$, $\forall X \in \mathcal{U}_q$. It follows that the subspace $\mathcal{H}_I$ of $\mathcal{U}_q$-invariant vectors in $\mathcal{H}$ is stable by A, i.e. $A(\mathcal{H}_I) \subset \mathcal{H}_I$. Thus $(\mathcal{H}_I, A)$ is an N-differential subspace of the N-differential vector space $(\mathcal{H}, A)$ and it turns out that the "interesting object" (the physical space) is the generalized homology of $(\mathcal{H}_I, A)$. We would like to avoid the restriction to the invariant subspace $\mathcal{H}_I$ that is, in complete analogy with the BRS methods, we would like to define an extended N-differential space in such a way that the $\mathcal{U}_q$-invariance is captured by its N-differential in the sense that it has the same generalized homology as $(\mathcal{H}_I, A)$. The most natural thing to do is to try to construct a nilpotent endomorphism Q of $\mathcal{H}$ with $Q^N = 0$ such that its generalized homology coincides with the one of A on $\mathcal{H}_I$ i.e. such that one has $H_{(n)}(\mathcal{H}, Q) = H_{(n)}(\mathcal{H}_I, A)$, $\forall n \in \{1, \ldots, N-1\}$. It turns out that this is impossible in general. Indeed in the above case (for the $SU(2)$ WZNW model) $\mathcal{H}$ is finite dimensional and then Proposition 4 (see in Section 5) imposes strong constraints connecting $\dim \mathcal{H}$ and the $\dim H_{(n)}(\mathcal{H}, Q) = \dim H_{(n)}(\mathcal{H}_I, A)$ for $n \in \{1, \cdots, N-1\}$ which are not satisfied [**23**]. This is not astonishing since in the usual BRS methods one has to add the ghost sector (see e.g. in last section or in Section 4).

We first present an abstract optimal construction in which the Hopf algebra $\mathcal{U}_q$ plays no role. We assume that $(\mathcal{H}, A)$ is an N-differential vector space, that there is a subspace $\mathcal{H}_I$ of $\mathcal{H}$ stable by A and we shall construct an N-differential vector space $(\mathcal{H}^\bullet, Q)$ with $\mathcal{H} \subset \mathcal{H}^\bullet$ such that $H_{(n)}(\mathcal{H}^\bullet, Q) = H_{(n)}(\mathcal{H}_I, A)$ for all $n \in \{1, \cdots, N-1\}$. Throughout the following q^2 is still a primitive N-th root of unity ($q^N = -1$). Let us define the graded vector space $\mathcal{H}^\bullet = \oplus_{n \geq 0} \mathcal{H}^n$ by $\mathcal{H}^0 = \mathcal{H}$, $\mathcal{H}^n = \mathcal{H}/\mathcal{H}_I$ for $1 \leq n \leq N-1$ and $\mathcal{H}^n = 0$ for $n \geq N$. One then defines an endomorphism d of degree 1 of $\mathcal{H}^\bullet$ by setting $d = \pi : \mathcal{H}^0 \to \mathcal{H}^1$ where $\pi : \mathcal{H} \to \mathcal{H}/\mathcal{H}_I$ is the canonical projection, $d = \mathrm{Id} : \mathcal{H}^n \to \mathcal{H}^{n+1}$ for $1 \leq n \leq N-2$ where Id

is the identity mapping of $\mathcal{H}/\mathcal{H}_I$ onto itself and $d = 0$ on $\mathcal{H}^n$ for $n \geq N - 1$. One has $d^N = 0$ and therefore $(\mathcal{H}^\bullet, d)$ is an N-complex, so its generalized (co)homology is graded $H_{(k)}(\mathcal{H}^\bullet, d) = \underset{n \geq 0}{\oplus} H_{(k)}^n(\mathcal{H}^\bullet, d)$. It is given by the following easy lemma.

LEMMA 12. *One has* $H_{(k)}^n(\mathcal{H}^\bullet, d) = 0$ *for* $n \geq 1$ *and* $H_{(k)}^0(\mathcal{H}^\bullet, d) = \mathcal{H}_I$, $\forall k \in \{1, \ldots, N - 1\}$.

It is worth noticing here that given the vector space $\mathcal{H}$ together with the subspace $\mathcal{H}_I$, the N-complex $(\mathcal{H}^\bullet, d)$ is characterized (uniquely up to an isomorphism) by the following universal property (the proof of which is straightforward).

LEMMA 13. *Any linear mapping* $\alpha : \mathcal{H} \to \mathcal{C}^0$ *of* $\mathcal{H}$ *into the subspace* $\mathcal{C}^0$ *of elements of degree* 0 *of an N-complex* $(\mathcal{C}^\bullet, d)$ *which satisfies* $d \circ \alpha(\mathcal{H}_I) = 0$ *extends uniquely as a homomorphism* $\bar{\alpha} : (\mathcal{H}^\bullet, d) \to (\mathcal{C}^\bullet, d)$ *of N-complexes.*

By using this universal property one can extend A to $\mathcal{H}^\bullet$ in the following manner.

LEMMA 14. *The endomorphism A of $\mathcal{H} = \mathcal{H}^0$ has a unique extension to $\mathcal{H}^\bullet$, again denoted by A, as a homogeneous endomorphism of degree 0 satisfying $Ad - q^2\, dA = 0$. On $\mathcal{H}^\bullet$, one has $A^N = 0$ and $(d + A)^N = 0$.*

Thus $Q = d + A$ is an N-differential on $\mathcal{H}^\bullet$ and we have the following result.

THEOREM 5. *The generalized Q-homology of $\mathcal{H}^\bullet$ coincides with the generalized A-homology of $\mathcal{H}_I$, i.e. one has $H_{(k)}(\mathcal{H}^\bullet, Q) = H_{(k)}(\mathcal{H}_I, A)$ for $1 \leq k \leq N - 1$.*

Notice that $(\mathcal{H}^\bullet, Q)$ is only an N-differential vector space and not an N-complex since $d + A = Q$ is inhomogeneous.

In the problem of the zero modes of the $SU(2)$ WZNW model, $\mathcal{H}_I$ is the invariant subspace of $\mathcal{H}$ by the action of the quantum group (i.e. the Hopf algebra) $\mathcal{U}_q$ which plays the role of a gauge group, or more precisely of the universal enveloping algebra of the Lie algebra of a gauge group, so (in view of Theorem 1) it is natural to produce a construction where $\mathcal{U}_q$ and its (Hochschild) cohomology enter as in the usual BRS construction for gauge theory in order to get a similar "geometrico-physical" interpretation. This is the aim of the end of this section. Since the above construction based on universal property is quite minimal, one cannot be astonished that it occurs as an N-differential subspace of the following one.

By definition $\mathcal{H}_I$ is the set of $\Psi \in \mathcal{H}$ such that $X\Psi = \Psi\varepsilon(X)$ for any $X \in \mathcal{U}_q$, where ε denotes the counit of $\mathcal{U}_q$. This means that if one considers $\mathcal{H}$ as a $(\mathcal{U}_q, \mathcal{U}_q)$-bimodule by equipping it with the trivial right action given by the counit, $\mathcal{H}_I$ identifies with the $\mathcal{H}$-valued Hochschild cohomology in degree 0 of $\mathcal{U}_q$, i.e. $\mathcal{H}_I = H^0(\mathcal{U}_q, \mathcal{H})$. The idea of the construction is to mix the Hochschild differential with A in a similar way as the mixing of δ_0 with $d_\mathcal{F}$ in last section. However, A is an N-differential whereas the Hochschild differential is an ordinary differential i.e. a 2-differential. Fortunately the next lemma shows that for the description of $\mathcal{H}_I$ one can replace the Hochschild differential by the N-differential d_1 of Section 6 with the replacement of q by q^2 since here it is q^2 which is a primitive N-th root of unity. To simplify the notations, this N-differential d_1 on $C(\mathcal{U}_q, \mathcal{H})$ will be denoted by d. That is the N-differential d is defined by

$$\begin{aligned}d(\omega)(X_0,\ldots,X_n) &= X_0\omega(X_1,\ldots,X_n)\\ &+ \sum_{k=1}^n q^{2k}\omega(X_0,\ldots,(X_{k-1}X_k),\ldots,X_n)\\ &- q^{2n}\omega(X_0,\ldots,X_{n-1})\varepsilon(X_n).\end{aligned}$$

for $\omega \in C^n(\mathcal{U}_q,\mathcal{H})$, $X_i \in \mathcal{U}_q$. One has the following lemma.

LEMMA 15. *Let $\Psi \in \mathcal{H} = C^0(\mathcal{U}_q,\mathcal{H})$; the following conditions (i), (ii) and (iii) are equivalent*
(i) $d^k(\Psi) = 0$ *for some k with $1 \leq k \leq N-1$*
(ii) $\Psi \in \mathcal{H}_I$
(iii) $d^n(\Psi) = 0$ *for any $n \in \{1,\ldots,N-1\}$.*

Observe first that $d(= d_1)$ coincides in degree 0 with the Hochschild differential. Then the result is a consequence of the following formula which one proves by induction on n.

$$d^n\Psi(\mathbf{1},\ldots,\mathbf{1},X) = (1+q^2)\ldots(1+q^2+\cdots+q^{2(n-1)})d\Psi(X)$$

for $\Psi \in C^0(\mathcal{U}_q,\mathcal{H})$ and for any $n \geq 1$, $X \in \mathcal{U}_q$ where $\mathbf{1}$ is the unit of $\mathcal{U}_q$.

This lemma implies : $H^0_{(k)}(C(\mathcal{U}_q,\mathcal{H}),d) = H^0(\mathcal{U}_q,\mathcal{H})$, $\forall k \in \{1,\ldots,N-1\}$. This is a special case of Theorem 2 of Section 6. As an easy consequence, one obtains the following result.

PROPOSITION 8. *The N-complex $(\mathcal{H}^\bullet,d)$ can be canonically identified with the N-subcomplex of $(C(\mathcal{U}_q,\mathcal{H}),d)$ generated by $\mathcal{H}$.*

Thus one has $\mathcal{H}^\bullet \subset C(\mathcal{U}_q,\mathcal{H})$ and the N-differential d of $C(\mathcal{U}_q,\mathcal{H})$ extends the one of $\mathcal{H}^\bullet$; we now extend A to $C(\mathcal{U}_q,\mathcal{H})$.

LEMMA 16. *Let us extend A to $C(\mathcal{U}_q,\mathcal{H})$ as a homogeneous endomorphism $\omega \mapsto (A\omega)$ of degree 0 by setting*

$$(A\omega)(X_1,\ldots,X_n) = q^{2n}A\omega(X_1,\ldots,X_n)$$

for $\omega \in C^n(\mathcal{U}_q,\mathcal{H})$ and $X_i \in \mathcal{U}_q$. On $C(\mathcal{U}_q,\mathcal{H})$ one has $Ad - q^2 dA = 0$, $A^N = 0$ and $(d+A)^N = 0$.

We have now extended to $C(\mathcal{U}_q,\mathcal{H})$ the whole structure defined previously on $\mathcal{H}^\bullet$. Indeed the uniqueness in Lemma 14 implies that A defined on $C(\mathcal{U}_q,\mathcal{H})$ in last lemma is an extension of A defined on $\mathcal{H}^\bullet$ in Lemma 14. One then extends to $C(\mathcal{U}_q,\mathcal{H})$ the definition of Q by setting again $Q = d + A$.

As explained in Section 6, Theorem 2 (1), the spaces $H^n_{(k)}(C(\mathcal{U}_q,\mathcal{H}),d)$ can be computed in terms of the Hochschild cohomology $H(\mathcal{U}_q,\mathcal{H})$. In particular, one sees that $H^n_{(k)}(C(\mathcal{U}_q,\mathcal{H}),d)$ does not generally vanish for $n \geq 1$. This implies that one cannot expect for the generalized homology of Q on $C(\mathcal{U}_q,\mathcal{H})$ such a simple result as the one given by Theorem 5 for the generalized homology of Q on $\mathcal{H}^\bullet$. Nevertheless, in view of Lemma 15, one has $H^0_{(k)}(C(\mathcal{U}_q,\mathcal{H}),d) = \mathcal{H}_I = H^0_{(k)}(\mathcal{H}^\bullet,d)$ and therefore one may expect $H^0_{(k)}(C(\mathcal{U}_q,\mathcal{H}),Q) = H_{(k)}(\mathcal{H}_I,A) (= H_{(k)}(\mathcal{H}^\bullet,Q))$. In fact, this is essentially true. However some care must be taken because Q is not homogeneous so $H_{(k)}(C(\mathcal{U}_q,\mathcal{H}),Q)$ is not a graded vector space. Instead of a graduation, one has an increasing filtration $F^n H_{(k)}(C(\mathcal{U}_q, \mathcal{H}), Q)$, $(n \in \mathbb{Z})$, with $F^n H_{(k)}(C(\mathcal{U}_q, \mathcal{H}), Q) = 0$ for $n < 0$ and where, for $n \geq 0$, $F^n H_{(k)}(C(\mathcal{U}_q, \mathcal{H}), Q)$ is the canonical image in $H_{(k)}(C(\mathcal{U}_q, \mathcal{H}), Q)$ of $\text{Ker}(Q^k) \cap \bigoplus_{r=0}^{r=n} C^r(\mathcal{U}_q, \mathcal{H})$. There is an associated graded vector space

$$^{\text{gr}} H_{(k)}(C(\mathcal{U}_q, \mathcal{H}), Q) = \bigoplus_n F^n H_{(k)}(C(\mathcal{U}_q, \mathcal{H}), Q)/F^{n-1} H_{(k)}(C(\mathcal{U}_q, \mathcal{H}), Q)$$

which here is $\mathbb{N}$-graded. One has $F^0 H_{(k)}(C(\mathcal{U}_q, \mathcal{H}), Q) = ^{\text{gr}} H_{(k)}^0(C(\mathcal{U}_q, \mathcal{H}), Q)$ and it is this space which is the correct version of the $H_{(k)}^0(C(\mathcal{U}_q, \mathcal{H}), Q)$ above in order to identify $H_{(k)}(\mathcal{H}_I, A)$ in the generalized homology of Q on $C(\mathcal{U}_q, \mathcal{H})$.

THEOREM 6. *The inclusion $\mathcal{H}^\bullet \subset C(\mathcal{U}_q, \mathcal{H})$ induces the isomorphisms*

$$H_{(k)}(\mathcal{H}^\bullet, Q) \simeq F^0 H_{(k)}(C(\mathcal{U}_q, \mathcal{H}), Q) \text{ for } 1 \leq k \leq N - 1.$$

In particular, with obvious identifications, one has

$$F^0 H_{(k)}(C(\mathcal{U}_q, \mathcal{H}), Q) = H_{(k)}(\mathcal{H}_I, A), \quad \forall k \in \{1, \ldots, N-1\}.$$

The proof is not difficult, for it as well as for complete proofs of all the results of this section we refer to [**23**].

If one compares this construction involving Hochschild cochains with the preceeding one, what has been gained here besides the explicit occurrence of the quantum gauge aspect is that the extended space $C(\mathcal{U}_q, \mathcal{H})$ is a tensor product $\mathcal{H} \otimes \mathcal{H}'$ of the original space $\mathcal{H}$ with the tensor algebra $\mathcal{H}' = T(\mathcal{U}_q^*)$ of the dual space of $\mathcal{U}_q$. The factor $\mathcal{H}'$ can thus be interpreted as the state space for some generalized ghost. What has been lost is the minimality of the generalized homology, i.e. besides the "physical" $H_{(k)}(\mathcal{H}_I, A)$, the generalized homology of Q on $\mathcal{H} \otimes \mathcal{H}'$ contains some other non trivial subspace in contrast to what happens on $\mathcal{H}^\bullet$. In the usual homological (BRS) methods however such a "non minimality" also occurs. Indeed, as explained in last section, in the homological approach to constrained classical systems, the relevant homology contains besides the functions on the reduced phase space the whole cohomology of longitudinal forms. The same is true for the BRS cohomology of gauge theory [**2**], [**4**].

In the usual situations where one applies the BRS construction (gauge theory, constrained systems) one has a Lie algebra $\mathfrak{g}$ (the Lie algebra of infinitesimal gauge transformations) acting on some space $\mathcal{H}$ and what is really relevant at this stage is the Lie algebra cohomology $H(\mathfrak{g}, \mathcal{H})$ of $\mathfrak{g}$ acting on $\mathcal{H}$. The extended space is then the space of $\mathcal{H}$-valued Lie algebra cochains of $\mathfrak{g}$, $C(\mathfrak{g}, \mathcal{H})$. This extended space is thus also a tensor product $\mathcal{H} \otimes \mathcal{H}'$ but now $\mathcal{H}'$ is the exterior algebra $\mathcal{H}' = \Lambda \mathfrak{g}^*$ of the dual space of $\mathfrak{g}$. That is why this factor can be interpreted (due to antisymmetry) as a fermionic state space; indeed that is the reason why one gives a fermionic character to the ghost [**2**], [**4**]. There is however another way to proceed in these situations which is closer to what has been done in our case here. To understand it, we recall that any representation of $\mathfrak{g}$ in $\mathcal{H}$ is also a representation of the enveloping algebra $U(\mathfrak{g})$ in $\mathcal{H}$. Thus $\mathcal{H}$ is a left $U(\mathfrak{g})$-module. Since $U(\mathfrak{g})$ is a Hopf algebra, one can convert as above $\mathcal{H}$ into a bimodule for $U(\mathfrak{g})$ by taking as right action the trivial representation given by the counit. It turns out that as explained in Section 3, Theorem 1, the $\mathcal{H}$-valued Hochschild cohomology of $U(\mathfrak{g})$, $H(U(\mathfrak{g}), \mathcal{H})$, coincides with the $\mathcal{H}$-valued Lie algebra cohomology of $\mathfrak{g}$, $H(\mathfrak{g}, \mathcal{H})$. Since it is the latter space which is relevant one can as well take as extended space the space of

$\mathcal{H}$-valued Hochschild cochains of $U(\mathfrak{g})$, $C(U(\mathfrak{g}), \mathcal{H})$, and then compute its cohomology. Again this space is a tensor product $\mathcal{H} \otimes \mathcal{H}'$ but now $\mathcal{H}' = T(U(\mathfrak{g})^*)$ is a tensor algebra as in our case.

Acknowledgments

I thank Robert Coquereaux and Raymond Stora for their constructive critical reading of the manuscript.

Appendix A. Remarks on tensor products

Let $\wedge\{d\}$ be the associative unital $\mathbf{k}$-algebra generated by an element d satisfying $d^2 = 0$. As a $\mathbf{k}$-algebra $\wedge\{d\} = \mathbf{k}\mathbf{1} \oplus \mathbf{k}d$ is the exterior algebra (over $\mathbf{k}$) of the free $\mathbf{k}$-module of rank one. It is clear that a $\wedge\{d\}$-module is the same thing as a differential module (as defined in Section 2). Given two differential modules E and F there is a canonical structure of $\wedge\{d\} \otimes \wedge\{d\}$-module on $E \otimes F$, where the first factor (resp. the second factor) corresponds to the structure of $\wedge\{d\}$-module of E (resp. of F). To say that for any such E and F there is a canonical differential on $E \otimes F$ (i.e. a canonical structure of $\wedge\{d\}$-module on $E \otimes F$) which only depends on the differentials of E and F (i.e. on their $\wedge\{d\}$-module structures) is the same thing as to say that one has a coproduct Δ on $\wedge\{d\}$, that is a homomorphism of unital $\mathbf{k}$-algebras $\Delta : \wedge\{d\} \to \wedge\{d\} \otimes \wedge\{d\}$. One must have $\Delta(\mathbf{1}) = \mathbf{1} \otimes \mathbf{1}$ so Δ is fixed by giving a $\Delta(d) \in \wedge\{d\} \otimes \wedge\{d\}$ satisfying $(\Delta(d))^2 = 0$. One has $\Delta(d) = \alpha \mathbf{1} \otimes \mathbf{1} + \beta \mathbf{1} \otimes d + \gamma d \otimes \mathbf{1} + \delta d \otimes d$ with $\alpha, \beta, \gamma, \delta \in \mathbf{k}$ and $(\Delta(d))^2 = 0$ implies $\alpha^2 = 0, 2\alpha\beta = 0, 2\alpha\gamma = 0$ and $2(\alpha\delta + \beta\gamma) = 0$.

Let us now assume that $\mathbf{k}$ is a field of characteristic different from 2. Then the above conditions imply $\alpha = 0$ and $\beta\gamma = 0$ i.e. either $\Delta(d) = \beta \mathbf{1} \otimes d + \delta d \otimes d$ or $\Delta(d) = \gamma d \otimes \mathbf{1} + \delta d \otimes d$. It is already clear that generically the differential $\beta \mathbf{1} \otimes d + \delta d \otimes d$ (resp. $\gamma d \otimes \mathbf{1} + \delta d \otimes d$) on $E \otimes F$ will lead to a homology $H(E \otimes F)$ for $E \otimes F$ different from $H(E) \otimes H(F)$. Notice that if one imposes the natural requirement of coassociativity for Δ one is led to the only 3 possibilities $\mathbf{1} \otimes d, d \otimes \mathbf{1}$ or $d \otimes d$ for the differential on the tensor products.

Let us come back to a general ring $\mathbf{k}$. Consider the associative unital $\mathbf{k}$-algebra $\mathcal{D}_{-1}$ generated by two elements d and Γ satisfying $d^2 = 0$, $\Gamma^2 = \mathbf{1}$ and $\Gamma d = -d\Gamma$. This algebra is a Hopf algebra for the counit ε, the antipode S and the coproduct Δ given by : $\varepsilon(d) = 0$, $\varepsilon(\Gamma) = 1$, $S(d) = -\Gamma d$, $S(\Gamma) = \Gamma$, $\Delta(d) = d \otimes \mathbf{1} + \Gamma \otimes d$ and $\Delta(\Gamma) = \Gamma \otimes \Gamma$. The Hopf algebra $\mathcal{D}_{-1}$ can be understood as a version of the universal enveloping algebra of the super Lie algebra with only one odd element d such that $[d, d] = 0$. Let $E = E^0 \oplus E^1$ be a $\mathbb{Z}_2$-complex then E is a $\mathcal{D}_{-1}$-module if d is represented by the differential of E and if Γ is represented by the multiplication by $(-1)^i$ on E^i for $i \in \{0, 1\}$. One verifies easily that the tensor product of $\mathbb{Z}_2$-complexes defined in Section 3 corresponds to the above structure, i.e. that it is induced by the coproduct Δ. Thus one can understand the tensor product of complexes in terms of a Hopf algebra. We now show that the same is true for N-complexes.

Let $q \in \mathbf{k}$ be such that Assumption (A_1) of Section 5 is satisfied and let us consider the associative unital $\mathbf{k}$-algebra $\mathcal{D}_q$ generated by two elements d and Γ

satisfying $d^N = 0$, $\Gamma^N = \mathbf{1}$ and $\Gamma d = qd\Gamma$. Again $\mathcal{D}_q$ is a Hopf algebra for the counit ε, the antipode S and the coproduct Δ given by : $\varepsilon(d) = 0$, $\varepsilon(\Gamma) = 1$, $S(d) = -\Gamma^{N-1}d$, $S(\Gamma) = \Gamma^{N-1}$, $\Delta(d) = d \otimes \mathbf{1} + \Gamma \otimes d$ and $\Delta\Gamma = \Gamma \otimes \Gamma$. Let $E = E^0 \oplus \cdots \oplus E^{N-1}$ be a $\mathbb{Z}_N$-complex (see in Section 6) then E is a $\mathcal{D}_q$-module if d is represented by the N-differential of E and Γ is represented by the multiplication by q^i on E^i for $i \in \{0, \ldots, N-1\}$. Again one verifies easily that the (q) tensor product of $\mathbb{Z}_N$-complexes (or of N-complexes) defined in Section 8 (introduced originally in [**40**]) is induced by the coproduct of $\mathcal{D}_q$.

Finally it is worth noticing here that instead of $\mathcal{D}_{-1}$ one can use the exterior algebra of the free module of rank one $\wedge^\bullet\{d\} = \wedge \mathbf{k} = \mathbf{k}\mathbf{1} \oplus \mathbf{k}d$ considered as a graded Hopf algebra. That is, as an associative algebra $\wedge^\bullet\{d\}$ is isomorphic to $\wedge\{d\}$ but it is a $\mathbb{Z}_2$-graded algebra with $\wedge^0\{d\} = \mathbf{k}\mathbf{1}$, $\wedge^1\{d\} = \mathbf{k}d$ and furthermore it is a graded Hopf algebra for the counit ε, the antipode S and the coproduct Δ given by $\varepsilon(d) = 0$, $S(d) = -d$ and $\Delta d = d \otimes \mathbf{1} + \mathbf{1} \otimes d$ where now Δ is a homomorphism of graded algebras of $\wedge^\bullet\{d\}$ into $\wedge^\bullet\{d\} \underline{\otimes} \wedge^\bullet\{d\}$ with $\wedge^\bullet\{d\} \underline{\otimes} \wedge^\bullet\{d\}$ being the (twisted) tensor product of graded algebras defined in Section 8. A $\mathbb{Z}_2$-complex is canonically a graded $\wedge^\bullet\{d\}$-module and the tensor product of complexes can be also defined by using the above graded coproduct. Of course $\wedge^\bullet\{d\}$ is also a version (which is graded) of the universal enveloping algebra of the super Lie algebra with one odd generator d satisfying $[d,d] = 0$. The advantage of $\mathcal{D}_{-1}$ is that it generalizes as $\mathcal{D}_q$ for N-complexes as explained above and that it is an ordinary Hopf algebra (in fact Γ plays the role of the graduation).

Appendix B. Longitudinal forms

Let $\mathcal{A}$ be an associative unital algebra over $\mathbb{R}$ or $\mathbb{C}$ (here $\mathbf{k} = \mathbb{R}$ or $\mathbb{C}$) and let us denote by $Z(\mathcal{A})$ the center of $\mathcal{A}$ that is the commutative unital subalgebra of $\mathcal{A}$ defined by $Z(\mathcal{A}) = \{z \in \mathcal{A} | za = az, \forall a \in \mathcal{A}\}$. One has $\mathcal{A} = Z(\mathcal{A})$ if and only if $\mathcal{A}$ is commutative. Recall that a derivation of $\mathcal{A}$ is a linear mapping $X : \mathcal{A} \to \mathcal{A}$ such that one has (Leibniz rule) $X(ab) = X(a)b + aX(b)$, $\forall a, b \in \mathcal{A}$. If X and Y are derivations of $\mathcal{A}$, their composition XY (product in $\mathrm{End}(\mathcal{A})$) is not a derivation but the commutator $[X, Y] = XY - YX$ is again a derivation of $\mathcal{A}$. On the other hand, if X is a derivation of $\mathcal{A}$ and if z is in the center of $\mathcal{A}$ then zX (defined by $(zX)(a) = zX(a)$, $\forall a \in \mathcal{A}$) is again a derivation of $\mathcal{A}$. Thus the vector space $\mathrm{Der}(\mathcal{A})$ of all derivations of $\mathcal{A}$ is a Lie algebra (for $[\cdot, \cdot]$) and also a $Z(\mathcal{A})$-module, both structures being connected through the identity

(DZ) $[X, zY] = z[X, Y] + X(z)Y$

for any derivation X and Y and for any $z \in Z(\mathcal{A})$; one verifies easily that the center is stable by derivation i.e. that one has $X(z) \in Z(\mathcal{A})$ for any $X \in \mathrm{Der}(\mathcal{A})$ and $z \in Z(\mathcal{A})$. Let $\mathcal{F} \subset \mathrm{Der}(\mathcal{A})$ be a $Z(\mathcal{A})$-submodule which is also a Lie subalgebra of $\mathrm{Der}(\mathcal{A})$. The graded space $C_\wedge(\mathcal{F}, \mathcal{A})$ of (Chevalley-Eilenberg) $\mathcal{A}$-valued cochains of the Lie algebra $\mathcal{F}$ (see in Section 3) is canonically a graded algebra and, since $\mathcal{F}$ operates by derivation on $\mathcal{A}$, the corresponding Chevalley-Eilenberg differential d is an antiderivation of $C_\wedge(\mathcal{F}, \mathcal{A})$. Thus $C_\wedge(\mathcal{F}, \mathcal{A})$ is a graded differential algebra (see in Section 8). It follows from the above identity (DZ) that the graded subalgebra $\underline{\Omega}_\mathcal{F}(\mathcal{A})$ of cochains which are $Z(\mathcal{A})$-multilinear is stable by d so $\underline{\Omega}_\mathcal{F}(\mathcal{A})$ is a graded differential algebra.

Let V be a smooth manifold and let $\mathcal{F}$ be a foliation of V (see in Section 9), then the graded differential algebra $\underline{\Omega}_{\mathcal{F}}(C^\infty(V))$ is referred to as the graded differential algebra of *longitudinal forms* and is denoted by $\Omega(V, \mathcal{F})$; its elements are called *longitudinal forms*. This is the graded differential algebra considered in Section 9. Notice that when $\mathcal{F}$ coincides with the module $\text{Der}(C^\infty(V))$ of all vector fields on V then $\underline{\Omega}_{\text{Der}}(C^\infty(V)) = \underline{\Omega}_{\text{Der}(C^\infty(V))}(C^\infty(V))$ is the graded differential algebra $\Omega(V)$ of differential forms on V. This is why $\underline{\Omega}_{\text{Der}}(\mathcal{A}) = \underline{\Omega}_{\text{Der}(\mathcal{A})}(\mathcal{A})$ is a noncommutative generalization of the graded differential algebra of differential forms when $\mathcal{A}$ is noncommutative; there are other noncommutative generalizations of differential forms (see e.g. in [16]).

References

[1] H. Araki, *Indecomposable representations with invariant inner product*, Commun. Math. Phys.**97** (1985) 149-159.

[2] C. Becchi, A. Rouet, R. Stora, *Renormalization models with broken symmetries* in "Renormalization Theory", Erice 1975, G. Velo, A.S. Wightman Eds, Reidel 1976.

[3] C. Becchi, A. Rouet, R. Stora, *Renormalization of gauge theories*, Ann. Phys. **98** (1976) 287-321.

[4] L. Bonora, P. Cotta-Ramusino, *Some remarks on BRS transformations, anomalies and the cohomology of the Lie algebra of the group of gauge transformations*, Commun. Math. Phys. **87** (1983) 589-603.

[5] S. Boukraa, *The BRS algebra of a free minimal differential algebra*, Nucl. Phys. **B303** (1988) 237-259.

[6] H. Cartan, *Notion d'algèbre différentielle; application aux groupes de Lie et aux variétés où opère un groupe de Lie* and *La trangression dans un groupe de Lie et dans un espace fibré principal*, Colloque de topologie (Bruxelles 1950), Paris, Masson 1951.

[7] H. Cartan, S. Eilenberg, *Homological algebra*, Princeton University Press 1973.

[8] R. Coquereaux, *Noncommutative geometry and theoretical physics*, J. Geom. Phys. **6** (1989) 425-490.

[9] T. Damour, S. Deser, *Geometry of spin 3 gauge theories*, Ann. Inst. H. Poincaré **47** (1987) 277-307.

[10] B. de Wit, D.Z. Freedman, *Systematics of higher-spin gauge fields*, Phys. Rev.**D21** (1980) 358-367.

[11] M. Dubois-Violette, *The Weil-BRS algebra of a Lie algebra and the anomalous terms in gauge theory*, J. Geom. Phys., **3** (1987) 525-565.

[12] M. Dubois-Violette, *Systèmes dynamiques contraints : l'approche homologique*, Ann. Inst. Fourier Grenoble **37** (1987) 45-57.

[13] M. Dubois-Violette, *Generalized differential spaces with $d^N = 0$ and the q-differential calculus*, Czech J. Phys.**46** (1997) 1227-1233.

[14] M. Dubois-Violette, $d^N = 0$: *Generalized homology*, K-Theory **14** (1998) 371-404.

[15] M. Dubois-Violette, *Generalized homologies for $d^N = 0$ and graded q-differential algebras*, Contemporary Mathematics **219** (1998) 69-79.

[16] M. Dubois-Violette, *Lectures on graded differential algebras and noncommutative geometry*, math.QA/9912017.

[17] M. Dubois-Violette, M. Henneaux, *Generalized cohomology for irreducible tensor fields of mixed Young symmetry type*, Lett. Math. Phys. **49** (1999) 245-252.

[18] M. Dubois-Violette, M. Henneaux, *Tensor fields of mixed Young symmetry type and N-complexes*, in preparation.

[19] M. Dubois-Violette, M. Henneaux, M. Talon, C.M. Viallet, *General solution of the consistency equation*, Phys. Lett. **B289** (1992) 361-367.

[20] M. Dubois-Violette, R. Kerner, *Universal q-differential calculus and q-analog of homological algebra*, Acta Math. Univ. Comenian. **65** (1996) 175-188.

[21] M. Dubois-Violette, M. Talon, C.M. Viallet, *B.R.S. algebras. Analysis of consistency equations in gauge theory*, Commun. Math. Phys. **102** (1985) 105-122.

[22] M. Dubois-Violette, I.T. Todorov, *Generalized cohomology and the physical subspace of the $SU(2)$ WZNW model*, Lett. Math. Phys. **42** (1997) 183-192.

[23] M. Dubois-Violette, I.T. Todorov, *Generalized homology for the zero mode of the $SU(2)$ WZNW model*, Lett. Math. Phys. **48** (1999) 323-338.

[24] H. Epstein, V. Glaser, *The role of locality in perturbation theory*, Ann. Inst. Henri Poincaré **A19** (1973) 211-295.

[25] L.D. Faddeev, V.N. Popov, *Feynman diagrams for the Yang-Mills field*, Phys. Lett. **25B** (1967) 29-30.

[26] J.M.L. Fisch, M. Henneaux, *Homological perturbation theory and the algebraic structure of the antifield-antibracket formalism for gauge theories*, Commun. Math. Phys. **128** (1990) 627-640.

[27] J.M.L. Fisch, M. Henneaux, J. Stasheff, C. Teitelboim, *Existence, uniqueness and cohomology of the classical BRST charge with ghosts of ghosts*, Commun. Math. Phys. **120** (1989) 379-407.

[28] E.S. Fradkin, G.A. Vilkovisky, *Quantization of relativistic systems with constraints*, Phys. Lett. **B55** (1975) 224-226.

[29] C. Fronsdal, *Massless fields with integer spins*, Phys. Rev. **D 18** (1978) 3624-3629.

[30] W. Fulton, *Young tableaux*, Cambridge University Press 1997.

[31] P. Furlan, L.K. Hadjiivanov, I.T. Todorov, *Operator realization of the $SU(2)$ WZNW model*, Nucl. Phys. **B474** (1996) 497-511.

[32] P. Furlan, L.K. Hadjiivanov, I.T. Todorov, *A quantum gauge group approach to the 2D $SU(N)$ WZNW model*, Int. J. Mod. Phys. **A12** (1997) 23-32.

[33] J. Gasqui, *Sur les structures de courbure d'ordre 2 dans $\mathbb{R}^n$*, J. Differential Geometry **12** (1977) 493-497.

[34] H.S. Green, *A generalized method of field quantization*, Phys. Rev. **90** (1953) 270-273.

[35] W. Greub, *Linear algebra*, Springer-Verlag 1963.

[36] W. Greub, S. Halperin, R. Vanstone, *Connections, curvature, and cohomology*, Vol. III, Academic Press 1976.

[37] M. Henneaux, *Hamiltonian form of the path integral for theories with a gauge freedom*, Physics Reports **126** (1985) 1-66.

[38] M. Henneaux, C. Teitelboim, *Quantization of gauge systems*, Princeton University Press 1992.

[39] C. Itzykson, J.B. Zuber, *Quantum field theory*, McGraw-Hill Inc., 1980.

[40] M.M. Kapranov, *On the q-analog of homological algebra*, Preprint Cornell University 1991, q-alg/9611005.

[41] M. Karoubi, *Homologie cyclique des groupes et algèbres*, C.R. Acad. Sci. Paris **297**, Série I, (1983) 381-384.

[42] M. Karoubi, *Homologie cyclique et K-théorie*, Astérisque **149** (SMF), 1987.

[43] C. Kassel, M. Wambst, *Algèbre homologique des N-complexes et homologies de Hochschild aux racines de l'unité* Publ. RIMS, Kyoto Univ. **34** (1998) 91-114.

[44] B. Kostant, S. Sternberg, *Symplectic reduction, BRS cohomology and infinite-dimensional Clifford algeras*, Ann. Physics **176** (1987) 49-113.

[45] J.L. Koszul, *Homologie et cohomologie des algèbres de Lie*, Bull. Soc. Math. Fr. **78** (1950) 65-127.

[46] J.P. Labesse, *Champs libres: Cas de l'électron de Dirac et du photon*. Exposés à Dijon 1980.

[47] J.-L. Loday, *Cyclic homology*, Springer-Verlag, New York 1992.

[48] D. McMullan, *Constraints and B.R.S. symmetry*, Imperial College Preprint TP 83-84/21 (1984).
A.D. Browning, D. McMullan, *The Batalin-Fradkin-Vilkovsky formalism for higher-order theories*, J. Math. Phys. **28** (1987) 438-444.

[49] W. Mayer, *A new homology theory I, II*, Annals of Math. **43** (1942) 370-380 and 594-605.

[50] N. Nakanishi, I. Ojima, *Covariant operator formalism of gauge theories and quantum gravity*, World Scientific 1990.

[51] G. Rideau, *Noncompletely reducible representations of the Poincaré group associated with the generalized Lorentz gauge*, J. Math. Phys. **19** (1978) 1627-1634.

[52] M. Rosso, *Espace des phases réduit et cohomologie B.R.S.*, Adv. Ser. Math. Phys. **4** (1988) 263-269.

[53] B. Schroer, *A course on: An algebraic approach to nonperturbative quantum field theory*, CBPF, Rio de Janeiro 1998, hep-th 9707230 revised.

[54] L.P.S. Singh, C.R. Hagen, *Lagrangian formulation for arbitrary spin. 1. The boson case*, Phys. Rev. **D 9** (1974) 898-909.
[55] A. Sitarz, *On the tensor product construction for q-differential algebras*, Lett. Math. Phys. **44** (1998) 17-21.
[56] J.D. Stasheff, *Constrained hamiltonians: a homological approach*, Suppl. Rendiconti del Circ. Mat. di Palermo, Proc. Winter School on Geometry and Physics (1987) 239-252.
[57] J.D. Stasheff, *Homological (ghost) methods in mathematical physics* in "Infinite dimensional geometry, non commutative geometry, operator algebras, fundamental interactions", Saint François - Guadeloupe 1993, R. Coquereaux, M. Dubois-Violette, P. Flad Eds, World Scientific 1995.
[58] R. Stora, *Algebraic structure of chiral anomalies* in "New perspectives in quantum field theories", Jaca (Spain) 1985, World Scientific 1986.
[59] R. Stora, *De la fixation de jauge considérée comme un des beaux arts et de la symétrie de Slavnov qui s'ensuit*, hep-th/9611115.
[60] R. Stora, *Exercises in equivariant cohomology* in "Quantum fields and quantum space-time", Cargèse 1996, G. 't Hooft, A. Jaffe, G. Mack, P.K. Mitter, R. Stora Eds, Plenum Press 1997.
[61] F. Strocchi, *Locality and covariance in qed and gravitation. General proof of Gupta-Bleuler type formulations*, Lectures in Theoretical Physics, **Vol. XIV B**, Boulder, Colorado 1973.
[62] F. Strocchi, A.S. Wightman, *Proof of the charge superselection rule in local relativistic quantum field theory*, J. Math. Phys. **15** (1974) 2198-2224.
[63] D. Sullivan, *Infinitesimal computations in topology*, Publ. IHES **47** (1977) 269-331.
[64] C.-M. Viallet, *The geometry of the space of fields in Yang-Mills theory* in "Fields and geometry", Karpacz (Poland) 1986, A. Jadczyk Ed., World Scientific 1986.
[65] R. M. Wald, *Spin-two fields and general covariance*, Phys. Rev. **D 33** (1986) 3613-3625.
[66] C.A. Weibel, *An introduction to homological algebra*, Cambridge University Press 1994.
[67] S. Weinberg, *The quantum theory of fields, I, II*, Cambridge University Press 1996.

LABORATOIRE DE PHYSIQUE THÉORIQUE[1], BÂTIMENT 210, UNIVERSITÉ PARIS XI, F-91405 ORSAY CEDEX
E-mail address: patricia@osiris.th.u-psud.fr

---

[1]Unité Mixte de Recherche du Centre National de la Recherche Scientifique - UMR 8627

Modular Invariants from Subfactors

Jens Böckenhauer and David E. Evans

ABSTRACT. In these lectures we explain the intimate relationship between modular invariants in conformal field theory and braided subfactors in operator algebras. A subfactor with a braiding determines a matrix Z which is obtained as a coupling matrix comparing two kinds of braided sector induction ("α-induction"). It has non-negative integer entries, is normalized and commutes with the S- and T-matrices arising from the braiding. Thus it is a physical modular invariant in the usual sense of rational conformal field theory. The algebraic treatment of conformal field theory models, e.g. $SU(n)_k$ models, produces subfactors which realize their known modular invariants. Several properties of modular invariants have so far been noticed empirically and considered mysterious such as their intimate relationship to graphs, as for example the A-D-E classification for $SU(2)_k$. In the subfactor context these properties can be rigorously derived in a very general setting. Moreover the fusion rule isomorphism for maximally extended chiral algebras due to Moore-Seiberg, Dijkgraaf-Verlinde finds a clear and very general proof and interpretation through intermediate subfactors, not even referring to modularity of S and T. Finally we give an overview on the current state of affairs concerning the relations between the classifications of braided subfactors and two-dimensional conformal field theories. We demonstrate in particular how to realize twisted (type II) descendant modular invariants of conformal inclusions from subfactors and illustrate the method by new examples.

1. Introduction and overview

A subfactor in its simplest guise arises from a group action $M^G \subset M$, the fixed point algebra M^G in the ambient von Neumann algebra M where a group G acts upon. If say the group is finite and acts outerly on M (equivalently $(M^G)' \cap M = \mathbb{C}1$, where the prime denotes the commutant) and both the group and the algebra are amenable, then we can recover both the group and the action from the inclusion $M^G \subset M$. (If M is not amenable, i.e. hyperfinite, one may recover the group but not the action as in free group factors in free probability theory). However we will concentrate on (infinite-dimensional) hyperfinite von Neumann algebras M which are inductive limits of finite dimensional algebras and are factors i.e. have trivial

---

1991 *Mathematics Subject Classification.* Primary 81T40, 46L37; Secondary 46L60, 81T05, 81R10, 22E67, 82B23, 18D10.

This project was supported by the EU TMR Network in Non-Commutative Geometry.

Lectures given by the second author at "Quantum Symmetries in Theoretical Physics and Mathematics", 10–21 January 2000, Bariloche, Argentine.

center $M' \cap M = \mathbb{C}\mathbf{1}$. A subfactor $N \subset M$ is then an inclusion of one factor in another, which is thought to represent a deformation of a group, for us we will restrict to the case where we only think of those inclusions which are deviants of finite groups. (Cf. [29] as a general reference.)

Rather than a group of $*$-automorphisms of a von Neumann algebra M, we will more generally consider a system Δ of $*$-endomorphisms which is closed under composition

$$\lambda \circ \mu = \bigoplus_{\nu \in \Delta} N_{\lambda,\mu}^{\nu} \nu$$

for a suitable notion of addition of endomorphisms (for which we will need infinite von Neumann factors and consider endomorphisms up to inner equivalence, i.e. as sectors [61]) and non-negative integral coefficients $N_{\lambda,\mu}^{\nu}$. In our relationship with modular invariant partition functions in conformal field theory, our starting point will be a system of endomorphisms labelled by vertices of graphs as e.g. given in Fig. 1. Each $\lambda \in \Delta$ defines a matrix $N_\lambda = [N_{\lambda,\mu}^{\nu}]_{\mu,\nu}$ of multiplication by λ so that

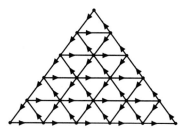

FIGURE 1. Fusion graphs of fundamental generators $\square$ of systems for $SU(2)_{10}$ and $SU(3)_5$

in the above setting the graph of $N_\square$ where $\square$ is the fundamental generator is as described in the figures. For example in the case of the Dynkin diagram A_3, as in Fig. 2. Here we labelled the vertices by b, s, v, and the graph represents the 'fusion'

FIGURE 2. Dynkin diagram A_3 as fusion graph

by s, and so the multiplication by s gives the sum of nearest neighbors:

$$\text{s} \cdot \text{b} = \text{s}, \qquad \text{s} \cdot \text{s} = \text{b} \oplus \text{v}, \qquad \text{s} \cdot \text{v} = \text{s}.$$

(Here and in general it is understood that an unoriented edge represents an arrow in both directions.)

These are the well-known fusion rules of the conformal Ising model. A treatment of the Ising model in the framework of local quantum physics realizing these fusion rules in terms of endomorphisms on von Neumann factors was carried out in

[4], building on [63]. The transfer matrix formalism allows one to study classical statistical mechanical models via non-commutative operator algebras. A study of the Ising model in this framework was carried out in [1, 27, 15]. The fundamental example of this non-commutative framework for understanding the Ising model was the driving force towards the present work on understanding modular invariant partition functions via non-commutative operator algebras (cf. the lecture by the second author at the CBMS meeting in Eugene, Oregon, September 1993).

Using associativity of the fusion product one obtains for the fusion matrices

$$N_\lambda N_\mu = \sum_\nu N_{\lambda,\mu}^\nu N_\nu,$$

i.e. the matrices N_λ themselves give a ("regular") representation of the fusion rules of Δ. Usually the system Δ will be closed under a certain conjugation $\lambda \mapsto \bar\lambda$ (generalizing the notion of inverse and conjugate representation in a group and group dual, respectively) which is anti-multiplicative and additive. This will mean that the transpose of N_λ is $N_{\bar\lambda}$. If we start with a system obeying commutative fusion rules (which will not always be the case), the collection $\{N_\lambda\}_{\lambda \in \Delta}$ will therefore constitute a family of normal commuting matrices, and hence be simultaneously diagonalizable, with spectra $\text{spec}(N_\lambda) = \{\gamma_\rho^\lambda\}_\rho$. In fact their spectra will be labelled naturally by the entire set Δ itself, i.e. we will have $\rho \in \Delta$. In this diagonalization we have

(1.1) $$\gamma_\rho^\lambda \gamma_\rho^\mu = \sum_\nu N_{\lambda,\mu}^\nu \gamma_\rho^\nu,$$

i.e. the eigenvalues provide one-dimensional representations of the fusion rules. The matrix γ_ρ^λ is invertible and we can invert Eq. (1.1) to obtain the Verlinde formula [81]

(1.2) $$N_{\lambda,\mu}^\nu = \sum_\rho \frac{S_{\lambda,\rho}}{S_{0,\rho}} S_{\mu,\rho} S_{\nu,\rho}^*.$$

Here we write the eigenvalues of N_λ as $\gamma_\rho^\lambda = S_{\lambda,\rho}/S_{0,\rho}$, where the label "0" refers to the distinguished identity element ("vacuum") of the fusion rules, and $S_{0,\rho} = (\sum_\lambda |\gamma_\rho^\lambda|^2)^{1/2}$. (See [36] for fusion rules in the context of conformal field theory.)

In our subfactor approach to modular invariants we will have representations of the Verlinde fusion rules appearing naturally, with spectrum a proper subset of Δ and with multiplicities $Z_{\lambda,\lambda}$, $\lambda \in \Delta$, given by the diagonal part of a modular invariant. The representation matrices can be interpreted a adjacency matrices of graphs associated with modular invariants.

Modular invariant partition functions arise as continuum limits in statistical mechanics and play a fundamental role in conformal field theory. Recall that a modular invariant partition function is of the form (cf. Zuber's lectures, or see [21, 35, 54, 20, 41] for more details on these matters)

$$Z(\tau) = \sum_{\lambda,\mu} Z_{\lambda,\mu} \chi_\lambda(\tau) \chi_\mu(\tau)^*.$$

Here $\chi_\lambda = \text{tr}(q^{L_0 - c/24})$, $q = e^{2\pi i \tau}$, is the trace in the irreducible representation of a chiral algebra, which for us will be a positive energy representation of a loop group with the conformal Hamiltonian L_0 being the infinitesimal generator of the rotation group on the circle. (More typically we would take un-specialized characters in order to have linearly independent characters. See for example [18] or [29, Sect. 8.3] for explicit computations with corner transfer matrices and derivations of the Virasoro characters in the context of the Ising model.) Then the action of the modular group $SL(2;\mathbb{Z})$ on $q = e^{2\pi i \tau}$ via $\mathcal{S} = \begin{pmatrix} 0 & -1 \\ 1 & 0 \end{pmatrix}$: $\tau \mapsto -1/\tau$, and $\mathcal{T} = \begin{pmatrix} 1 & 1 \\ 0 & 1 \end{pmatrix}$: $\tau \mapsto \tau + 1$, transforms the family of characters $\{\chi_\lambda\}$ linearly. More precisely, there are matrices S and T such that

$$\chi_\lambda(-1/\tau) = \sum_\mu S_{\lambda,\mu} \chi_\mu(\tau), \qquad \chi_\lambda(\tau+1) = \sum_\mu T_{\lambda,\mu} \chi_\mu(\tau).$$

Note first that what is remarkable about the Verlinde formula, Eq. (1.2), is that the matrix which diagonalizes the fusion rules is the same as the modular matrix S which transforms the characters (e.g. the Kac-Peterson matrix for current algebra models, see [54, 35]). It is also remarkable that this matrix is symmetric: $S_{\lambda,\mu} = S_{\mu,\lambda}$.

From physical considerations we will require solutions to the matrix equations $ZS = SZ$, $ZT = TZ$, subject to the constraint $Z_{0,0} = 1$ ("uniqueness of the vacuum") and the "coupling matrix" Z having only non-negative integer entries (from multiplicities of the representations). There will always be at least one solution, the diagonal partition function

$$Z = \sum_\lambda |\chi_\lambda|^2$$

(or $Z_{\lambda,\mu} = \delta_{\lambda,\mu}$), or more generally there may be permutation invariants

$$Z = \sum_\lambda \chi_\lambda \chi^*_{\omega(\lambda)},$$

whenever ω is a permutation of the labels which preserves the fusion rules, the vacuum, and the "conformal dimensions". Moore and Seiberg argue in [66] (see also [24]) that after a "maximal extension of the chiral algebra" (the hardest part is to make this mathematically precise) the partition function of a RCFT is at most a permutation matrix $Z^{\text{ext}}_{\tau,\tau'} = \delta_{\tau,\omega(\tau')}$, where τ, τ' label the representations of the extended chiral algebra and now ω denotes a permutation of these with analogous invariance properties. Decomposing the extended characters χ^{ext}_τ in terms of the original characters χ_λ, we have $\chi^{\text{ext}}_\tau = \sum_\lambda b_{\tau,\lambda} \chi_\lambda$ for some non-negative integral branching coefficients $b_{\tau,\lambda}$. The maximal extension yields the coupling matrix expression

$$Z_{\lambda,\mu} = \sum_\tau b_{\tau,\lambda} b_{\omega(\tau),\mu}.$$

There is a distinction [23] between so-called type I invariants which arise from the diagonal invariant of the maximal extension, i.e. for which ω is trivial, and type II invariants corresponding to non-trivial automorphisms of the extended fusion rules. The coupling matrix of a type I invariant is in particular symmetric whereas type II invariants need not be so but still the "vacuum coupling" is symmetric: $Z_{0,\lambda} = Z_{\lambda,0}$ for all labels λ. To allow more generally for possibly non-symmetric vacuum coupling, $Z_{0,\lambda} \neq Z_{\lambda,0}$, one may need different extensions for the left and right chiral algebra [9] (see also [58] where this possibility is explicitly addressed in the context of simple current extensions), and then the distinction between type I and type II modular invariants does no longer make sense.[1]

A simple argument of Gannon [38] shows that there are at most finitely many solutions to our modular invariant problem. Since $d_\lambda = S_{\lambda,0}/S_{0,0}$ will be positive and at least 1 (the d_λ's will be the Perron-Frobenius weights of the graphs as in

---

[1] Surprisingly enough, all known modular invariants of $SU(n)_k$ models are entirely symmetric. Nevertheless there are known modular invariants of other models with non-symmetric ("heterotic") vacuum coupling — see Section 7.

Fig. 1, or indeed d_λ will be the statistical dimension of λ as a sector in the von Neumann algebra theory or the square root of the Jones index [52]), we obtain from $SZS^* = Z$ that $\sum_{\lambda,\mu} Z_{\lambda,\mu} \leq \sum_{\lambda,\mu} d_\lambda Z_{\lambda,\mu} d_\mu = 1/S_{0,0}^2$. Consequently each integer $Z_{\lambda,\mu}$ must be bounded by $1/S_{0,0}^2 = \sum_\lambda d_\lambda^2$ (from unitarity of the S-matrix), so that there are only finitely many solutions. Note that this bound will be our "global index" w, and this suggests a strong relation between Gannon's argument and Ocneanu's rigidity theorem (presented at a conference in January 1997 in Madras, India), the latter implying the finiteness of the number of subequivalent paragroups for a given paragroup.

Gannon's estimate can even be refined to the inequality

(1.3) $$Z_{\lambda,\mu} \leq d_\lambda d_\mu$$

for each individual entry of a modular invariant coupling matrix as follows. As by Verlinde's formula, Eq. (1.2), the eigenvalues of the non-negative fusion matrices N_λ are given by $\gamma_\rho^\lambda = S_{\lambda,\rho}/S_{0,\rho}$, Perron-Frobenius theory tells us that $|\gamma_\rho^\lambda|$ is bounded by the Perron-Frobenius eigenvalue γ_0^λ, so that $|S_{\lambda,\rho}| \leq S_{\lambda,0} S_{0,\rho}/S_{0,0}$ (cf. [41]). Commutativity of Z with the unitary S then yields

$$Z_{\lambda,\mu} = \sum_{\rho,\nu} S_{\lambda,\rho} Z_{\rho,\nu} S^*_{\nu,\mu} \leq \sum_{\rho,\nu} |S_{\lambda,\rho}| Z_{\rho,\nu} |S_{\nu,\mu}| \leq d_\lambda Z_{0,0} d_\mu,$$

which provides Eq. (1.3) by the normalization $Z_{0,0} = 1$.

2. Operator algebraic input

We will study the classification of modular invariants and construction of maximal extensions through subfactors, in particular starting with braided systems of endomorphisms on loop group factors which are purely infinite factors with no traces, or more precisely type III_1. Recall that a factor is type I when there is a trace on the algebra taking discrete values on projections, type II when there is a trace that takes continuous values. (We hope that the classification of factors into types I, II and III will not be confused with the distinction of type I and type II modular invariants — it has nothing to do with it.) A trace on a von Neumann algebra M is a (possibly unbounded) linear functional τ satisfying $\tau(ab) = \tau(ba)$, $a, b \in M$, where the algebra or trace is finite if $\tau(\mathbf{1}) < \infty$, or infinite otherwise. Thus a finite type I factor is (isomorphic to) $\text{Mat}(n) = \text{End}(\mathbb{C}^n)$, the $n \times n$ complex matrices, the infinite factor is $B(H)$, the bounded linear operators on an infinite-dimensional Hilbert space. A factor is of type III (or purely infinite) otherwise, there is no trace and every non-zero projection p is equivalent to the unit in the sense that there is a partial isometry v in the algebra such that $v^*v = \mathbf{1}$ and $vv^* = p$. The factors relevant for RCFT are amenable, in the sense that they are hyperfinite, the completions of unions of finite-dimensional algebras. Murray and von Neumann showed that there is an unique hyperfinite II_1 (i.e. finite type II) factor which can be realized as for example the infinite tensor product of matrix algebras (arbitrarily chosen as long as they are non-commutative) completed with respect to the trace, i.e. use the trace τ (constructed as the tensor product of traces over the matrices) to define an inner product on $\mathcal{M}$ (the algebraic tensor product) $\langle a, b \rangle = \tau(b^*a)$. Letting $\Omega = \mathbf{1}$ regarded as a vector in the completion $\mathcal{H}$ of $\mathcal{M}$ with respect to this inner product, we can let $\mathcal{M}$ act on $\mathcal{H}$ by the induced left action of $\mathcal{M}$ on itself, and the hyperfinite II_1 factor is the von Neumann algebra generated by $\mathcal{M}$ in this representation. There is by Connes [16] an unique hyperfinite II_∞

(i.e. infinite type II factor) which is $R \otimes B(H)$ where R is the unique hyperfinite II_1 factor and $B(H)$ is type I_∞. There is a finer classification of type III factors into III_λ, $0 \leq \lambda \leq 1$. For each $\lambda \in (0,1]$ there is by Connes an unique hyperfinite III_λ factor (the analysis completed by Haagerup [47] in the case $\lambda = 1$). The type III_0 factors are classified by their flow of weights.

In the semi-finite case (I or II) where there is a trace τ, we can define a conjugation J, a conjugate linear map of the Hilbert space $\mathcal{H}$ of the trace, by $J : a \mapsto a^*$, $a \in M \subset \mathcal{H}$, or $Ja\Omega = a^*\Omega$. Then J is isometric because τ is a trace and interchanges left and right multiplication, indeed $JMJ = M'$. Thus M and M' are of comparable size. If $M = \mathrm{Mat}(n) = \mathbb{C}^n \otimes \overline{\mathbb{C}^n}$ is finite dimensional, then acting on itself (regarded as a Hilbert space) M becomes $M \otimes \mathbf{1}$ with commutant $\mathbf{1} \otimes M$. In general represent a factor M on a Hilbert space H with vector $\Phi \in H$ cyclic for M, i.e. $H = \overline{M\Phi}$, which is also cyclic for M', $H = \overline{M'\Phi}$. (Take a faithful normal state φ on M and the associated Hilbert space.) Then we can define $S : a\Phi \mapsto a^*\Phi$, $a \in M$, and take the polar decomposition $S = J\Delta^{1/2}$, where J is a conjugation and Δ the (possibly unbounded) Tomita-Takesaki modular operator. Then Tomita-Takesaki theory [79] tells us that $JMJ = M'$, and $\sigma_t = \mathrm{Ad}(\Delta^{it})$ defines a one-parameter automorphism group of M which describes how far the vector state $\varphi(\cdot) = \langle \cdot \Phi, \Phi \rangle$ is from being a trace, $\varphi(ab) = \varphi(b\sigma_i(a))$ for analytic $a, b \in M$. In the case of a semi-finite algebra, and if φ is a trace, then $S = J$, $\Delta = \mathbf{1}$, and $\sigma_t = \mathrm{id}$, whilst for other choices of cyclic and separating vectors Φ, the Tomita-Takesaki modular group σ_t is at least inner.

Now consider the case of an infinite subfactor $N \subset M$, i.e. both factors N and M are infinite which means that they contain isometries with range projections being different from the identity. Then we can represent M on a Hilbert space H where there is a vector Φ which is cyclic and separating for both N and M. Taking the corresponding Tomita-Takesaki modular conjugations J_N and J_M where $J_N N J_N = N'$, $J_M M J_M = M'$, we define

$$\gamma = \mathrm{Ad}(J_N J_M)|_M : M \to M' \subset N' \to N$$

called the canonical endomorphism [60] from M into N. Different choices of Hilbert spaces and cyclic and separating vectors only amount to a change $\gamma \to \mathrm{Ad}(u) \circ \gamma$ by a unitary $u \in N$, i.e. the N-M sector determined by γ is well-defined. (If $\rho \in \mathrm{Mor}(A, B)$ is a unital morphism from A to B, the B-A sector $[\rho]$ is the equivalence class of ρ where $\rho' \simeq \rho$ iff $\rho' = \mathrm{Ad}(u) \circ \rho$ for unitaries $u \in B$.) We then have an inclusion of factors:

(2.1) $$\gamma(N) \subset \gamma(M) \subset N \subset M \subset M_1 = \mathrm{Ad}(J_M J_N)(N).$$

We can continue upwards (called the Jones tower) or downwards (called the Jones tunnel) but the sequence is of period two, e.g. the inclusion $\gamma(N) \subset \gamma(M)$ is isomorphic to $N \subset M$, and $\gamma(M) = \mathrm{Ad}(J_N)(M') \subset N$ is isomorphic to $M = \mathrm{Ad}(J_M)(M') \subset \mathrm{Ad}(J_M J_N)(N) = M_1$. This periodicity reduces to that between a group and its dual $G \leftrightarrow \hat{G}$ in the case of a group subfactor tower $M^G \subset M \subset M \rtimes G$. So there are basically two canonical endomorphisms, $\gamma \in \mathrm{End}(M)$ and $\theta \in \mathrm{End}(N)$, where $\theta = \gamma|_N$. We call γ the canonical endomorphism, and θ the dual canonical endomorphism for $N \subset M$.

The tower can be identified with the Jones extensions, in the case of finite index obtained by adjoining a sequence of projections satisfying the Temperley-Lieb relations. We could define the Jones index using the Pimsner-Popa inequality as follows. If $E : M \to N$ is a conditional expectation (a projection of norm one), then let $\mathrm{Ind}(E)$ be the best constant ξ such that $E(x^*x) \geq \xi^{-1} x^* x$ for all $x \in M$. Then the (Jones) index $[M : N]$ is the infimum of $\mathrm{Ind}(E)$ over all expectations E, and there is an unique expectation called the minimal expectation which realizes the index.

For us, all the relative commutants $N' \cap M_j$, $M' \cap M_j$, in the tower will be finite-dimensional and moreover $N' \cap M_j \subset N' \cap M_{j+1}$, $M' \cap M_j \subset M' \cap M_{j+1}$, will be described by *finite* graphs. Due to the periodicity of the tower, only two graphs appear here, the principal and dual principal graph. The finiteness of the graphs (equivalent to the finiteness in RCFT) will imply finite index and the Jones index will be the square of the norm of either graph.

One question which will engage us will be whether a particular endomorphism of a factor N should be a dual canonical endomorphism (of some subfactor $N \subset M$ without any a priori knowledge of what M should be). For example if Z is a modular invariant, we can consider $\bigoplus_{\lambda \in \Delta} Z_{0,\lambda}[\lambda]$, $\bigoplus_{\lambda \in \Delta} Z_{\lambda,0}[\lambda]$, $\bigoplus_{\lambda \in \Delta} Z_{\lambda,\mu}[\lambda \otimes \mu^{\mathrm{opp}}]$ as candidates for (the sectors of) dual canonical endomorphisms (on N, N, $N \otimes N^{\mathrm{opp}}$, respectively, if Δ is a system of endomorphisms of N).

Before we go any further let us formalize the notion of algebraic operations and sectors. Let A and B be type III von Neumann factors. A unital $*$-homomorphism $\rho : A \to B$ is called a B-A morphism, and we write $\rho \in \mathrm{Mor}(A, B)$. The positive number $d_\rho = [B : \rho(A)]^{1/2}$ is called the statistical dimension of ρ; here $[B : \rho(A)]$ is the Jones index of the subfactor $\rho(A) \subset B$. Now if $\sigma \in \mathrm{Mor}(B, C)$ (with C being another type III factor) then the multiplication or "fusion"

$$[\sigma][\rho] = [\sigma \rho]$$

is well defined on sectors. (We usually abbreviate $\sigma \rho \equiv \sigma \circ \rho$.) For $\tau_1, \tau_2 \in \mathrm{Mor}(A, B)$ take isometries $t_1, t_2 \in B$ such that $t_1 t_1^* + t_2 t_2^* = \mathbf{1}$ which we can find by infiniteness of B. Then define the sum

$$[\tau_1] \oplus [\tau_2] = [\tau], \quad \text{where} \quad \tau(a) = t_1 \tau_1(a) t_1^* + t_2 \tau_2(a) t_2^*, \quad a \in A.$$

This is well-defined as if $s_1, s_2 \in B$ is another choice of isometries satisfying $s_1 s_1^* + s_2 s_2^* = \mathbf{1}$ then $u = s_1 t_1^* + s_2 t_2^*$ is a unitary in B, intertwining τ and τ' where $\tau'(a) = s_1 \tau_1(a) s_1^* + s_2 \tau_2(a) s_2^*$ for all $a \in A$. This notion of a sum is basically writing $[\tau_1] \oplus [\tau_2]$ as a 2×2 matrix $\begin{pmatrix} \tau_1(\cdot) & 0 \\ 0 & \tau_2(\cdot) \end{pmatrix}$ in B using the infiniteness of B to achieve the matrix decomposition. If ρ and σ are B-A morphisms with finite statistical dimensions, then the vector space of intertwiners

$$\mathrm{Hom}(\rho, \sigma) = \{t \in B : t\rho(a) = \sigma(a)t, \; a \in A\}$$

is finite-dimensional, and we denote its dimension by $\langle \rho, \sigma \rangle$. Note that for τ, τ_1, τ_2 and t_1, t_2 as above we have e.g. $t_1 \in \mathrm{Hom}(\tau_1, \tau)$. The impossibility of decomposing some $\rho \in \mathrm{Mor}(A, B)$ as $[\rho] = [\rho_1] \oplus [\rho_2]$ for some $\rho_1, \rho_2 \in \mathrm{Mor}(A, B)$, or irreducibility is then equivalent to the subfactor $\rho(A) \subset B$ being irreducible, i.e. $\rho(A)' \cap B = \mathbb{C}\mathbf{1}$.

For groups (and group duals) we have a notion of a conjugate of λ, namely the inverse λ^{-1} of λ (respectively the conjugate representation). There is a similar notion for sectors. For an irreducible $\lambda \in \mathrm{Mor}(A, B)$, an irreducible morphism $\bar{\lambda} \in \mathrm{Mor}(B, A)$ is a representative of the conjugate sector if $[\lambda \bar{\lambda}]$ or $[\bar{\lambda} \lambda]$ contain the identity sector ($[\mathrm{id}_A]$ or $[\mathrm{id}_B]$, respectively), and the multiplicity is then automatically one for both cases [**48**]. More generally, for an arbitrary morphism $\rho \in \mathrm{Mor}(A, B)$ of finite statistical dimension d_ρ, an A-B morphism $\bar{\rho}$ is a conjugate morphism if there are isometries $r_\rho \in \text{Hom}(\text{id}_A, \bar\rho\rho)$ and $\bar r_\rho \in \text{Hom}(\text{id}_B, \rho\bar\rho)$ such that

(2.2) $\qquad \rho(r_\rho)^* \bar r_\rho = d_\rho^{-1} \mathbf{1}_B \qquad$ and $\qquad \bar\rho(\bar r_\rho)^* r_\rho = d_\rho^{-1} \mathbf{1}_A$.

Recall the tower-tunnel of Eq. (2.1). Suppose $E : M \to N$ is a conditional expectation and φ is a faithful normal state on N, and set $\omega = \varphi \circ E$. Then ω is a faithful normal state on M such that $\omega \circ E = \omega$. Take the GNS Hilbert space H of this state on M, with cyclic and separating vector Ω. We can identify this space with our previous Hilbert space (where there is vector Φ being cyclic and separating for both N and M) and the actions coincide. However Φ is not identified with Ω, as $\overline{N\Omega}$ is a proper subspace if $N \neq M$, with orthogonal Jones projection $e_N : \overline{M\Omega} \to \overline{N\Omega}$ such that $m\Omega \mapsto E(m)\Omega$, $m \in M$. Define $v' : n\Phi \mapsto n\Omega$, $n \in N$, on H so that $v' \in N'$, and $v'v'^* = e_N$. Then $v_1 = \text{Ad}(J_M)(v')$ is an isometry in M_1, and also $v_1 v_1^* = e_N$. It is easily checked starting from $J_M v' = v' J_N$ that v_1 is an intertwiner in $\text{Hom}(\text{id}_{M_1}, \gamma_1)$, where $\gamma_1 = \text{Ad}(J_{M_1} J_M)$ is the canonical endomorphism of $M \subset M_1$. Thus, by translating in the tunnel-tower the canonical and dual canonical endomorphism contain the identity sector.

Denoting by $\iota : N \hookrightarrow M$ the inclusion homomorphism we put $\bar\iota : M \to N$ as $\bar\iota(m) = \gamma(m)$, $m \in M$. Then $\gamma = \iota\bar\iota$ and $\theta = \bar\iota\iota$ both contain the identity sector so that $\bar\iota$ is in fact a conjugate morphism for ι. Similarly, if $\lambda \in \text{End}(N)$, we can take $\gamma_\lambda, \theta_\lambda$ to be the canonical and dual canonical endomorphisms of the inclusion $\lambda(N) \subset N$. Then we can set $\bar\lambda = \lambda^{-1}\gamma_\lambda$, which is well defined so that $\lambda\bar\lambda = \gamma_\lambda$. In the group case, say if we have an outer action $\alpha : G \to \text{Aut}(M)$ of a finite group G on a type III factor M and let N be the corresponding fixed point algebra, $N = M^G$, then γ decomposes as a sector into the group elements, $[\gamma] = \bigoplus_{g \in G}[\alpha_g]$, whereas the decomposition of θ is according to the group dual $\hat G$, i.e. $[\theta] = \bigoplus_{\pi \in \hat G} d_\pi[\rho_\pi]$, with multiplicities given by the dimensions d_π of the irreducible representations π of G.

Sometimes it is useful to use graphical expressions for formulae involving intertwiners. Roughly speaking, for an intertwiner $t \in \text{Hom}(\rho, \sigma)$ we draw a picture as in Fig. 3, i.e. we represent morphisms by oriented "wires" and intertwiners by boxes. Reversing an arrow means replacing a label ρ by its conjugate $\bar\rho$, and taking ad-

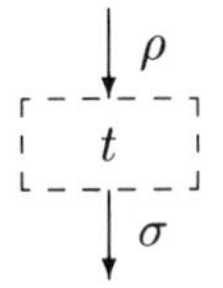

FIGURE 3. An intertwiner $t \in \text{Hom}(\rho, \sigma)$

joints then corresponds to vertical reflection of the picture together with reversing all arrows. As $\text{Hom}(\rho, \sigma) \subset \text{Hom}(\rho\tau, \sigma\tau)$ we are allowed to add or remove straight wires on the right, i.e. we are free to pass from Fig. 3 to Fig. 4. On the other hand, the intertwiner $\mu(t)$ is in $\text{Hom}(\mu\rho, \mu\sigma)$ and is represented graphically as in Fig. 5. With the convention that the identity morphism (of some factor) is labelled by "the invisible wire", the isometries $r_\rho, \bar r_\rho$ and $r_\rho^*, \bar r_\rho^*$ are represented as caps and cups, respectively, with different orientations of the wire labelled by ρ. Then, with certain

FIGURE 4. An intertwiner $t \in \mathrm{Hom}(\rho, \sigma) \subset \mathrm{Hom}(\rho\tau, \sigma\tau)$

FIGURE 5. The intertwiner $\mu(t) \in \mathrm{Hom}(\mu\rho, \mu\sigma)$

normalization procedures taken care of in [**10**] (where the graphical framework is worked out in full detail – but see also [**65, 57, 84, 34, 32, 53**]), the relations of Eq. (2.2) become topological moves as in Fig. 6. The minimal conditional expecta-

FIGURE 6. A topological invariance

tion is obtained as follows. First, the map $\phi_\rho : B \to A$, $b \mapsto r_\rho^* \overline{\rho}(b) r_\rho$, is the unique standard left inverse for ρ (as $\phi_\rho \circ \rho = \mathrm{id}_A$) and then $E_\rho = \rho \circ \phi_\rho : B \to \rho(A)$ is the minimal conditional expectation for the subfactor $\rho(A) \subset B$. In the graphical framework, Jones projections in the tunnel which were translates of $v_1 v_1^*$ appear as in Fig. 7. The Pimsner-Popa bound in the Kosaki-Jones index is realized by such

FIGURE 7. Jones projections in the tunnel

Jones projections so that the constant d_ρ in Eq. (2.2) is identified with $[B : \rho(A)]^{1/2}$, the square root of the Jones index.

Returning to our original subfactor $N \subset M$ with inclusion homomorphism $\iota : N \hookrightarrow M$, $\gamma = \iota\overline{\iota}$, $\theta = \overline{\iota}\iota$, where $\overline{\iota}$ is a conjugate for ι, we have isometries $w \equiv r_\iota \in \mathrm{Hom}(\mathrm{id}_N, \theta)$ and $v \equiv \overline{r}_\iota \in \mathrm{Hom}(\mathrm{id}_N, \gamma)$ satisfying the consistency relations

$w^*v = w^*\gamma(v) = d_\iota^{-1}\mathbf{1}$ with $d_\iota^2 = [M:N] = d_\gamma = d_\theta$. Note that we have pointwise equality $M = Nv$ as $m = [M:N]^{1/2}w^*\gamma(m)v$ where $[M:N]^{1/2}w^*\gamma(m) \in N$, $m \in M$, which means that v is a basis element for M as an N-module. The previous characterization of conjugates can be used to characterize which endomorphisms arise as a canonical endomorphism.

If $\gamma \in \mathrm{End}(M)$ where M is an infinite factor, then γ is a canonical endomorphism of some subfactor $N \subset M$ if and only if there exist isometries $v \in \mathrm{Hom}(\mathrm{id}_M, \gamma)$ and $w \in \mathrm{Hom}(\gamma, \gamma^2)$ such that

(2.3) $\qquad\qquad w^*\gamma(w) = ww^*, \qquad \gamma(w)w = w^2,$

(2.4) $\qquad\qquad v^*w = w^*\gamma(v) = d^{-1}\mathbf{1}, \quad d > 0.$

Note that if $v = \bar{r}_\iota$ and $w = r_\iota$ as before, then $w \in \mathrm{Hom}(\mathrm{id}_N, \theta) \subset \mathrm{Hom}(\gamma, \gamma^2)$. Conversely, if Eq. (2.3) and Eq. (2.4) hold then we can define $N = \{x \in M : wx = \gamma(x)w, \ wx^* = \gamma(x^*)w\}$, and then $E : M \to N$ defined by $E(x) = w^*\gamma(x)w$, $x \in M$, is a conditional expectation.

3. Subfactors arising from loop groups

We now turn to the actual algebras which we will use to describe our modular invariants, arising from loop groups. The loop group $LSU(n)$ consists of smooth maps $f : S^1 \to SU(n)$, the product being pointwise multiplication. The representations of interest will be projective representations of $LSU(n)$ which extend to positive energy representations of $LSU(n) \rtimes \mathrm{Rot}(S^1)$ where the rotation group acts on the maps of S^1 in a natural way so that the "Hamiltonian" or infinitesimal generator L_0 is positive. The ones of particular interest, the irreducible unitary positive energy representations are classified as follows. First there is a level k, a positive integer describing a cocycle because we are dealing with projective representations. The projective representation restricts to a genuine irreducible representation of the constant loops, identified with $SU(n)$ itself, the multiplier becomes irrelevant now since we are dealing with simply connected groups. In order to obtain positive energy, only finitely many irreducible representations are admissible, namely the vertices of (i.e. integrable weights in) the Weyl alcove $\mathcal{A}^{(n,k)}$. The (adjacency matrices of the) graphs N_λ, such as $N_\square$ itself, describe the fusion of positive energy representations.

Restricting to loops concentrated on an interval $I \subset S^1$ (proper, i.e. $I \neq S^1$ and non-empty), the corresponding subgroup denoted by

$$L_I SU(n) = \{f \in LSU(n) : f(z) = 1, \ z \notin I\},$$

one finds that in each positive energy representation π_λ the sets of operators $\pi_\lambda(L_I SU(n))$ and $\pi_\lambda(L_{I^c} SU(n))$ commute where I^c is the complementary interval of I, again using that $SU(n)$ is simply connected. In turn we obtain a subfactor

(3.1) $\qquad\qquad \pi_\lambda(L_I SU(n))'' \subset \pi_\lambda(L_{I^c} SU(n))',$

involving hyperfinite type III$_1$ factors (see [83]). In the vacuum representation, labelled by $\lambda = 0$, we have Haag duality in that the inclusion collapses to a single factor $N(I) = N(I)$, and in general we obtain a subfactor. The level 1 representations of $LSU(n)$ are realized through the Fock state of the Hardy space projection P on $L^2(S^1; \mathbb{C}^n)$. Since $[f, P]$ is Hilbert-Schmidt for $f \in LSU(n)$ acting naturally on $L^2(S^1; \mathbb{C}^n)$, we have that $LSU(n)$ is implemented in the corresponding Fock space giving a positive energy representation.

The vacuum representation π_0 gives a clear geometric picture of the Tomita-Takesaki modular group action σ and modular conjugation J on the Fock vacuum vector, cyclic and separating for say $\pi_0(L_I SU(n))''$ for I being the upper half circle. The Tomita-Takesaki modular group is induced by the second quantization of the geometric action of $SU(1,1)$ on S^1, which is seen to be ergodic. Consequently the algebra must be type III_1 as the action σ is never ergodic otherwise (see e.g. [**3**, Cor. 1.10.8]). Similarly the conjugation J acts by flipping the circle, taking I into the complementary interval:

$$\pi_0(L_I SU(n))'' = J\pi_0(L_I SU(n))'J = \pi_0(L_{I^c} SU(n))',$$

so that Haag duality holds in the vacuum representation and is a consequence of Tomita-Takesaki theory. More generally the inclusion Eq. (3.1) can be read as providing an endomorphism λ (by abuse of notation denoted by the same symbol as the label) of the local algebra $N(I)$ such that Eq. (3.1) is isomorphic to $\lambda(N(I)) \subset N(I)$. By A. Wassermann's work [**83**] we obtain this way a system of endomorphisms $\Delta = \{\lambda\}$, the morphisms being labelled by the Weyl alcove $\mathcal{A}^{(n,k)}$, which is closed under sector fusion, and the fusion coefficients $N_{\lambda,\mu}^\nu$ match exactly the loop group fusion. Similar results have been obtained for minimal models [**59**] and (partially) $LSpin(2n)$ models [**80**]. (That the DHR morphisms of net of a conformal field theory model obey exactly the Verlinde fusion rules from the conformal character transformations was conjectured in [**33**]. Proofs for special cases can be found in [**83, 59, 80, 4, 5**]. Antony Wassermann has informed us that he has computed fusion for all simple, simply connected loop groups; and with Toledano-Laredo all but E$_8$ using a variant of the Dotsenko-Fateev differential equation considered in his thesis.) If we take the relative commutants for the tunnel

$$\cdots \subset \lambda\bar{\lambda}\lambda(N) \subset \lambda\bar{\lambda}(N) \subset \lambda(N) \subset N,$$

we are decomposing products $\lambda\bar{\lambda}\lambda \cdots$ into irreducible components obtaining the same relative commutants as for the Jones-Wenzl type II_1 $SU(n)_k$ subfactors. More precisely, if λ is the fundamental representation $\square$ and $A \subset B$ is the hyperfinite type II_1 subfactor

(3.2) $$\{g_i : i = 1, 2, 3, \ldots\}'' \subset \{g_i : i = 0, 1, 2, \ldots\}'',$$

where the g_i's are the Hecke algebra generators obtained as explained below, then (using Popa [**69**]) the loop group subfactor $\lambda(N) \subset N$ is isomorphic to $A \otimes N \subset B \otimes N$.

The statistical mechanical models of [**17**] are generalizations of the Ising model. The configuration space of the Ising model, distributions of symbols "+" and "−" on the vertices of the square lattice $\mathbb{Z}^2$, can be thought of edges on the Dynkin diagram A$_3$ on the edges of a square lattice, where the end vertices are labelled by "+" and "−". This model can be generalized by replacing A$_3$ by other graphs Γ such as other Dynkin diagrams or indeed the Weyl alcove $\mathcal{A}^{(n,k)}$. A configuration is then a distribution of the edges of Γ over $\mathbb{Z}^2$, and associated to each local configuration is a Boltzmann weight satisfying the Yang-Baxter equation. The justification of $SU(n)$ models is as follows. By Weyl duality, the representation of the permutation group on $\bigotimes \text{Mat}(n)$ is the fixed point algebra of the product action of $SU(n)$. Deforming this, there is a representation of the Hecke algebra in $\bigotimes \text{Mat}(n)$, whose commutant is a representation of a deformation of $SU(n)$, the quantum group $SU(n)_q$ [**51**]. The Boltzmann weights lie in this algebra representation, and at critically reduce to the natural braid generators g_i, so that the Yang-Baxter equation satisfied by the Boltzmann weights reduces to the braid relation $g_i g_{i+1} g_i = g_{i+1} g_i g_{i+1}$. When $q = e^{2\pi i/(n+k)}$ is a root of unity, the irreducible representations of the corresponding Hecke algebra are labelled precisely by $\mathcal{A}^{(n,k)}$.

The graph Γ generates a von Neumann algebra by considering larger and larger matrices generated by larger and larger partition functions. A subfactor can be obtained by the adjoint action, placing the initial Boltzmann weights on the boundary. For $SU(n)_q$ subfactors this amounts to Eq. (3.2) because of the braid relations $\mathrm{Ad}(g_1 g_2 \cdots)(g_i) = g_{i+1}$. The principal graph of these inclusions is not the entire graph $N_\square$ (corresponding to $\mathcal{A}^{(n,k)}$ as in Fig. 1) but merely its zero-one part (the first two colours) as in Fig. 8. Nevertheless the entire graphs do have a meaning in

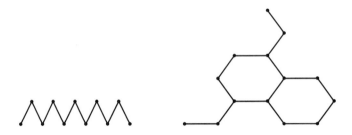

FIGURE 8. Colour zero-one part of the graphs in Fig. 1

subfactor theory simply as graphs encoding the fusion rules of associated systems of bimodules or sectors. Moreover, the center of $SU(n)$, namely $\mathbb{Z}_n$, induces an action on these subfactors and one may construct crossed product or orbifold subfactors $A^{\mathbb{Z}_n} \subset B^{\mathbb{Z}_n}$ [28, 85] which will in turn produce "orbifold" sector systems and graphs. As such graphs have been noticed to label certain modular invariants, this can be seen as a first indication that there is a relation between modular invariant partition functions and subfactors. Another strong indication is the special role of the Dynkin diagrams D_odd and E_7: In the classification of $SU(2)_k$ modular invariants [13, 14, 55], the Dynkin diagrams A, D_even, E_6 and E_8 label the type I invariants whereas the invariants labelled by D_odd and E_7 are type II, i.e. involve a non-trivial "twist". In subfactor theory it turned out that it is precisely the diagrams A, D_even, E_6 and E_8 which appear as principal graphs whereas D_odd and E_7 are not allowed (see [56] and references therein).

4. Braiding, α-induction, and all that

The geometry on the circle together with Haag duality in the vacuum induces a braiding on the endomorphisms. The endomorphisms λ appearing above can be thought of as being defined on a global algebra $\mathcal{N}$ generated by the $N(J)$'s where J varies in the proper intervals on S^1, neither touching nor containing a fixed distinguished "point at infinity" $\zeta \in S^1$. Then λ will be localized on I in the sense that $\lambda(a) = a$ whenever $a \in N(J)$ with $J \cap I = \emptyset$, and transportable in the sense that for each interval J there is a unitary $u \in \mathcal{N}$ such that $\mathrm{Ad}(u) \circ \lambda$ is localized in J. Then if λ and μ are localized on disjoint intervals then they commute: $\lambda\mu = \mu\lambda$. If however λ and μ are localized in the same interval I, then we may choose a relatively disjoint interval J (whose closure does not contain ζ as well) and a unitary u such that $\mathrm{Ad}(u) \circ \mu$ is localized in J. Then λ and $\mathrm{Ad}(u) \circ \mu$ commute and in turn $\varepsilon_u(\lambda, \mu) = u^*\lambda(u)$ is a unitary intertwining $\lambda\mu$ and $\mu\lambda$. It turns out that this unitary is entirely independent on the choice of J and u, except that it may depend on the choice of J lying in the left or right connected complement of I with respect to the point at infinity ζ. (See e.g. [46, 30, 31, 6] for more detailed discussions of such matters.) Therefore we have in fact only two "statistics" or braiding operators $\varepsilon^+(\lambda, \mu)$ and $\varepsilon^-(\lambda, \mu)$, according to this choice. Indeed we have $\varepsilon^-(\lambda, \mu) = \varepsilon^+(\mu, \lambda)^*$, but $\varepsilon^+(\lambda, \mu)$ and $\varepsilon^-(\lambda, \mu)$ can be different. The statistics operators obey a couple of consistency equations which are called braiding fusion relations: Whenever $t \in \mathrm{Hom}(\lambda, \mu\nu)$ one has

$$\rho(t)\varepsilon^\pm(\lambda, \rho) = \varepsilon^\pm(\mu, \rho)\mu(\varepsilon^\pm(\nu, \rho))t,$$
$$t\varepsilon^\pm(\rho, \lambda) = \mu(\varepsilon^\pm(\rho, \nu))\varepsilon^\pm(\rho, \mu)\rho(t).$$

This in turn implies the braid relation (or "Yang-Baxter equation")

$$\rho(\varepsilon^\pm(\lambda, \mu))\varepsilon^\pm(\lambda, \rho)\lambda(\varepsilon^\pm(\mu, \rho)) = \varepsilon^\pm(\mu, \rho)\mu(\varepsilon^\pm(\lambda, \rho))\varepsilon^\pm(\lambda, \mu).$$

These equations turn our system Δ of endomorphisms into a "braided C^*-tensor category" (cf. [26]).

The braiding operators can be nicely incorporated in our graphical intertwiner calculus. Namely, for $\varepsilon^+(\lambda, \mu)$ and $\varepsilon^-(\lambda, \mu)$ we draw over- and undercrossings, respectively, of wires λ and μ as in Fig. 9. Then the consistency relations are

FIGURE 9. Braiding operators $\varepsilon^+(\lambda, \mu)$ and $\varepsilon^-(\lambda, \mu)$ as over- and undercrossings

translated into some kind of topological moves for the pictures, as e.g. the second braiding fusion relation for overcrossings is drawn graphically as in Fig. 10 whereas

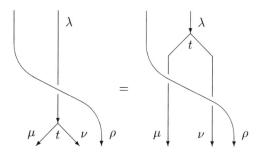

FIGURE 10. The second braiding fusion equation for over-crossings

the braid relation becomes a vertical Reidemeister move of type III, presented in Fig. 11. We would like to obtain generators of the modular group $SL(2;\mathbb{Z})$ (up to normalization) from the Hopf link and the twist, which is in fact possible if and only if the braiding is subject to a certain maximality condition, called "non-degeneracy", basically stating that $+$ and $-$ braiding operators are as different as

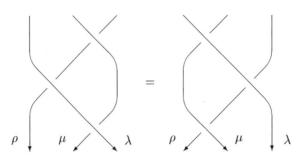

FIGURE 11. The braid relation as a vertical Reidemeister move of type III

possible [**72**]. For ρ irreducible we find $\varepsilon^+(\rho,\bar\rho)^* \bar r_\rho = \omega_\rho r_\rho$ for some scalar $\omega_\rho \in \mathbb{T}$, thanks to the uniqueness of isometries in the one-dimensional $\mathrm{Hom}(\mathrm{id}, \bar\rho\rho)$.

We will need net versions of canonical and dual canonical endomorphisms to handle inclusions $N(I) \subset M(I)$, where $N(I)$ are local, and which are standard in the sense that there is a single vector $\Omega \in H$ being cyclic and separating for all $M(I)$ on H and all $N(I)$ on a subspace $H_0 \subset H$, and such that there is a consistent family of conditional expectations $E_I : M(I) \to N(I)$ preserving Ω. In this case for each interval $I_\circ$, there is an endomorphism γ of the global algebra $\mathcal{M}$ associated to the net $\{M(I)\}$ such that $\gamma|_{M(I)}$ is a canonical endomorphism for $N(I) \subset M(I)$ whenever $I \supset I_\circ$. Moreover, the restricted $\theta = \gamma|_{\mathcal{N}}$ is localized and transportable and we have

$$(4.1) \qquad \pi^0 \simeq \pi_0 \circ \gamma, \qquad \pi^0|_{\mathcal{N}} \simeq \pi_0 \circ \theta,$$

for π^0 denoting the defining representation of $\mathcal{M}$ on H and π_0 the representation of $\mathcal{N}$ on $\overline{\mathcal{N}\Omega}$. There is an isometry v intertwining the identity and γ, and then we have $\mathcal{M} = \mathcal{N}v$, indeed $M(I) = N(I)v$ whenever $I \supset I_\circ$. It is crucial to note that, though the net $\{N(I)\}$ satisfies locality by assumption, the net $\{M(I)\}$ is not local in general. In fact the latter is local if and only if the chiral locality condition holds

$$\varepsilon^+(\theta,\theta)v^2 = v^2,$$

(see the original work [**62**] as an excellent mathematical reference for these matters and [**76**] for a more physical discussion of local extensions) and locality of the extended net $\{M(I)\}$ is extremely constraining, e.g. this automatically implies that all the inclusions $N(I) \subset M(I)$ are irreducible, as shown in [**6**].

The statistics phase ω_ρ (ρ again irreducible) can also be obtained using the left inverse $\phi_\rho(\varepsilon^+(\rho,\rho)) = \omega_\rho/d_\rho$. Such formulae in algebraic quantum field theory (see [**46**] and references therein) predate subfactor theory. Graphically ω_ρ can be displayed as in Fig. 12. By a conformal spin and statistics theorem [**32**, **31**, **45**] one can identify

$$\omega_\rho = e^{2\pi i h_\rho}$$

where h_ρ is the lowest eigenvalue of the Hamiltonian L_0 in the superselection sector $[\rho]$. This will ensure that the statistics phase (and the modular T-matrix) in our subfactor context coincide with that in conformal field theory. Now note that for μ, ν irreducible the expression $d_\mu d_\nu \phi_\mu(\varepsilon(\nu,\mu)\varepsilon(\mu,\nu))^*$ must be a scalar (as it is in $\mathrm{Hom}(\nu,\nu)$) which we will denote by $Y_{\mu,\nu}$ and which is given graphically as in

FIGURE 12. Statistics phase ω_ρ as a "twist"

Fig. 13. In case we are dealing with a closed system Δ of braided endomorphisms

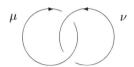

FIGURE 13. Matrix element $Y_{\lambda,\mu}$ of Rehren's Y-matrix as a "Hopf link"

it turns out [**72, 32, 31**] that

$$(4.2) \qquad Y_{\mu,\nu} = \sum_{\lambda \in \Delta} \frac{\omega_\mu \omega_\nu}{\omega_\lambda} N^\lambda_{\mu,\nu} d_\lambda, \qquad \mu,\nu \in \Delta.$$

Normalizing the matrix Y will yield the (modular) S-matrix. Then from Eq. (4.2) it follows that if the ω's and N's coincide then so does the modular matrix S in the subfactor context and that in conformal field theory. Next define $z = \sum_{\lambda \in \Delta} d_\lambda^2 \omega_\lambda$. If $z \neq 0$ put $c = 4\arg(z)/\pi$, the central charge which is defined modulo 8, and set

$$S_{\lambda,\mu} = |z|^{-1} Y_{\lambda,\mu}, \qquad T_{\lambda,\mu} = \mathrm{e}^{-\pi i c/12} \omega_\lambda \delta_{\lambda,\mu}.$$

Then the matrices S and T obey the partial Verlinde modular algebra

$$TSTST = S, \quad CTC = T, \quad CSC = S, \quad T^*T = \mathbf{1},$$

where C is the conjugation matrix, i.e. $C_{\lambda,\mu} = \delta_{\lambda,\bar{\mu}}$. Moreover, the following conditions are equivalent [**72**]:

- The braiding is non-degenerate, i.e. $\varepsilon^+(\lambda,\mu) = \varepsilon^-(\lambda,\mu)$ for all $\mu \in \Delta$ only if $\lambda = \mathrm{id}$.
- We have $|z|^2 = w$ (recall that $w = \sum_{\lambda \in \Delta} d_\lambda^2$ is the global index of the system Δ) and S is invertible so that S and T obey the full Verlinde modular algebra, in particular $(ST)^3 = S^2 = C$, and S diagonalizes the fusion rules, i.e. the Verlinde formula of Eq. (1.2) holds.

In our setting of a subfactor $N \subset M$ with a system $_N\mathcal{X}_N$ of braided endomorphisms of N we will show how to induce endomorphisms of M. This method corresponds to Mackey-induction in the group-subgroup subfactor. The standard subfactor induction $\lambda \mapsto \iota\lambda\bar{\iota}$ will not be multiplicative on sectors as e.g. the statistical dimension is multiplied by d_θ — so that in some sense we need to divide out by θ. This is achieved by the notion of α-induction which goes back to Longo and Rehren [**62**] in the (nets of) subfactor setting, and it was studied in [**6, 7, 8, 10, 11, 9**] and in a similar framework (the relation is explained in [**87**]) also in [**86**].

Note that $\gamma(v) \in \mathrm{Hom}(\theta, \theta^2)$ as $v \in \mathrm{Hom}(\mathrm{id}, \gamma)$. Therefore the braiding fusion relations can be applied to obtain

$$\varepsilon^\pm(\lambda, \theta) \lambda \gamma(v) \varepsilon^\pm(\lambda, \theta)^* = \theta(\varepsilon^\pm(\lambda, \theta)^*) \gamma(v),$$

and as

$$\varepsilon^\pm(\lambda, \theta) \lambda \gamma(n) \varepsilon^\pm(\lambda, \theta)^* = \theta \lambda(n), \qquad n \in N,$$

we find by $M = Nv$ that $\mathrm{Ad}(\varepsilon^\pm(\lambda, \theta)) \circ \lambda \gamma$ maps M into $\gamma(M)$, and so, in a more stream-lined notation,

$$\alpha_\lambda^\pm = \bar{\iota}^{-1} \circ \mathrm{Ad}(\varepsilon^\pm(\lambda, \theta)) \circ \lambda \circ \bar{\iota}$$

is a well-defined endomorphism of M such that $\alpha_\lambda^\pm(v) = \varepsilon^\pm(\lambda, \theta)^* v$ and $\alpha_\lambda^\pm|_N = \lambda$. The maps α^+ and α^- are well-defined on sectors and are multiplicative, additive and preserve conjugates:

$$\alpha_{\lambda\mu}^\pm = \alpha_\lambda^\pm \alpha_\mu^\pm, \qquad \overline{\alpha_\lambda^\pm} = \alpha_{\bar\lambda}^\pm, \qquad [\alpha_\nu^\pm] = [\alpha_{\nu_1}^\pm] \oplus [\alpha_{\nu_2}^\pm]$$

for $[\nu] = [\nu_1] \oplus [\nu_2]$. In particular the sectors $[\alpha_\lambda^\pm]$ commute; indeed

$$\alpha_\mu^\pm \alpha_\lambda^\pm = \mathrm{Ad}(\varepsilon^\pm(\lambda, \mu)) \circ \alpha_\lambda^\pm \alpha_\mu^\pm.$$

In the restriction direction we write

$$\sigma_\beta = \bar{\iota} \circ \beta \circ \iota \equiv \gamma \circ \beta|_N$$

for β an endomorphism of M. Now σ is additive on sectors and preserves conjugates but it is not multiplicative (as e.g. $\sigma_{\mathrm{id}} = \theta$). In general we have

$$\langle \alpha_\lambda^\pm, \beta \rangle \leq \langle \lambda, \sigma_\beta \rangle,$$

with equality in the case of chiral locality. One has to be careful though for which endomorphisms β of M one is considering in this formula. The inequality is true for any subsector of $[\alpha_\lambda^\pm]$, $\lambda \in {}_N\mathcal{X}_N$.

To help compute such subsectors and their fusion rules, one has the relation

$$\langle \alpha_\lambda^\pm, \alpha_\mu^\pm \rangle \leq \langle \theta \lambda, \mu \rangle,$$

again with equality in the case of chiral locality. Note that we really have divided out by θ, as in the case of standard sector induction $\lambda \mapsto \iota \lambda \bar{\iota}$ we would have $\langle \iota \lambda \bar{\iota}, \iota \mu \bar{\iota} \rangle = \langle \theta^2 \lambda, \mu \rangle$ by Frobenius reciprocity [49].

We may also compare the two different "chiral" inductions α^+ and α^-. Then $\alpha_\lambda^+ = \alpha_\lambda^-$ is equivalent to the monodromy being trivial, i.e. $\varepsilon^+(\lambda, \theta) \varepsilon^+(\theta, \lambda) = \mathbf{1}$. Moreover, whenever chiral locality holds then we even have that the chiral induced sectors coincide, $[\alpha_\lambda^+] = [\alpha_\lambda^-]$, if and only if the monodromy is trivial [6]. Nevertheless one has quite generally that

$$\alpha_\mu^- \alpha_\lambda^+ = \mathrm{Ad}(\varepsilon^+(\lambda, \mu)) \circ \alpha_\lambda^+ \alpha_\mu^-,$$

so that the sectors $[\alpha_\mu^-]$ and $[\alpha_\lambda^+]$ clearly commute. Indeed even their subsectors commute and this gives rise to a relative braiding symmetry between the chiral induced sectors [8].

5. Modular invariants, graphs and α-induction

The A-D-E classification of [13, 14, 55] associates a Dynkin diagram to each $SU(2)$ modular invariant in such a way that the multiplicities of the eigenvalues $S_{1,\lambda}/S_{0,\lambda}$ of the associated graphs match the diagonal entries $Z_{\lambda,\lambda}$ of the modular invariant. Here S is the modular S-matrix for $SU(2)$ at level k, and λ just takes the values in the $SU(2)_k$ spins $\lambda \in \{0, 1, 2, ..., k\}$. For $SU(3)$, Di Francesco and Zuber [22, 23, 20] sought graphs to describe the modular invariants in an analogous way, guided partly by the principle that the affine A-D-E diagrams correspond to the finite subgroups of $SU(2)$, and so began with fusion or McKay graphs [64] of finite subgroups of $SU(3)$ and sought truncations with the correct eigenvalues — a science essentially based on trial and error. Nevertheless they found a lot of interesting and puzzling relations between graphs, fusion rules and coupling matrices, giving the impetus to further research. We illustrate our subfactor approach through analyzing one of the exceptional $SU(2)$ modular invariants which occurs at level $k = 10$. The modular invariant is

$$(5.1) \qquad Z_{E_6} = |\chi_0 + \chi_6|^2 + |\chi_4 + \chi_{10}|^2 + |\chi_3 + \chi_7|^2 .$$

This invariant was labelled by the Dynkin diagram E_6 by [13] since the diagonal part $\{\lambda : Z_{\lambda,\lambda} \neq 0\}$ of the invariant is $\{0, 3, 4, 6, 7, 10\}$ in this case are the Coxeter exponents of E_6, i.e. the eigenvalues of the incidence (or adjacency) matrix of E_6 are precisely $\{S_{1,\lambda}/S_{0,\lambda} = 2\cos((\lambda+1)\pi/12) : \lambda = 0, 3, 4, 6, 7, 10\}$. The E_6 modular invariant can be obtained from the conformal embedding $SU(2)_{10} \subset SO(5)_1$, i.e. an inclusion of $SU(2)$ in $SO(5)$ such that the level 1 positive energy representations of $LSO(5)$ decompose into the level 10 representations of $LSU(2)$, with finite multiplicity. The loop group $LSO(5)$ has three level 1 representations, the basic (b), vector (v) and spinor (s) representation, with characters χ_b, χ_v and χ_s, respectively, decomposing as

$$\chi_b = \chi_0 + \chi_6, \quad \chi_v = \chi_4 + \chi_{10}, \quad \chi_s = \chi_3 + \chi_7,$$

on $LSU(2)$. The diagonal invariant $|\chi_b|^2 + |\chi_v|^2 + |\chi_s|^2$ of $SO(5)_1$ then immediately produces the exceptional E_6 invariant of $SU(2)_{10}$ of Eq. (5.1). The positive energy representations of b, v, s of $SO(5)_1$ satisfy the Ising fusion rules with b being the identity and in particular fusion by s corresponds to the Dynkin diagram A_3 as in Fig. 2. Analogous to what we discussed for $LSU(n)$, they give rise to three endomorphisms in the loop group subfactor setting of $LSO(5)$ with the same Ising fusion rules [5]. The conformal embedding $SU(2)_{10} \subset SO(5)_1$ then gives in the vacuum representation a net of subfactors $\pi^0(L_I SU(2))'' \subset \pi^0(L_I SO(5))''$ or $N(I) \subset M(I)$. Over the net $\{N(I)\}$ we have a system of braided endomorphisms $\{\lambda_j\}$ labelled by vertices (enumerated by $j = 0, 1, ..., 10$, $\lambda_0 = \text{id}$) of the Dynkin diagram A_{11}, and braided endomorphisms $\{\tau_b = \text{id}, \tau_v, \tau_s\}$ over $\{M(I)\}$ corresponding to the vertices of A_3, where the graphs A_{11} and A_3 represent fusion by λ_1 and τ_s.

We can put the Ising A_3 system to one side for the time being and focus on the A_{11} system of $\{N(I)\}$. Then (cf. [62, 76]) the dual canonical endomorphism θ of $\mathcal{N}$ is as a sector the sum $[\lambda_0] \oplus [\lambda_6]$ coming from the vacuum block — this is basically Eq. (4.1). We can thus perform first our α^+-induction to obtain 11 endomorphisms $\{\alpha_j^+ : j = 0, 1, ..., 10\}$ (we abbreviate $\alpha_j^+ \equiv \alpha_{\lambda_j}^+$, but after decomposition into irreducible sectors, we only find six sectors $[\alpha_0^+] = [\text{id}]$, $[\alpha_1^+]$, $[\alpha_2^+]$, $[\alpha_9^+]$, $[\alpha_{10}^+]$ and $[\varsigma]$, the latter appearing as a subsector of $[\alpha_3^+]$ which decomposes as $[\alpha_3^+] = [\varsigma] \oplus [\alpha_9^+]$.

The graph E$_6$ appears as fusion graph of $[\alpha_1^+]$ as in Fig. 14. We now turn to our

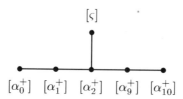

FIGURE 14. $SU(2)_{10} \subset SO(5)_1$: Fusion graph of $[\alpha_1^+]$ on the chiral system E$_6^+$

original braided system A$_3$ on $\{M(I)\}$. Reading off from the blocks of the modular invariant we find for σ-restriction:

$$[\sigma_\text{b}] = [\lambda_0] \oplus [\lambda_6] \equiv [\theta], \quad [\sigma_\text{v}] = [\lambda_4] \oplus [\lambda_{10}], \quad [\sigma_\text{s}] = [\lambda_3] \oplus [\lambda_7],$$

because σ-restriction reflects the restriction of (DHR) representations [62], which is basically Eq. (4.1) again. Since the net of loop group factors $M(I) = \pi^0(L_I SO(5))''$ satisfies locality we have $\alpha\sigma$-reciprocity

$$\langle \alpha_\lambda^\pm, \beta \rangle = \langle \lambda, \sigma_\beta \rangle$$

for λ in the A$_{11}$ system and β representing any subsector of the induced system $\{[\alpha_j^+]\}$, here E$_6$. Then as σ-restriction takes us back into the A$_{11}$ system, the sectors $[\tau_\text{b}]$, $[\tau_\text{v}]$, $[\tau_\text{s}]$ must lie amongst the six E$_6$ sectors. They are identified as $[\alpha_0^+] = [\tau_\text{b}]$, $[\alpha_{10}^+] = [\tau_\text{v}]$, $[\varsigma] = [\tau_\text{s}]$, and indeed satisfy the Ising fusion rules.

Other conformal inclusions and also simple current extension invariants (often also called orbifold invariants) can be handled similarly, the latter are realized by so-called crossed product subfactors using the simple current groups which represent the center $\mathbb{Z}_n$ of $SU(n)$ amongst the $SU(n)_k$ fusion rules [7, 8]. (See also Section 8.)

So far we have only considered "positive" α^+-induction, arising from the braiding ε^+. The same way we can use the opposite braiding ε^-, giving α^--induction with "negative" chirality. In either case, say for the conformal inclusion $SU(2)_{10} \subset SO(5)_1$, we have two induced systems E$_6^+$ and E$_6^-$ of sectors on $M(I)$, but at least they intersect on the Ising sectors b, v, s of $LSO(5)$ at level 1, symbolically: E$_6^+ \cap$ E$_6^- \supset$ A$_3$. In fact they only coincide on these "marked vertices", in the terminology of Di Francesco and Zuber, b, v, s of E$_6^\pm$. Di Francesco and Zuber [22, 23, 20] had already empirically observed that the graphs which they sought to describe the diagonal part of a given modular invariant carried in the type I case fusion rule algebras with certain distinguished marked vertices forming fusion rule subalgebras describing the extended fusion rules. This now finds a clear explanation in terms of the game of induction and restriction of sectors.

More generally, in the case of conformal embedding subfactors, the following were shown to be equivalent [8]:

- $\mathcal{V}^+ \cap \mathcal{V}^- = \mathcal{T}$,
- $Z_{\lambda,\mu} = \langle \alpha_\lambda^+, \alpha_\mu^- \rangle$,
- The irreducible subsectors of $[\gamma]$ all lie in $\mathcal{V}^\alpha = \mathcal{V}^+ \vee \mathcal{V}^-$,
- $\sum_{\beta \in \mathcal{V}^\alpha} d_\beta^2 = w$.

Here $\mathcal{V}^\pm$ are the two chiral systems of induced irreducible sectors, $\mathcal{T} \subset \mathcal{V}^\pm$ is the subsystem of neutral or "ambichiral" sectors, arising from either induction and corresponding to the marked vertices, and finally $\mathcal{V}^\alpha$ is the system of irreducible sectors generated by products of the different chiral systems or equivalently obtained by decomposing sectors $[\alpha_\lambda^+ \alpha_\mu^-]$ into irreducibles. The second condition gives a nice interpretation of the modular invariant matrix Z as counting the coupling of the two chiral inductions. Note that it immediately produces the upper bound of Eq. (1.3) because the largest possible coupling occurs when $[\alpha_\lambda^+]$ and $[\alpha_\mu^-]$ both purely decompose into multiples of one and the same irreducible sector and the multiplicities are bounded by the statistical dimensions. In fact the bound $\langle \alpha_\lambda^+, \alpha_\mu^- \rangle \leq d_\lambda d_\mu$ even holds for degenerate braidings (i.e. with non-unitary S-matrices). The third and hence all completeness properties could be verified in a case by case analysis for e.g. all $SU(2)$ and $SU(3)$ conformal inclusion subfactors, and by a general proof for all simple current extensions of $SU(n)$ all levels in [8]. By adopting a graphical argument of Ocneanu [68] from the bimodule sectors to the sector framework, the generating property was proven in [10] to hold quite generally, provided the braiding is non-degenerate. (And this is the case for $SU(n)_k$ due to unitarity of the S-matrix.)

A more careful analysis in [9] using algebraic instead of graphical techniques shows that the "α-global index" $w_\alpha = \sum_{\beta \in \mathcal{V}^\alpha} d_\beta^2$ is in fact given by

$$w_\alpha = \frac{w}{\sum_\lambda \deg Z_{0,\lambda} d_\lambda}$$

with summation over degenerate elements λ for which $\varepsilon^+(\lambda, \mu) = \varepsilon^-(\lambda, \mu)$ for all μ. Thus the generating property can hold even for some degenerate systems. (An example is the conformal inclusion subfactor $SU(2)_{10} \subset SO(5)_1$ if we start only with the smaller system A_{11}^{even} of even spins.) Moreover, the methods of [10, 11] allow us to handle type II modular invariants as well as conformal embedding and simple current invariants.

We now turn to the general framework of [10, 11, 9]. We take a subfactor $N \subset M$ and a system $_N\mathcal{X}_N \subset \text{End}(N)$ of endomorphisms by which we mean a collection of irreducible endomorphisms of finite statistical dimension, containing the identity morphism and closed under conjugation and irreducible decomposition of products. Then for $\iota : N \hookrightarrow M$ being the inclusion homomorphism and $\theta = \bar{\iota}\iota$ and $\gamma = \iota\bar{\iota}$ the dual canonical endomorphism and canonical endomorphism, respectively, we assume that θ lies in $\Sigma(_N\mathcal{X}_N)$, the set of morphisms representing sector sums corresponding to the irreducibles in $_N\mathcal{X}_N$ — but make no assumption on γ. Moreover we assume that the system $_N\mathcal{X}_N$ is braided. We let $_M\mathcal{X}_M \subset \text{End}(M)$ denote a system of endomorphisms consisting of a choice of representative of each irreducible subsector of sectors $[\iota \lambda \bar{\iota}]$, $\lambda \in {}_N\mathcal{X}_N$. We define $_M\mathcal{X}_M^\alpha \subset {}_M\mathcal{X}_M$ to be the subsystem of those endomorphisms which are representatives of some subsectors of $[\alpha_\lambda^+ \alpha_\mu^-]$, $\lambda, \mu \in {}_N\mathcal{X}_N$. (Note that by $\alpha_\lambda^\pm \iota = \iota \lambda$, any subsector of $[\alpha_\lambda^+ \alpha_\mu^-]$ will automatically be a subsector of $[\iota \lambda \mu \bar{\iota}]$ since $[\gamma]$ contains the identity sector.) Then we similarly define the chiral induced systems as the subsystems $_M\mathcal{X}_M^\pm \subset {}_M\mathcal{X}_M$ of irreducible sectors arising from positive/negative $\alpha^\pm$-induction, and the neutral system $_M\mathcal{X}_M^0 = {}_M\mathcal{X}_M^+ \cap {}_M\mathcal{X}_M^-$. Their global indices, i.e. sums over squares of statistical dimensions, are denoted by w, w_α, $w_\pm$, and w_0 (it follows from the assumptions that $_N\mathcal{X}_N$ and $_M\mathcal{X}_M$ have the same global index w) and fulfill $1 \leq w_0 \leq w_\pm \leq w_\alpha \leq w$.

Defining now a "coupling matrix" Z by setting

$$Z_{\lambda,\mu} = \langle \alpha_\lambda^+, \alpha_\mu^- \rangle, \qquad \lambda, \mu \in {}_N\mathcal{X}_N,$$

turns out to commute [10] with matrices Ω and Y, where $\Omega_{\lambda,\mu} = \delta_{\lambda,\mu}\omega_\lambda$ and Y is defined as in Eq. (4.2). (When the braiding is non-degenerate, we thus have a physical modular invariant Z which commutes with the modular S- and T-matrices, being the normalized matrices Y and Ω.) Moreover, the relative sizes of the various systems are encoded in Z, namely we have [11]

(5.2) $$w_+ = \frac{w}{\sum_{\lambda \in {}_N\mathcal{X}_N} d_\lambda Z_{\lambda,0}} = \frac{w}{\sum_{\lambda \in {}_N\mathcal{X}_N} Z_{0,\lambda} d_\lambda} = w_-$$

as well as [9]

$$w_\alpha = \frac{w}{\sum_{\lambda \in {}_N\mathcal{X}_N^{\deg}} Z_{0,\lambda} d_\lambda}, \qquad w_0 = \frac{w_+^2}{w_\alpha},$$

where ${}_N\mathcal{X}_N^{\deg} \subset {}_N\mathcal{X}_N$ denotes the subsystem of degenerate morphisms (i.e. ${}_N\mathcal{X}_N^{\deg} = \{\text{id}\}$ in the non-degenerate case). Here the equality $\sum_\lambda d_\lambda Z_{\lambda,0} = \sum_\lambda Z_{0,\lambda} d_\lambda$ is due to the invariance $YZ = ZY$.

Although the original system ${}_N\mathcal{X}_N$ is braided, the induced systems ${}_M\mathcal{X}_M$ or even ${}_M\mathcal{X}_M^\pm$ need not even be commutative. Indeed if we complexify the fusion rules of ${}_M\mathcal{X}_M^\pm$ to obtain finite-dimensional C^*-algebras $\mathcal{Z}^\pm$ we find [11] (assuming non-degeneracy of the braiding)

(5.3) $$\mathcal{Z}^\pm \simeq \bigoplus_{\tau \in {}_M\mathcal{X}_M^0} \bigoplus_{\lambda \in {}_N\mathcal{X}_N} \text{Mat}(b_{\tau,\lambda}^\pm),$$

where $b_{\tau,\lambda}^\pm = \langle \tau, \alpha_\lambda^\pm \rangle$ are the chiral branching coefficients for $\lambda \in {}_N\mathcal{X}_N$ and a neutral morphism $\tau \in {}_M\mathcal{X}_M^0$ — a marked vertex of Di Francesco and Zuber.

In particular, the chiral systems are commutative only when $b_{\tau,\lambda}^\pm \leq 1$ for all τ, λ. This explains the non-commutativity discovered by Feng Xu [86] with direct computations of some fusion rules for the conformal embedding $SU(4)_4 \subset SU(15)_1$ (and which lead to conceptual problems in the partially systematic approach to graphs from modular invariants of [70] based on certain assumptions) and provides now a whole series of non-commutative chiral fusion rules for $SU(n)_n \subset SU(n^2-1)_1$, $n \geq 4$. Moreover, by counting dimensions we find for the cardinality $\#{}_M\mathcal{X}_M^\pm$ of ${}_M\mathcal{X}_M^\pm$ that

$$\#{}_M\mathcal{X}_M^\pm = \sum_{\tau,\lambda} (b_{\tau,\lambda}^\pm)^2 = \text{tr}({}^tb^\pm b^\pm).$$

In the matrix algebra $\text{Mat}(b_{\tau,\lambda}^\pm)$, the induced $[\alpha_\nu^\pm]$ is scalar, being $S_{\lambda,\nu}/S_{\lambda,0}$. However, even in the degenerate case, the neutral elements always possess a braiding (hence have commutative fusion) arising as restriction of the relative braiding, and this braiding is non-degenerate if the original braiding on ${}_N\mathcal{X}_N$ is. In that case we have "extended" S- and T-matrices S^{ext} and T^{ext} from the neutral system, and in $\text{Mat}(b_{\tau,\lambda}^\pm)$ a neutral sector $[\tau']$ ($\tau' \in {}_M\mathcal{X}_M^0$) acts as a central element since it commutes with all subsectors of $[\alpha_\nu^\pm]$, as $S_{\tau,\tau'}^{\text{ext}}/S_{\tau,0}^{\text{ext}}$.

Even if the chiral systems are commutative, the full system ${}_M\mathcal{X}_M^\alpha = {}_M\mathcal{X}_M^+ \vee {}_M\mathcal{X}_M^-$ may not be. Although ${}_M\mathcal{X}_M^+$ and ${}_M\mathcal{X}_M^-$ relatively commute (thanks to the relative braiding), it may happen that "mixed" products of elements of ${}_M\mathcal{X}_M^+$ and

${}_M\mathcal{X}_M^-$ decompose into non-commuting irreducibles. Indeed (cf. Eq. (5.3) for the chiral fusion rules), if we complexify the fusion rules of ${}_M\mathcal{X}_M^\alpha = {}_M\mathcal{X}_M$ in the non-degenerate case to obtain a finite-dimensional C^*-algebra $\mathcal{Z}$, then we find [**10**]

$$\mathcal{Z} \simeq \bigoplus_{\lambda,\mu \in {}_N\mathcal{X}_N} \mathrm{Mat}(Z_{\lambda,\mu}).$$

(This particular decomposition has also been claimed by Ocneanu in his lectures [**68**] in case of A-D-E graphs and $SU(2)$ modular invariants.) Moreover, in the matrix algebra $\mathrm{Mat}(Z_{\lambda,\mu})$, the induced $[\alpha_\nu^+]$ and $[\alpha_\nu^-]$ are scalars, being $S_{\lambda,\nu}/S_{\lambda,0}$ and $S_{\mu,\nu}/S_{\mu,0}$, respectively. We have seen in the case of chiral locality (which holds e.g. for conformal embeddings) that we can obtain graphs with spectrum corresponding to the diagonal part of the modular invariant through the fusion graphs of $[\alpha_\lambda^\pm]$ on ${}_M\mathcal{X}_M^\pm$. In the general case where chiral locality may not necessarily hold, we instead look at the action of ${}_M\mathcal{X}_M \supset {}_M\mathcal{X}_M^\pm$ on the system ${}_M\mathcal{X}_N$ of M-N sectors. Here the system ${}_M\mathcal{X}_N$ is a choice of representatives of irreducible subsectors of the sectors $[\iota\lambda]$, $\lambda \in {}_N\mathcal{X}_N$. As M-N sectors cannot be multiplied among themselves there is no associated fusion rule algebra to decompose. (Nevertheless, when chiral locality does holds, ${}_M\mathcal{X}_N$ can be canonically identified with either ${}_M\mathcal{X}_M^\pm$ by $\beta \mapsto \beta \circ \iota$, $\beta \in {}_M\mathcal{X}_M^\pm$.) However, the left action of ${}_M\mathcal{X}_M$ on ${}_M\mathcal{X}_N$ defines a representation ϱ of the M-M fusion rule algebra, with matrix elements $[\varrho([\beta])]_{\xi,\xi'} = \langle \xi, \beta\xi' \rangle$, $\xi, \xi' \in {}_M\mathcal{X}_N$, and decomposes as [**10, 11**]

$$\varrho \simeq \bigoplus_{\lambda \in {}_N\mathcal{X}_N} \pi_{\lambda,\lambda},$$

where $\pi_{\lambda,\lambda}$ is the irreducible representation corresponding to the matrix block $\mathrm{Mat}(Z_{\lambda,\lambda})$, so that $\pi_{\lambda,\lambda}([\alpha_\nu^\pm]) = S_{\lambda,\nu}/S_{\lambda,0}\mathbf{1}_{Z_{\lambda,\lambda}}$. In particular the spectrum is determined by the diagonal part of the modular invariant. Thus it is precisely this representation ϱ which provides an automatic connection between the modular invariant and fusion graphs (e.g. the representation matrix of some fundamental generator $\Box$ corresponding to the left multiplication of $[\alpha_\Box^\pm]$ on the M-N sectors) in such a way that (the multiplicities in) their spectra are canonically given by the diagonal entries of the coupling matrix. In fact, evaluation of ϱ on the $[\alpha_\lambda^\pm]$'s yields a "nimrep" of the original N-N fusion rules, i.e. a matrix representation where all the matrix entries are non-negative integers. Finally, by counting dimensions we see that $\#{}_M\mathcal{X}_N = \mathrm{tr}(Z)$.

We can illustrate this with the E_7 modular invariant of $SU(2)$:

$$Z_{\mathrm{E}_7} = |\chi_0 + \chi_{16}|^2 + |\chi_4 + \chi_{12}|^2 + |\chi_6 + \chi_{10}|^2$$
$$+ |\chi_8|^2 + (\chi_2 + \chi_{14})\chi_8^* + \chi_8(\chi_2 + \chi_{14})^*.$$

Instead of simply extending to a diagonal invariant, as in the E_6 case, we also insert a twist on the blocks. This is an example of the setting of Moore and Seiberg [**66**] (see also Dijkgraaf and Verlinde [**24**]) that taking a maximal extension of the "chiral algebra" $\mathcal{A} \subset \mathcal{B}$, a modular invariant of $\mathcal{A}$ is the restriction of some permutation invariant $Z_{\tau,\tau'}^{\mathrm{ext}} = \delta_{\tau,\omega(\tau')}$ where ω is a permutation of the sectors of the extended theory $\mathcal{B}$, defining an automorphism of their fusion rules and preserving the extended vacuum sector, $\omega(0) = 0$. The E_7 invariant is a twist of the D_{10} invariant, the latter we can realize from a subfactor with the dual canonical endomorphism sector decomposing as $[\lambda_0] \oplus [\lambda_{16}]$, being a simple current extension [**7**]. As shown in [11], the E_7 modular invariant appears for a subfactor with dual canonical endomorphism sector $[\lambda_0] \oplus [\lambda_8] \oplus [\lambda_{16}]$. For either invariant we find $\mathrm{tr}({}^t b^\pm b^\pm) = 10$ ($= \mathrm{tr}(Z_{D_{10}})$) so that indeed in either case the fusion graph of the generator $[\alpha_1^\pm]$ on ${}_M\mathcal{X}_M^\pm$ is D_{10}. However, $\mathrm{tr}(Z_{E_7}) = 7$ and the fusion graph of $[\alpha_1^\pm]$ on ${}_M\mathcal{X}_N$ is E_7.

6. Type II modular invariants, extended fusion rule automorphisms, and all that

Now in our general setting we have

$$Z_{\lambda,\mu} = \langle \alpha_\lambda^+, \alpha_\mu^- \rangle = \sum_{\tau \in {}_M\mathcal{X}_M^0} b_{\tau,\lambda}^+ b_{\tau,\mu}^-,$$

with chiral branching coefficients $b_{\tau,\lambda}^\pm = \langle \tau, \alpha_\lambda^\pm \rangle$. To write this in Moore-Seiberg form we would need $b_{\tau,\lambda}^- = b_{\omega(\tau),\lambda}^+$ for a permutation of the extended system, being identified as the neutral system ${}_M\mathcal{X}_M^0$, so that

$$Z_{\lambda,\mu} = \sum_{\tau \in {}_M\mathcal{X}_M^0} b_{\tau,\lambda}^+ b_{\omega(\tau),\mu}^+.$$

Note that by $\omega(0) = 0$ and $b_{\tau,0}^\pm = \delta_{\tau,0}$ (do not worry that we denote both the original and the extended "vacuum" i.e. identity morphism by the same symbol "0") we are automatically forced to have symmetric vacuum coupling $Z_{\lambda,0} = Z_{0,\lambda}$. To cover more general cases, which do occur as we shall see, we should consider instead of one maximal extension $\mathcal{A} \subset \mathcal{B}$ of the chiral algebra $\mathcal{A}$, but two different extensions $\mathcal{A} \subset \mathcal{B}_\pm$, yielding different labelling sets of extended fusion rules so that the extended modular invariant is

$$Z_{\tau_+,\tau_-}^{\mathrm{ext}} = \delta_{\tau_+,\omega(\tau_-)},$$

where ω now is an isomorphism between the two sets of extended fusion rules, still subject to $\omega(0) = 0$. Note that when we have two different labelling sets it makes no sense to ask whether a coupling matrix is symmetric or not.

When chiral locality does hold then

$$b_{\beta,\lambda}^\pm = \langle \alpha_\lambda^\pm, \beta \rangle = \langle \lambda, \sigma_\beta \rangle,$$

whenever $\beta \in {}_M\mathcal{X}_M^\pm$. In particular, when $\beta = \tau$ is neutral, i.e. lies in the intersection ${}_M\mathcal{X}_M^0 = {}_M\mathcal{X}_M^+ \cap {}_M\mathcal{X}_M^-$, then

$$b_{\tau,\lambda}^+ = b_{\tau,\lambda}^- \equiv b_{\tau,\lambda},$$

and we have a block decomposition or "type I" modular invariant

$$Z_{\lambda,\mu} = \sum_{\tau \in {}_M\mathcal{X}_M^0} b_{\tau,\lambda} b_{\tau,\mu}.$$

Permutation invariants can be classified as follows. The following conditions are equivalent [11]:

- $Z_{\lambda,\mu} = \delta_{\lambda,\omega(\mu)}$ with ω a permutation of ${}_N\mathcal{X}_N$ with $\omega(0) = 0$ and defining a fusion rule automorphism,
- $Z_{\lambda,0} = \delta_{\lambda,0}$,
- $Z_{0,\lambda} = \delta_{\lambda,0}$,
- $w_\pm = w$.

In this case the two inductions $\alpha^\pm$ are isomorphisms (i.e. each $[\alpha_\lambda^\pm]$ is irreducible) and $\omega = (\alpha^+)^{-1} \circ \alpha^-$. This result does not rely on non-degeneracy of the braiding.

We would like to decompose a modular invariant into its two parts, a type I part together with a twist, and in order to take care of heterotic vacuum coupling we will need to implement such a twist by an isomorphism rather than an automorphism. First we characterize chiral locality. If chiral locality holds, i.e. $\varepsilon^+(\theta,\theta)v^2 = v^2$, then $Z_{\lambda,0} = \langle \alpha_\lambda^+, \alpha_0^- \rangle = \langle \alpha_\lambda^+, \mathrm{id} \rangle = \langle \lambda, \theta \rangle$, and similarly $Z_{0,\lambda} = \langle \lambda, \theta \rangle$. Indeed the following conditions are equivalent [9]:

- We have $Z_{\lambda,0} = \langle \theta, \lambda \rangle$ for all $\lambda \in {}_N\mathcal{X}_N$.
- We have $Z_{0,\lambda} = \langle \theta, \lambda \rangle$ for all $\lambda \in {}_N\mathcal{X}_N$.
- Chiral locality holds: $\varepsilon^+(\theta,\theta)v^2 = v^2$.

Thus chiral locality holds if and only if

$$[\theta] = \bigoplus_{\lambda \in {}_N\mathcal{X}_N} \langle \theta, \lambda \rangle [\lambda] = \bigoplus_{\lambda \in {}_N\mathcal{X}_N} Z_{\lambda,0}[\lambda] = \bigoplus_{\lambda \in {}_N\mathcal{X}_N} Z_{0,\lambda}[\lambda].$$

In general we define sectors

$$[\theta_+] = \bigoplus_{\lambda \in {}_N\mathcal{X}_N} Z_{\lambda,0}[\lambda], \qquad [\theta_-] = \bigoplus_{\lambda \in {}_N\mathcal{X}_N} Z_{0,\lambda}[\lambda].$$

Note that $d_{\theta_+} = \sum_\lambda d_\lambda Z_{\lambda,0} = \sum_\lambda Z_{0,\lambda} d_\lambda = d_{\theta_-}$ (due to $(YZ)_{0,0} = (ZY)_{0,0}$) but in general $[\theta_+]$ and $[\theta_-]$ may be different. Using results on intermediate subfactors [50] it was shown in [9] that, starting with an arbitrary subfactor $N \subset M$ subject to our assumptions, both $[\theta_+]$ and $[\theta_-]$ are dual canonical endomorphism sectors of N, corresponding to intermediate subfactors

$$N \subset M_\pm \subset M,$$

and that $N \subset M_\pm$ satisfy chiral locality. We then can form the $\tilde{\alpha}_\delta^\pm$-inductions ($\lambda \mapsto \tilde{\alpha}_{\delta;\lambda}^\pm$), on $N \subset M_\delta$, $\delta = \pm$, and consider then the symmetric type I modular invariants $Z^\pm$,

$$Z_{\lambda,\mu}^+ = \langle \tilde{\alpha}_{+;\lambda}^+, \tilde{\alpha}_{+;\mu}^- \rangle, \qquad Z_{\lambda,\mu}^- = \langle \tilde{\alpha}_{-;\lambda}^+, \tilde{\alpha}_{-;\mu}^- \rangle.$$

From the definition of $[\theta_\pm]$ we have $Z_{\lambda,0}^+ = \langle \theta_+, \lambda \rangle = Z_{0,\lambda}^+$ as $N \subset M_+$ satisfies chiral locality, and so $Z_{\lambda,0}^+ = Z_{0,\lambda}^+ = Z_{\lambda,0}$ and similarly $Z_{\lambda,0}^- = Z_{0,\lambda}^- = Z_{0,\lambda}$. For $Z = Z_{E_7}$ the E_7 modular invariant of $SU(2)$, $Z^\pm$ will both be $Z_{D_{10}}$ but it is possible as we shall see that $Z^+ \neq Z^-$.

Next we argue that we can canonically identify ${}_M\mathcal{X}_M^+$ with ${}_{M_+}\mathcal{X}_{M_+}^+$ and ${}_M\mathcal{X}_M^-$ with ${}_{M_-}\mathcal{X}_{M_-}^-$. To do this it will be enough to find an injective map ${}_M\mathcal{X}_M^+ \to {}_{M_+}\mathcal{X}_{M_+}^+$ (and ${}_M\mathcal{X}_M^- \to {}_{M_-}\mathcal{X}_{M_-}^-$) because the global indices w^+ are the same thanks to Eq. (5.2) and $Z_{\lambda,0} = Z_{\lambda,0}^+$. One can show that $\mathrm{Hom}(\mathrm{id}, \alpha_\nu^\pm) = \mathrm{Hom}(\mathrm{id}, \tilde{\alpha}_{\pm;\nu}^\pm)$ which in turn implies $\mathrm{Hom}(\alpha_\lambda^\pm, \alpha_\mu^\pm) = \mathrm{Hom}(\tilde{\alpha}_{\pm;\lambda}^\pm, \tilde{\alpha}_{\pm;\mu}^\pm)$. In particular $\mathrm{Hom}(\alpha_\lambda^\pm, \alpha_\mu^\pm) \subset M_\pm$. We can then move from intertwiners to endomorphisms. If $\beta \in {}_M\mathcal{X}_M^+$ represents a subsector of $[\alpha_\lambda^+]$, so that there is a $t \in M$ such that $\alpha_\lambda^+(\cdot) = t\beta(\cdot)t^* + \ldots$, then $tt^* \in \mathrm{Hom}(\alpha_\lambda^\pm, \alpha_\lambda^\pm) = \mathrm{Hom}(\tilde{\alpha}_{\pm;\lambda}^\pm, \tilde{\alpha}_{\pm;\lambda}^\pm)$. We can then construct an endomorphism $\tilde{\beta} \in \mathrm{End}(M_+)$ representing a subsector of $[\tilde{\alpha}_{\pm;\lambda}^\pm]$ and such that $\beta|_{M_+} = \tilde{\beta}$. In this way we construct bijections $\vartheta_\pm : {}_M\mathcal{X}_M^\pm \to {}_{M_\pm}\mathcal{X}_{M_\pm}^\pm$ which [9]:

- preserve chiral branching rules $\langle \beta, \alpha_\lambda^\pm \rangle = \langle \vartheta(\beta), \tilde{\alpha}_{\pm;\lambda}^\pm \rangle$, $\beta \in {}_M\mathcal{X}_M^\pm$,
- preserve chiral fusion rules,

- and restrict to bijections of the neutral systems ${}_M\mathcal{X}_M^0 \to {}_{M_\pm}\mathcal{X}_{M_\pm}^0$.

This means that ${}_M\mathcal{X}_M^0$ can be used (rather than ${}_{M_\pm}\mathcal{X}_{M_\pm}^0$) to decompose the type I coupling matrices

$$Z_{\lambda,\mu}^\pm = \sum_{\tau \in {}_M\mathcal{X}_M^0} b_{\tau,\lambda}^\pm b_{\tau,\mu}^\pm$$

with chiral branching coefficients $b_\lambda^\pm = \langle \tau, \alpha_\lambda^\pm \rangle$, $\tau \in {}_M\mathcal{X}_M^0$, $\lambda \in {}_N\mathcal{X}_N$. If the two intermediate subfactors happen to be identical, $M_+ = M_-$ (so that the "parent" coupling matrices coincide, $Z^+ = Z^-$), then we can write

$$Z_{\lambda,\mu} = \sum_{\tau \in {}_M\mathcal{X}_M^0} b_{\tau,\lambda}^+ b_{\omega(\tau),\mu}^+ .$$

for the (generically type II) coupling matrix Z. Here the permutation $\omega = \vartheta_+^{-1} \circ \vartheta_-$, satisfying $\omega(0) = 0$ clearly defines an automorphism of the neutral fusion rules.

In general when $M_+ \neq M_-$ we would write the extended coupling matrix as

$$Z_{\tau_+,\tau_-}^{\text{ext}} = \delta_{\tau_-, \vartheta(\tau_+)},$$

where $\tau_\pm \in {}_{M_\pm}\mathcal{X}_{M_\pm}^0$ and $\vartheta = \vartheta_- \circ \vartheta_+^{-1} : {}_{M_+}\mathcal{X}_{M_+}^0 \to {}_{M_-}\mathcal{X}_{M_-}^0$ is a bijection defining an isomorphism of the chiral fusion rules. We will illustrate that such heterotic situations do exist, in fact examples are already provided by certain $SO(n)_k$ current algebra models. We will deal with the simplest case at level $k = 1$ in Section 7.

What is the connection between the two chiral inductions and the picture of left- and right-chiral algebras in conformal field theory? An appropriate notion of chiral algebras in the setting of algebraic quantum field theory are "chiral observables" [74], and one can show that our coupling matrices describe in fact a Hilbert space decomposition of the vacuum sector of a two-dimensional quantum field theory upon restriction to the action of a tensor product of left- and right-chiral observables [75]. Suppose that our factor N is obtained as a local factor $N = N(I_\circ)$ of a quantum field theoretical net of factors $\{N(I)\}$ indexed by proper intervals $I \subset \mathbb{R}$ on the real line, and that the system ${}_N\mathcal{X}_N$ is obtained as restrictions of DHR-morphisms (cf. [46]) to N. This is in fact the case in our examples arising from conformal field theory where the net is defined in terms of local loop groups in the vacuum representation. Taking two copies of such a net and placing the real axes on the light cone, then this defines a local net $\{A(\mathcal{O})\}$, indexed by double cones $\mathcal{O}$ on two-dimensional Minkowski space (cf. [74] for such constructions). Given a subfactor $N \subset M$, determining in turn two subfactors $N \subset M_\pm$ obeying chiral locality, will provide two local nets of subfactors $\{N(I) \subset M_\pm(I)\}$ as a local subfactor basically encodes the entire information about the net of subfactors [62]. Arranging $M_+(I)$ and $M_-(J)$ on the two light cone axes defines a local net of subfactors $\{A(\mathcal{O}) \subset A_{\text{ext}}(\mathcal{O})\}$ in Minkowski space. Rehren has recently proven [75] (see also [12] for a different but less general derivation) that there is a (type III) factor B such that we have an irreducible inclusions $N \otimes N^{\text{opp}} \subset B$ such that the dual canonical endomorphism Θ of the inclusion $N \otimes N^{\text{opp}} \subset B$ decomposes as

$$[\Theta] = \bigoplus_{\lambda,\mu \in {}_N\mathcal{X}_N} Z_{\lambda,\mu} [\lambda \otimes \mu^{\text{opp}}].$$

(Here the superscript "opp" just denotes the opposite algebra, i.e. N^{opp} is N as a linear space, with reversed multiplication. There is a canonical way of identifying $N(I)^{\text{opp}}$ with the CPT reflection of $N(I)$ [44] which is involved in the two-dimensional construction.) Refining this result it has been shown [9] that our local extensions $M_\pm$ produce an intermediate subfactor

$$N \otimes N^{\text{opp}} \subset M_+ \otimes M_-^{\text{opp}} \subset B$$

such that moreover the dual canonical endomorphism Θ_{ext} of the inclusion $M_+ \otimes M_-^{\text{opp}} \subset B$ decomposes as

$$[\Theta_{\text{ext}}] = \bigoplus_{\tau \in {}_M\mathcal{X}_M^0} [\vartheta_+(\tau) \otimes \vartheta_-(\tau)^{\text{opp}}].$$

The embedding $M_+ \otimes M_-^{\text{opp}} \subset B$ gives rise to another net of subfactors $\{A_{\text{ext}}(\mathcal{O}) \subset B(\mathcal{O})\}$, and a condition which ensures that the net $\{B(\mathcal{O})\}$ obeys local commutation relations can be established. The existence of the local net was already proven in [75], and now the decomposition of $[\Theta_{\text{ext}}]$ tells us that the chiral extensions $N(I) \subset M_+(I)$ and $N(I) \subset M_-(I)$ for left and right chiral nets are indeed maximal (in the sense of [74]), following from the fact that the coupling matrix for $\{A_{\text{ext}}(\mathcal{O}) \subset B(\mathcal{O})\}$ is a bijection. This shows that the inclusions $N \subset M_\pm$ should in fact be regarded as the subfactor version of left- and right maximal extensions of the chiral algebra.

7. Heterotic examples

Let us now consider the $SO(n)$ loop group models at level 1, where n is a multiple of 16, $n = 16\ell$, $\ell = 1, 2, 3, \ldots$. These theories have four sectors, the basic (0), vector (v), spinor (s) and conjugate spinor (c) module, corresponding to highest weights 0, $\Lambda_{(1)}$, $\Lambda_{(r-1)}$ and $\Lambda_{(r)}$, respectively; here $r = n/2 = 8\ell$ is the rank of $SO(n)$. The conformal dimensions are given as $h_0 = 0$, $h_v = 1/2$, $h_s = h_c = \ell$, and the sectors obey $\mathbb{Z}_2 \times \mathbb{Z}_2$ fusion rules. The Kac-Peterson matrices are given explicitly as

$$(7.1) \quad S = \frac{1}{2}\begin{pmatrix} 1 & 1 & 1 & 1 \\ 1 & 1 & -1 & -1 \\ 1 & -1 & 1 & -1 \\ 1 & -1 & -1 & 1 \end{pmatrix}, \quad T = e^{-2\pi i \ell/3}\begin{pmatrix} 1 & 0 & 0 & 0 \\ 0 & -1 & 0 & 0 \\ 0 & 0 & 1 & 0 \\ 0 & 0 & 0 & 1 \end{pmatrix}.$$

It is easy to check that there are exactly six modular invariants, $Z = \mathbf{1}$, W, X_s, X_c, Q, ${}^tQ$. Here

$$W = \begin{pmatrix} 1 & 0 & 0 & 0 \\ 0 & 1 & 0 & 0 \\ 0 & 0 & 0 & 1 \\ 0 & 0 & 1 & 0 \end{pmatrix}, \quad X_s = \begin{pmatrix} 1 & 0 & 1 & 0 \\ 0 & 0 & 0 & 0 \\ 1 & 0 & 1 & 0 \\ 0 & 0 & 0 & 0 \end{pmatrix}, \quad Q = \begin{pmatrix} 1 & 0 & 0 & 1 \\ 0 & 0 & 0 & 0 \\ 1 & 0 & 0 & 1 \\ 0 & 0 & 0 & 0 \end{pmatrix},$$

and $X_c = W X_s W$. (Note that $Q = X_s W$ and ${}^tQ = W X_s$.) The matrix Q and its transpose ${}^tQ$ are two examples for modular invariants with non-symmetric vacuum coupling. Such "heterotic" invariants seem to be extremely rare and have not enjoyed particular attention in the literature, perhaps because they were erroneously dismissed as being spurious in the sense that they would not correspond to a physical partition function. Examples for truly spurious modular invariants were given in [**78, 82, 37**] and found to be "coincidental" linear combinations of proper physical invariants. Note that although there is a linear dependence here, namely

$$\mathbf{1} - W - X_s - X_c + Q + {}^tQ = 0,$$

we cannot express Q (or ${}^tQ$) alone as a linear combination of the four symmetric invariants. This may serve as a first indication that Q and ${}^tQ$ are not spurious. We will now demonstrate that they can be realized from subfactors.

The $\mathbb{Z}_2 \times \mathbb{Z}_2$ fusion rules for these models were proven in the DHR framework in [**5**], and together with the conformal spin and statistics theorem [**32, 31, 45**] we conclude that there is a net of type III factors on S^1 with a system $\{\mathrm{id}, \rho_v, \rho_s, \rho_c\}$ of localized and transportable, hence braided endomorphisms, such that the statistics S- and T-matrices are given by Eq. (7.1). Because the statistics phases are second roots of unity as $\omega_v = -1$ and $\omega_s = \omega_c = 1$, we can by [**73**] choose the morphisms in the system such that obey the $\mathbb{Z}_2 \times \mathbb{Z}_2$ fusion rules even by individual multiplication,

$$\rho_v^2 = \rho_s^2 = \rho_c^2 = \mathrm{id}, \qquad \rho_v \rho_s = \rho_s \rho_v = \rho_c.$$

This is enough to proceed with the DHR construction of the field net [**25**], as already carried out similarly for simple current extensions with cyclic groups in [**7, 8**]. In fact, all we need to do here is to pick a single local factor $N = N(I)$ such that the interval $I \subset S^1$ contains the localization region of the morphisms, and then we construct the cross product subfactor $N \subset N \rtimes (\mathbb{Z}_2 \times \mathbb{Z}_2)$. Then the corresponding dual canonical endomorphism θ decomposes as a sector as

$$[\theta] = [\mathrm{id}] \oplus [\rho_v] \oplus [\rho_s] \oplus [\rho_c].$$

Checking $\langle \iota\lambda, \iota\mu \rangle = \langle \theta\lambda, \mu \rangle = 1$ for $\lambda, \mu = \mathrm{id}, \rho_v, \rho_s, \rho_c$, we find that there is only a single M-N sector, namely $[\iota]$. From $\mathrm{tr}Z = \#_M\mathcal{X}_N$ we conclude that the modular invariant coupling matrix Z arising from this subfactor must fulfill $\mathrm{tr}Z = 1$. This leaves only the possibility that Z is Q or ${}^tQ$. We may and do assume that $Z = Q$, otherwise we exchange braiding and opposite braiding. It is easy to determine the intermediate subfactors $N \subset M_\pm \subset M$. Namely, we have $M_+ = N \rtimes_{\rho_s} \mathbb{Z}_2$ and $M_- = N \rtimes_{\rho_c} \mathbb{Z}_2$ with dual canonical endomorphism sectors $[\theta_+] = [\mathrm{id}] \oplus [\rho_s]$ and $[\theta_-] = [\mathrm{id}] \oplus [\rho_c]$, respectively. That both extensions are local also follows from $\omega_s = \omega_c = 1$. We therefore find $Z^+ = X_s$ and $Z^- = X_c$. Finally, the permutation invariant W is obtained from the non-local extension $M_v = N \rtimes_{\rho_v} \mathbb{Z}_2$.

8. Realization of modular invariants from subfactors

In our general setting, we have the following situation: For a given type III von Neumann factor N equipped with a braided system of endomorphism ${}_N\mathcal{X}_N$, any embedding $N \subset M$ of N in a larger factor M which is compatible with the system ${}_N\mathcal{X}_N$ (in the sense that the dual canonical endomorphism decomposes in ${}_N\mathcal{X}_N$) defines a coupling matrix Z through α-induction. This matrix Z commutes with the matrices Y and Ω arising from the braiding and in turn is a "modular invariant mass matrix" whenever the braiding is non-degenerate. Suppose we start with a system corresponding to the RCFT data of $SU(n)_k$. Then the following question is natural, but difficult to answer:

Can any physical modular invariant be realized from some subfactor $N \subset M$?

The first problem with this question is that one needs to specify what the term "physical" means. Quite often in the literature, any modular invariant matrix (i.e. $ZS = SZ$, $ZT = TZ$) subject to the constraint that all entries are non-negative integers and with normalization $Z_{0,0} = 1$ is called a physical invariant. Well, with this interpretation of "physical" the answer to the question is clearly negative. Namely, our general theory says that there is always some associate extended theory carrying another representation of the modular group $SL(2;\mathbb{Z})$ which is compatible with the chiral branching rules. As mentioned above, it is however known [**78, 82, 37**] that there are "spurious" modular invariants satisfying the above constraints but which do not admit an extended modular S-matrix. But even with this relatively simple specification we have another problem: Complete classifications of such modular invariant matrices are known only for very few models, not much more than $\mathbb{Z}_n$ conformal field theories [**19**], $SU(2)$ all levels [**14, 55**], $SU(3)$ all levels [**39**], and some classifications for affine partition functions at low levels [**40**].

Another specification of "physical" (but unfortunately mathematically harder to reach) would be that Z arises from "the existence of some 2D conformal field theory". A promising way of making this precise seems for us to be the concept of chiral observables as light-cone nets built in an observable net over 2D Minkowski space [**74**]. As mentioned in Section 6, Rehren has shown [**75**] that any subfactor $N \subset M$ of our kind which arises as an extension of a local factor $N = N(I_\circ)$ of a Möbius covariant net $\{N(I)\}$ over $\mathbb{R}$ (or equivalently $S^1 \setminus \zeta$) determines an entire 2D conformal field theory over Minkowski space. The converse direction, however, is an open problem: Does any 2D conformal field theory with chiral building blocks containing $\{N(I)\}$ determine a subfactor $N \subset M$ producing the modular invariant matrix Z which describes the coupling between left- and right-chiral sectors? (In particular in the case that the coupling matrix is type II.) Nevertheless there are partial answers to this question. First of all the trivial invariants, $Z_{\lambda,\mu} = \delta_{\lambda,\mu}$, are obtained from the trivial subfactor $N \subset M$ with $M = N$. Next, any conformal inclusion determines a subfactor which in turn produces a modular invariant, being the type I exceptional invariant which arises from the diagonal invariant of the extended theory, here the level 1 representation theory of the larger affine Lie algebra (e.g. of $SO(5)$ for the embedding $SU(2)_{10} \subset SO(5)_1$ as treated above). The situation is even better for simple current invariants, which in a sense produce the majority of non-trivial modular invariants. Simple currents [**77**] are primary fields with unit quantum dimension and appear in our framework as sectors with statistical dimension one, hence its representatives are automorphisms. They form a closed abelian group G under fusion which is hence a product of cyclic groups. Simple currents give rise to modular invariants, and all such invariants have been classified [**42, 58**].

If we take generators $[\sigma_i]$ for each cyclic subgroup $\mathbb{Z}_{n_i}$ then we can construct the crossed product subfactor $N \subset M = N \rtimes G$ whenever we can choose a representative σ_i in each such simple current sector such that we have exact cyclicity $\sigma_i^{n_i} = \mathrm{id}$ (and not only as sectors). As we are starting with a chiral quantum field theory (e.g. from loop groups), Rehren's lemma [**73**] applies which states that such a choice is possible if and only if the statistics phase is an n_i-th root of unity, or in the conformal context if and only if the conformal weight h_{σ_i} is an integer multiple of $1/n_i$. Sometimes this may only be possible for a simple current subgroup $H \subset G$, but any non-trivial subgroup ($H \neq \{0\}$) gives rise to a non-trivial subfactor and in turn to a modular invariant. In fact one can check by our methods that all simple current invariants are realized this way. For example, for $SU(n)_k$ the simple current group is just $\mathbb{Z}_n$, corresponding to weights $k\Lambda_{(j)}$, $j = 0, 1, ..., n-1$. The conformal dimensions are $h_{k\Lambda_{(j)}} = kj(n-j)/2n$ which allow for extensions except when n is even and k and j are odd. (This reflects the fact that e.g. for $SU(2)$ there are no D-invariants at odd levels.) An extension by a simple current subgroup $\mathbb{Z}_m \subset \mathbb{Z}_n$, i.e. m is a divisor of n, is moreover local, if the generating current (and hence all in the $\mathbb{Z}_m$ subgroup) has integer conformal weight, $h_{k\Lambda_{(q)}} \in \mathbb{Z}$, where $n = mq$. This happens exactly if $kq \in 2m\mathbb{Z}$ if n is even, or $kq \in m\mathbb{Z}$ if n is odd. For $SU(2)$ this corresponds to the D_{even} series whereas the D_{odd} series are non-local extensions. For $SU(3)$, there is a simple current extension at each level, but only those at $k \in 3\mathbb{Z}$ are local. Clearly, the cases with chiral locality match exactly the type I simple current modular invariants. Our results imply that the system ${}_M\mathcal{X}_M^0$ of neutral morphisms, which is obtained by decomposing $[\alpha_\lambda^\pm]$'s with colour zero mod m, carries a non-degenerate braiding. This nicely reflects a general fact about non-degenerate extensions of degenerate (sub-) systems conjectured by Rehren [72] and proven by Müger [67].

For the exceptional modular invariants arising from conformal inclusions, the corresponding subfactor comes (almost) for free. A conformal inclusion means that the level 1 representations of some loop group of a Lie group restrict in a finite manner to the positive energy representations of a certain embedded loop group of an embedded (simple) Lie group at some level. As discussed for the E_6 example, a subfactor is obtained by taking this embedding as a local subfactor in the vacuum representation. Since the embedding level one theory is always local, the modular invariant will necessarily be type I. For $SU(2)$, the modular invariants arising from conformal embeddings are, besides E_6, the E_8 and the D_4 ones, corresponding to embeddings $SU(2)_{28} \subset (G_2)_1$ and $SU(2)_4 \subset SU(3)_1$, respectively, the latter happens to be a simple current invariant at the same time. For $SU(3)$, the invariants from conformal embeddings are $\mathcal{D}^{(6)}$, $\mathcal{E}^{(8)}$, $\mathcal{E}^{(12)}$ and $\mathcal{E}^{(24)}$, corresponding to $SU(3)_3 \subset SO(8)_1$, $SU(3)_5 \subset SU(6)_1$, $SU(3)_9 \subset (E_6)_1$, $SU(3)_{21} \subset (E_7)_1$, respectively.

With these techniques we can obtain a huge amount of modular invariants from subfactors. Nevertheless we still do not have a systematic procedure to get all physical invariants. The more problematic cases are typically the exceptional type II invariants. We did realize the E_7 invariant of $SU(2)$ by some subfactor, namely we used the existence of a certain Goodman-de la Harpe-Jones subfactor [43] for this case, however, this method will not apply to general invariants of $SU(n)$. It seems to follow from Ocneanu's recent announcement (see his lectures) that there are subfactors realizing all $SU(3)$ modular invariants, but also his methods relying on the "$SU(3)$ wire model" (as well as on Gannon's classification of modular invariants) do not solve the general problem. Nevertheless a large class of exceptional type II invariants can be dealt with quite generally, namely those which are type II descendants of conformal embeddings. Since the embedding level 1 theories are typically (whenever simply laced Lie groups are worked with) $\mathbb{Z}_n$ theories, i.e. pure simple current theories, the subfactors producing their modular invariants can be constructed by simple current methods, and in turn we will obtain the relevant subfactors for the embedded theories, say $SU(n)$.

For a while we will be looking at the so-called $\mathbb{Z}_n$ conformal field theories as treated in [19], which have n sectors, labelled by $\lambda = 0, 1, 2, ..., n-1 \pmod{n}$, obeying $\mathbb{Z}_n$ fusion rules, and conformal dimensions of the form $h_\lambda = a\lambda^2/2n \pmod{1}$, where a is an integer mod $2n$, a and n coprime and a is even whenever n is odd. The modular invariants of such models have been classified [19]. They are labelled by the divisors δ of $\tilde{n}$, where $\tilde{n} = n$ if n is odd and $\tilde{n} = n/2$ if n is even. Explicitly, the modular invariants $Z^{(\delta)}$ are given by

$$Z^{(\delta)}_{\lambda,\mu} = \begin{cases} 1 & \text{if } \lambda, \mu = 0 \bmod \alpha \text{ and } \mu = \omega(\delta)\lambda \bmod n/\alpha, \\ 0 & \text{otherwise}, \end{cases}$$

where $\alpha = \gcd(\delta, \tilde{n}/\delta)$ so that there are numbers $r, s \in \mathbb{Z}$ such that $r\tilde{n}/\delta\alpha - s\delta/\alpha = 1$ and then $\omega(\delta)$ is defined as $\omega(\delta) = r\tilde{n}/\delta\alpha + s\delta/\alpha$. The trivial invariant corresponds to $\delta = \tilde{n}$, i.e. $Z^{(\tilde{n})} = \mathbf{1}$ and $\delta = 1$ gives the charge conjugation matrix, $Z^{(1)} = C$.

We now claim that

(8.1) $$Z^{(\delta)}_{\lambda,\lambda} = \begin{cases} 1 & \text{if } \lambda = 0 \bmod \tilde{n}/\delta, \\ 0 & \text{otherwise}. \end{cases}$$

Notice that $\omega(\delta) - 1 = 2s\delta/\alpha$. Assume first that $\lambda = x\tilde{n}/\delta$, $x \in \mathbb{Z}$. Then clearly $\lambda = 0 \bmod \alpha$ since α divides $\tilde{n}/\delta$, and we have $(\omega(\delta) - 1)\lambda = 2sx\tilde{n}/\alpha$, implying $\lambda = \omega(\delta)\lambda \bmod n/\alpha$, thus $Z^{(\delta)}_{\lambda,\lambda} = 1$. Conversely, assume $Z^{(\delta)}_{\lambda,\lambda} = 1$ so that $\lambda = y\alpha$ and $(\omega(\delta) - 1)\lambda = zn/\alpha$ with $y, z \in \mathbb{Z}$. This gives $2sy\delta = zn/\alpha$, hence $2sy = zn/\delta\alpha$. Now s is coprime to $\tilde{n}/\delta\alpha$, and therefore it follows that y is a multiple of $\tilde{n}/\delta\alpha$ (as we see that z must be even if n is odd) which implies in fact $\lambda = 0 \bmod \tilde{n}/\delta$.

From Eq. (8.1) we obtain the following trace property of $Z^{(\delta)}$:

$$\text{tr}(Z^{(\delta)}) = \epsilon\delta, \qquad \text{where} \quad \epsilon = \frac{n}{\tilde{n}} = \begin{cases} 2 & \text{if } n \text{ is even,} \\ 1 & \text{if } n \text{ is odd.} \end{cases}$$

Now suppose that for such a $\mathbb{Z}_n$ theory at hand we have corresponding braided endomorphisms ρ_λ of some type III factor N, such that their statistical phases are given by $e^{2\pi i h_\lambda}$ with conformal weights h_λ as above (as is the case for level 1 loop group theories). As we are dealing with $\mathbb{Z}_n$ fusion rules, all our morphisms ρ_λ will in fact be automorphisms. Note that if n is odd then we can always assume that $\rho_1^n = \text{id}$ as morphisms (and our system can be chosen as $\{\rho_1^\lambda\}_{\lambda=0}^{n-1}$). However, if n is even, then we cannot choose a representative of the sector $[\rho_1]$ such that its n-th power gives the identity, nevertheless we can always assume that $\rho_\epsilon^{\tilde{n}} = \text{id}$. Thus we have a simple current (sub-) group $\mathbb{Z}_{\tilde{n}}$, for which we can form the crossed product subfactor $N \subset M = N \rtimes \mathbb{Z}_{\tilde{n}/\delta}$ for any divisor δ of $\tilde{n}$. It is quite easy to see that $N \subset M = N \rtimes \mathbb{Z}_{\tilde{n}/\delta}$ indeed realizes $Z^{(\delta)}$: The crossed product by $\mathbb{Z}_{\tilde{n}/\delta}$ gives the dual canonical endomorphism sector $[\theta] = [\text{id}] \oplus [\rho_{\epsilon\delta}] \oplus [\rho_{\epsilon\delta}^2] \oplus \ldots \oplus [\rho_{\epsilon\delta}^{\tilde{n}/\delta-1}]$. The formula $\langle \iota\rho_\lambda, \iota\rho_\mu \rangle = \langle \theta\rho_\lambda, \rho_\mu \rangle$ then shows that the system of M-N morphisms is labelled by $\mathbb{Z}_n/\mathbb{Z}_{\tilde{n}/\delta} \simeq \mathbb{Z}_{\epsilon\delta}$, i.e. $\#_M \mathcal{X}_N = \epsilon\delta$. Therefore our general theory implies that the modular invariant arising from $N \subset M = N \rtimes \mathbb{Z}_{\tilde{n}/\delta}$ has trace equal to $\epsilon\delta$, and thus must be $Z^{(\delta)}$. Thus all modular invariants classified in [19] are realized from subfactors.

It is instructive to apply the above results to descendant modular invariants of conformal inclusions. Let us consider the conformal inclusion $SU(4)_6 \subset SU(10)_1$.

The associated modular invariant, which can be found in [**77**], reads
$$Z = \sum_{j \in \mathbb{Z}_{10}} |\chi^j|^2$$
with $SU(10)_1$ characters decomposing into $SU(4)_6$ characters as

$$\chi^0 = \chi_{0,0,0} + \chi_{0,6,0} + \chi_{2,0,2} + \chi_{2,2,2}, \quad \chi^5 = \chi_{0,0,6} + \chi_{6,0,0} + \chi_{0,2,2} + \chi_{2,2,0},$$
$$\chi^1 = \chi_{0,0,2} + \chi_{2,4,0} + \chi_{2,1,2}, \quad \chi^6 = \chi_{4,0,0} + \chi_{0,2,4} + \chi_{1,2,1},$$
$$\chi^2 = \chi_{0,1,2} + \chi_{2,3,0} + \chi_{3,0,3}, \quad \chi^7 = \chi_{3,0,1} + \chi_{1,2,3} + \chi_{0,3,0},$$
$$\chi^3 = \chi_{1,0,3} + \chi_{3,2,1} + \chi_{0,3,0}, \quad \chi^8 = \chi_{0,3,2} + \chi_{2,1,0} + \chi_{3,0,3},$$
$$\chi^4 = \chi_{0,0,4} + \chi_{4,2,0} + \chi_{1,2,1}, \quad \chi^9 = \chi_{2,0,0} + \chi_{0,4,2} + \chi_{2,1,2}.$$

As usual, this invariant can be realized from the conformal inclusion subfactor
$$N = \pi^0(L_I SU(4))'' \subset \pi^0(L_I SU(10))'' = M_+,$$
with π^0 denoting the level 1 vacuum representation of $LSU(10)$. The dual canonical endomorphism sector corresponds to the vacuum block,
$$[\theta_+] = [\lambda_{0,0,0}] \oplus [\lambda_{0,6,0}] \oplus [\lambda_{2,0,2}] \oplus [\lambda_{2,2,2}].$$

Proceeding with α-induction $\lambda_{p,q,r} \mapsto \alpha^\pm_{+;p,q,r} \in \mathrm{End}(M_+)$, it is a straightforward calculation that the graphs describing left multiplication by fundamental generators $[\alpha^\pm_{+;1,0,0}]$ and $[\alpha^\pm_{+;0,1,0}]$ (which is the same as right multiplication by $[\lambda_{1,0,0}]$ and $[\lambda_{0,1,0}]$, respectively) on the system of M_+-N sectors gives precisely the graphs found by Petkova and Zuber [**71**, Figs. 1 and 2] by their more empirical procedure to obtain graphs with spectrum matching the diagonal part of some given modular invariant. In our framework, the graph [**71**, Fig. 1] obtains the following meaning: Take the outer wreath, pick a vertex with 4-ality 0 and label it by $[\iota_+] \equiv [\tau_0 \iota_+]$, where $\iota_+ : N \hookrightarrow M_+$ denotes the injection homomorphism, as usual. Going around in a counter-clockwise direction the vertices will then be the marked vertices labelled by the $\mathbb{Z}_{10}$ sectors $[\tau_1 \iota_+], [\tau_2 \iota_+], \ldots, [\tau_9 \iota_+]$ of $SU(10)_1$. Passing to the next inner wreath the 4-ality 1 vertex adjacent to $[\iota_+]$ is then the sector $[\alpha^\pm_{+;1,0,0} \iota_+] = [\iota_+ \lambda_{1,0,0}]$, and the others its $\mathbb{Z}_{10}$ translates. Similarly the inner wreath consists of the $\mathbb{Z}_{10}$ translates of $[\iota_+ \lambda_{0,1,0}]$. The remaining two vertices in the center correspond to subsectors of the reducible $[\iota \lambda_{1,1,0}]$ and $[\iota \lambda_{0,1,1}]$. The graph itself then represents left (right) multiplication by $[\alpha^\pm_{+;1,0,0}]$ ($[\lambda_{1,0,0}]$).

As for $LSU(10)$ at level 1 we are in fact dealing with a $\mathbb{Z}_n$ conformal field theory, we have $n = 10$ and $\tilde{n} = 5$, we thus know that there are only two modular invariants: The diagonal one which in restriction to $LSU(4)$ gives exactly the above type I invariant $Z \equiv Z^{(5)}$, but there is also the charge conjugation invariant $Z^{(1)}$, written as
$$Z^{(1)} = \sum_{j \in \mathbb{Z}_{10}} \chi^j (\chi^{-j})^*.$$

Whereas $Z^{(5)}$ can be thought of as the trivial extension $M_+ \subset M_+$, the conjugation invariant $Z^{(1)}$ can be realized from the crossed product $M_+ \subset M = M_+ \rtimes \mathbb{Z}_5$ which has dual canonical endomorphism sector
$$[\theta^{\mathrm{ext}}] = [\tau_0] \oplus [\tau_2] \oplus [\tau_4] \oplus [\tau_6] \oplus [\tau_8].$$

So far we have considered the situation on the "extended level", but we may now descend to the level of $SU(4)_6$ sectors and characters. Namely we may consider the subfactor $N \subset M = M_+ \rtimes \mathbb{Z}_5$. Its dual canonical endomorphism sector $[\theta]$ is obtained by σ-restriction of $[\theta^{\text{ext}}]$ which can now be read off from the character decomposition,

$$[\theta] = \bigoplus_{j=0}^{4}[\sigma_{\tau_{2j}}] = [\lambda_{0,0,0}] \oplus [\lambda_{0,6,0}] \oplus [\lambda_{2,0,2}] \oplus [\lambda_{2,2,2}] \oplus [\lambda_{0,1,2}]$$
$$\oplus [\lambda_{2,3,0}] \oplus [\lambda_{3,0,3}] \oplus [\lambda_{0,0,4}] \oplus [\lambda_{4,2,0}] \oplus [\lambda_{1,2,1}] \oplus [\lambda_{4,0,0}] \oplus [\lambda_{0,2,4}]$$
$$\oplus [\lambda_{1,2,1}] \oplus [\lambda_{0,3,2}] \oplus [\lambda_{2,1,0}] \oplus [\lambda_{3,0,3}].$$

This subfactor produces the conjugation invariant $Z^{(1)}$ written in $SU(4)_6$ characters which is the same as taking the original $SU(4)_6$ conformal inclusion invariant and conjugating on the level of the $SU(4)_6$ characters. Note that this invariant has only 16 diagonal entries.

Also note that we will still have entries $Z_{\lambda,\mu} \geq 2$, for instance the diagonal entry corresponding to the weight $(2,1,2)$ is 2 as $|\chi_{2,1,2}|^2$ appears in $\chi^1(\chi^9)^*$ and in $\chi^9(\chi^1)^*$. Hence the system of M-M sectors will have non-commutative fusion rules (as had the M_+-M_+ system). When passing from M_+ to $M = M_+ \rtimes \mathbb{Z}_5$, the M_+-N system will change to the M-N system in such a way that all sectors which are translates by τ_{2j}, $j = 0,1,2,3,4$, have to be identified, and similarly fixed points split. Thus our new system of M-N morphisms will be some kind of orbifold of the old one. To see this, we first recall that all the irreducible M_+-N morphisms are of the form $\beta\iota_+$ with $\beta \in {}_{M_+}\mathcal{X}^{\pm}_{M_+}$. To such an irreducible M_+-N morphism $\beta\iota_+$ we can now associate an M-N morphism $\iota^{\text{ext}}\beta\iota_+$ which may no longer be irreducible; here ι^{ext} is the injection homomorphism $M_+ \hookrightarrow M$. Then the reducibility can be controlled by Frobenius reciprocity as we have

$$\langle \iota^{\text{ext}}\beta\iota_+, \iota^{\text{ext}}\beta'\iota_+ \rangle = \langle \theta^{\text{ext}}\beta\iota_+, \beta'\iota_+ \rangle,$$

and $\theta^{\text{ext}} = \bar\iota^{\text{ext}}\iota^{\text{ext}}$. Carrying out the entire computation we find that there are 16 M-N sectors, and the right multiplication by $[\lambda_{1,0,0}]$ is displayed graphically as in Fig. 15. Here the 4-alities 0,1,2,3 of the vertices are indicated by solid circles of decreasing size. The $[\iota]$ vertex (with $\iota = \iota^{\text{ext}}\iota_+$ denoting the injection homomorphism $N \hookrightarrow M$ of the total subfactor $N \subset M = M_+ \rtimes \mathbb{Z}_5$) is the 4-ality 0 vertex in the center of the picture, and the 4-ality 1 vertex above corresponds to $[\iota\lambda_{1,0,0}]$. Each group of five vertices on the top and the bottom of the picture arise from the splitting of the two central vertices of the graphs in [**71**] as they are $\mathbb{Z}_5$ fixed points. That our orbifold graph inherits the 4-ality of the original graph is due to the fact that all entries in $[\theta]$ have 4-ality zero which in turn comes from the fact that all even marked vertices (corresponding to the subgroup $\mathbb{Z}_5 \subset \mathbb{Z}_{10}$) of the graph of Petkova and Zuber have 4-ality zero. We also display the graph corresponding to the second fundamental representation, namely the right multiplication by $[\lambda_{0,1,0}]$ in Fig. 15.

The conformal inclusion $SU(3)_5 \subset SU(6)_1$ can be treated along the same lines. The associated $SU(3)_5$ modular invariant, i.e. the one which is the specialization of the diagonal $SU(6)_1$ invariant,

$$Z = \sum_{j \in \mathbb{Z}_6} |\chi^j|^2,$$

with $SU(6)_1$ characters decomposing in $SU(3)_5$ variables as

$$\chi^0 = \chi_{0,0} + \chi_{2,2}, \quad \chi^1 = \chi_{2,0} + \chi_{2,3}, \quad \chi^2 = \chi_{2,1} + \chi_{0,5},$$
$$\chi^3 = \chi_{3,0} + \chi_{0,3}, \quad \chi^4 = \chi_{1,2} + \chi_{5,0}, \quad \chi^5 = \chi_{0,2} + \chi_{3,2},$$

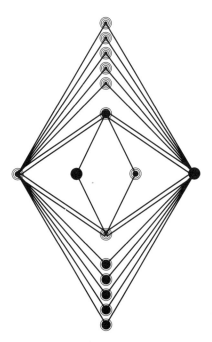

FIGURE 15. Graph G_1 associated to the conjugation invariant of the conformal inclusion $SU(4)_6 \subset SU(10)_1$

is labelled by the graph $\mathcal{E}^{(8)}$. Besides this diagonal invariant $Z \equiv Z^{(3)}$, the extended $SU(6)_1$ theory, being a $\mathbb{Z}_6$ theory, possesses only the conjugation invariant $Z^{(1)} = \sum_{j \in \mathbb{Z}_6} \chi^j (\chi^{-j})^*$, corresponding to the divisors 3 and 1 of 3, respectively. Writing again the conformal inclusion subfactor as $N \subset M_+$, the conjugation invariant can be realized from the extension $N \subset M = M_+ \rtimes \mathbb{Z}_3$ with canonical endomorphism sector

$$[\theta] = [\lambda_{0,0}] \oplus [\lambda_{2,2}] \oplus [\lambda_{2,1}] \oplus [\lambda_{0,5}] \oplus [\lambda_{1,2}] \oplus [\lambda_{5,0}],$$

which arises as $\theta = \sigma_{\theta^{\text{ext}}}$ where $[\theta^{\text{ext}}] = [\tau_0] \oplus [\tau_2] \oplus [\tau_4]$. Whereas the M_+-N system is labelled by the vertices of the graph $\mathcal{E}^{(8)}$ and can be given by $\{\beta \iota\}$ where β runs through the chiral M_+-M_+ system determined in [**7**, Subsect. 2.3 (iv)], the M-N system will now be obtained from this one by identification of all $\mathbb{Z}_3$ translations (corresponding to the vertices labelled by $[\alpha_{(0,0)}]$, $[\alpha_{(5,5)}]$ and $[\alpha_{(5,0)}]$ in [**7**, Fig. 11]). We have no fixed points here so that the 12 vertices of $\mathcal{E}^{(8)}$ collapse to 4 vertices, and it is easy to see that the new M-N fusion graph is exactly the graph $\mathcal{E}^{(8)*}$ in the list of Di Francesco and Zuber (see Zuber's lectures or [**2**]). Note that this time the orbifold graph $(\mathcal{E}^{(8)*})$ looses the triality of the original graph $(\mathcal{E}^{(8)})$ because the even marked vertices (corresponding to the subgroup $\mathbb{Z}_3 \subset \mathbb{Z}_6$) of $\mathcal{E}^{(8)}$ are not exclusively of colour zero.

This way we understand why the descendants of modular invariants of conformal inclusions (where the extended theory has $\mathbb{Z}_n$ fusion rules) are in fact labelled

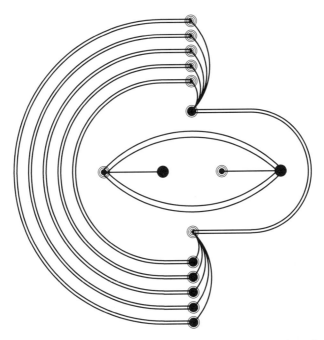

FIGURE 16. Graph G_2 associated to the conjugation invariant of the conformal inclusion $SU(4)_6 \subset SU(10)_1$

by orbifold graphs of the graph labelling the original, block-diagonal conformal inclusion invariant, and why the conjugation invariant corresponds to the maximal $\mathbb{Z}_{\tilde{n}}$ orbifold.

In the above examples, the trivial and conjugation invariant of the extended theory still remained distinct when written in terms of the $SU(4)_6$ characters. This need not be the case in general. Let us look at a familiar modular invariant of $SU(3)$ at level 9, namely

$$Z_{\mathcal{E}^{(12)}} = |\chi_{0,0} + \chi_{9,0} + \chi_{0,9} + \chi_{4,1} + \chi_{1,4} + \chi_{4,4}|^2 + 2|\chi_{2,2} + \chi_{5,2} + \chi_{2,5}|^2,$$

which arises from the conformal embedding $SU(3)_9 \subset (E_6)_1$. Now E_6 at level 1 gives a $\mathbb{Z}_3$ theory and in terms of the extended characters the above invariant is the trivial extended invariant

$$Z_{\mathcal{E}_1^{(12)}} = |\chi^0|^2 + |\chi^1|^2 + |\chi^2|^2,$$

using obvious notation. Here both the $(E_6)_1$ characters χ^1 and χ^2 specialize to $\chi_{2,2} + \chi_{5,2} + \chi_{2,5}$ in terms of $SU(3)_9$ variables. Let $N \subset M_+$ denote the conformal inclusion subfactor obtained by analogous means as in the previous example. It has been treated in [8] and produces the graph $\mathcal{E}_1^{(12)}$ of the list of Di Francesco and Zuber as chiral fusion graphs — and in turn as M_+-N fusion graph, thanks to chiral locality.

Corresponding to the two divisors 3 and 1 of 3, we know that besides the trivial there is only the conjugation invariant of our $\mathbb{Z}_3$ theory. It is given as

$$Z_{\mathcal{E}_2^{(12)}} = |\chi^0|^2 + \chi^1(\chi^2)^* + \chi^2(\chi^1)^*$$

but this distinct invariant restricts to the same invariant $Z_{\mathcal{E}^{(12)}}$ when specialized to $SU(3)_9$ variables. Nevertheless we will obtain a different subfactor $N \subset M$ since the conjugation invariant of our $\mathbb{Z}_3$ theory is realized from the extension $M_+ \subset M = M_+ \rtimes \mathbb{Z}_3$. In particular, the subfactor $N \subset M$ has dual canonical endomorphism sector

$$[\theta] = [\lambda_{0,0}] \oplus [\lambda_{9,0}] \oplus [\lambda_{0,9}] \oplus [\lambda_{4,1}] \oplus [\lambda_{1,4}] \oplus [\lambda_{4,4}] \oplus 2[\lambda_{2,2}] \oplus 2[\lambda_{5,2}] \oplus 2[\lambda_{2,5}],$$

determined by σ-restriction of

$$[\theta^{\text{ext}}] = [\tau_0] \oplus [\tau_1] \oplus [\tau_2].$$

As before, the M-N system can be obtained from the M_+-N system by dividing out the cyclic symmetry carried by $[\theta^{\text{ext}}]$. In terms of graphs, the cyclic $\mathbb{Z}_3$ symmetry corresponds to the three wings of the graph $\mathcal{E}_1^{(12)}$ which are transformed into each other by translation through the $[\tau_j]$'s, and dividing out this symmetry gives exactly the graph $\mathcal{E}_2^{(12)}$ as the wings are identified whereas each vertex on the middle axis splits into three nodes of identical Perron-Frobenius weight. This way we understand the graph $\mathcal{E}_2^{(12)}$ as the label for the conjugation invariant $Z_{\mathcal{E}_2^{(12)}}$ of $Z_{\mathcal{E}_1^{(12)}}$ which accidentally happens to be the same as the selfconjugate $Z_{\mathcal{E}^{(12)}}$ when specialized to $SU(3)_9$ variables.

Though here the same modular invariant, possessing a second interpretation as its own conjugation, gave rise to two different graphs, it often happens that an exceptional self-conjugate invariant is labelled by only one and the same graph which is its own orbifold. The very simplest case is the conformal inclusion $SU(2)_4 \subset SU(3)_1$, giving rise to the D_4 invariant which is self-conjugate for $SU(2)$ though the non-specialized diagonal $SU(3)_1$ invariant is not. We could proceed as above, passing from the conformal inclusion subfactor $N \subset M_+$ to $N \subset M = M_+ \rtimes \mathbb{Z}_3$, collapsing the M_+-N fusion graph D_4 into its $\mathbb{Z}_3$ orbifold. However, identifying the three external vertices and splitting the $\mathbb{Z}_3$ fixed point into 3 nodes gives us again D_4: The Dynkin diagram D_4 is its own $\mathbb{Z}_3$ orbifold.

Acknowledgement. We would like to thank J. Fuchs, T. Gannon, C. Schweigert and J.-B. Zuber for helpful comments on an earlier version of the manuscript.

References

[1] Araki, H., Evans, D.E.: *On a C^*-algebra approach to phase transition in the two dimensional Ising model.* Commun. Math. Phys. **91**, 489-503 (1983)

[2] Behrend, R.E., Pearce, P.A., Petkova, V.B., Zuber, J.-B.: *Boundary conditions in rational conformal field theories.* Nucl. Phys. **B570**, 525-589 (2000)

[3] Baumgärtel, H.: *Operatoralgebraic methods in quantum field theory.* Berlin: Akademie Verlag 1995

[4] Böckenhauer, J.: *Localized endomorphisms of the chiral Ising model.* Commun. Math. Phys. **177**, 265-304 (1996)

[5] Böckenhauer, J.: *An algebraic formulation of level one Wess-Zumino-Witten models.* Rev. Math. Phys. **8**, 925-947 (1996)

[6] Böckenhauer, J., Evans, D.E.: *Modular invariants, graphs and α-induction for nets of subfactors. I.* Commun. Math. Phys. **197**, 361-386 (1998)

[7] Böckenhauer, J., Evans, D.E.: *Modular invariants, graphs and α-induction for nets of subfactors. II.* Commun. Math. Phys. **200**, 57-103 (1999)

[8] Böckenhauer, J., Evans, D.E.: *Modular invariants, graphs and α-induction for nets of subfactors. III.* Commun. Math. Phys. **205**, 183-228 (1999)

[9] Böckenhauer, J., Evans, D.E.: *Modular invariants from subfactors: Type I coupling matrices and intermediate subfactors.* Preprint math.OA/9911239, to appear in Commun. Math. Phys.
[10] Böckenhauer, J., Evans, D.E., Kawahigashi, Y.: *On α-induction, chiral generators and modular invariants for subfactors.* Commun. Math. Phys. **208**, 429-487 (1999)
[11] Böckenhauer, J., Evans, D.E., Kawahigashi, Y.: *Chiral structure of modular invariants for subfactors.* Commun. Math. Phys. **210**, 733-784 (2000)
[12] Böckenhauer, J., Evans, D.E., Kawahigashi, Y.: *Longo-Rehren subfactors arising from α-induction.* Preprint math.OA/0002154
[13] Cappelli, A., Itzykson, C., Zuber, J.-B.: *Modular invariant partition functions in two dimensions.* Nucl. Phys. **B280**, 445-465 (1987)
[14] Cappelli, A., Itzykson, C., Zuber, J.-B.: *The A-D-E classification of minimal and $A_1^{(1)}$ conformal invariant theories.* Commun. Math. Phys. **113**, 1-26 (1987)
[15] Carey, A.L., Evans, D.E.: *The operator algebras of the two dimensional Ising model.* In: Birman, J. et al (eds.): *Braids.* Contemp. Math. **78**, 117-165 (1988)
[16] Connes, A.: *Classification of injective factors.* Ann. Math. **104**, 73-115 (1976)
[17] Date, E., Jimbo, M., Miwa, T., Okado, M.: *Solvable lattice models.* In: *Theta functions – Bowdoin 1987, Part 1.* Proc. Symp. Pure Math. **49**, Providence, R.I.: American Math. Soc., pp. 295-332 (1987)
[18] Davies, E.B., Pearce, P.A.: *Conformal invariance and critical spectrum of corner transfer matrices.* J. Phys. **A23**, 1295-1312 (1990)
[19] Degiovanni, P.: *Z/NZ Conformal field theories.* Commun. Math. Phys. **127**, 71-99 (1990)
[20] Di Francesco, P.: *Integrable lattice models, graphs and modular invariant conformal field theories.* Int. J. Mod. Phys. **A7**, 407-500 (1992)
[21] Di Francesco, P., Mathieu, P., Sénéchal, D.: *Conformal field theory.* New York: Springer-Verlag 1996
[22] Di Francesco, P., Zuber, J.-B.: *SU(N) lattice integrable models associated with graphs.* Nucl. Phys. **B338**, 602-646 (1990)
[23] Di Francesco, P., Zuber, J.-B.: *SU(N) lattice integrable models and modular invariance.* In: *Recent Developments in Conformal Field Theories.* Trieste 1989, Singapore: World Scientific 1990, pp. 179-215
[24] Dijkgraaf, R., Verlinde, E.: *Modular invariance and the fusion algebras.* Nucl. Phys. (Proc. Suppl.) **5B**, 87-97 (1988)
[25] Doplicher, S., Haag, R., Roberts, J.E.: *Fields, observables and gauge transformations. II.* Commun. Math. Phys. **15**, 173-200 (1969)
[26] Doplicher, S., Roberts, J.E.: *A new duality theory for compact groups.* Invent. Math. **98**, 157-218 (1989)
[27] Evans, D.E., Lewis, J.T.: *On a C^*-algebra approach to phase transition in the two dimensional Ising model. II.* Commun. Math. Phys. **102**, 521-535 (1986)
[28] Evans, D.E., Kawahigashi, Y.: *Orbifold subfactors from Hecke algebras.* Commun. Math. Phys. **165**, 445-484 (1994)
[29] Evans, D.E., Kawahigashi, Y.: *Quantum symmetries on operator algebras.* Oxford: Oxford University Press 1998
[30] Fredenhagen, K., Rehren, K.-H., Schroer, B.: *Superselection sectors with braid group statistics and exchange algebras. I.* Commun. Math. Phys. **125**, 201-226 (1989)
[31] Fredenhagen, K., Rehren, K.-H., Schroer, B.: *Superselection sectors with braid group statistics and exchange algebras. II.* Rev. Math. Phys. **Special issue**, 113-157 (1992)
[32] Fröhlich, J., Gabbiani, F.: *Braid statistics in local quantum theory.* Rev. Math. Phys. **2**, 251-353 (1990)
[33] Fröhlich, J., Gabbiani, F.: *Operator algebras and conformal field theory.* Commun. Math. Phys. **155**, 569-640 (1993)
[34] Fröhlich, J., King, C.: *Two-dimensional conformal field theory and three-dimensional topology.* Int. J. Mod. Phys. **A4**, 5321-5399 (1989)
[35] Fuchs, J.: *Affine Lie algebras and quantum groups.* Cambridge: Cambridge University Press 1992
[36] Fuchs, J.: *Fusion rules in conformal field theory.* Fortschr. Phys. **42**, 1-48 (1994)
[37] Fuchs, J., Schellekens, A.N., Schweigert, C.: *Galois modular invariants of WZW models.* Nucl. Phys. **B437**, 667-694 (1995)

[38] Gannon, T.: *WZW commutants, lattices and level–one partition functions.* Nucl. Phys. **B396**, 708-736 (1993)
[39] Gannon, T.: *The classification of affine $SU(3)$ modular invariants.* Commun. Math. Phys. **161**, 233-264 (1994)
[40] Gannon, T.: *The level two and three modular invariants of $SU(n)$.* Lett. Math. Phys. **39**, 289-298 (1997)
[41] Gannon, T.: *Monstrous moonshine and the classification of CFT.* Lectures given in Istambul, August 1998, math.QA/9906167
[42] Gato-Rivera, B., Schellekens A.N.: *Complete classification of modular invariants for RCFT's with a center $(\mathbf{Z}_p)^k$.* Commun. Math. Phys. **145**, 85-121 (1992)
[43] Goodman, F., de la Harpe, P., Jones, V.F.R.: *Coxeter graphs and towers of algebras.* MSRI publications 14, Berlin: Springer, 1989
[44] Guido, D., Longo, R.: *Relativistic invariance and charge conjugation in quantum field theory.* Commun. Math. Phys. **148**, 521-551 (1992)
[45] Guido, D., Longo, R.: *The conformal spin and statistics theorem.* Commun. Math. Phys. **181**, 11-35 (1996)
[46] Haag, R.: *Local Quantum Physics.* Berlin: Springer-Verlag 1992
[47] Haagerup, U.: *Connes' bicentralizer problem and the uniqueness of the injective factor of type III_1.* Acta Math. **158**, 95-148 (1987)
[48] Izumi, M.: *Application of fusion rules to classification of subfactors.* Publ. RIMS, Kyoto Univ. **27**, 953-994 (1991)
[49] Izumi, M.: *Subalgebras of infinite C^*-algebras with finite Watatani indices II: Cuntz-Krieger algebras.* Duke Math. J. **91**, 409-461 (1998)
[50] Izumi, M., Longo, R., Popa, S.: *A Galois correspondence for compact groups of automorphisms of von Neumann algebras with a generalization to Kac algebras.* J. Funct. Anal. **155**, 25-63 (1998)
[51] Jimbo, M.: *A q-analogue of $U(N+1)$, Hecke algebra and the Yang-Baxter equation.* Lett. Math. Phys. **11**, 247-252 (1986)
[52] Jones, V.F.R.: *Index for subfactors.* Invent. Math. **72**, 1-25 (1983)
[53] Jones, V.F.R.: *Planar algebras.* Preprint math.QA/9909027
[54] Kač, V.G.: *Infinite dimensional Lie algebras*, 3rd edition, Cambridge: Cambridge University Press, 1990
[55] Kato, A.: *Classification of modular invariant partition functions in two dimensions.* Mod. Phys. Lett. **A2**, 585-600 (1987)
[56] Kawahigashi, Y.: *On flatness of Ocneanu's connections on the Dynkin diagrams and classification of subfactors.* J. Funct. Anal. **127**, 63-107 (1995)
[57] Kirillov, A.N., Reshetikhin, N.Yu.: *Representations of the algebra $U_q(sl_2)$, q-orthogonal polynomials and invariants for links.* In: Kač, V.G. (ed.): *Infinite dimensional Lie algebras and groups.* Advanced Series in Mathematical Physics, Vol. **7** (1988) pp. 285-339
[58] Kreuzer, M., Schellekens A.N.: *Simple currents versus orbifolds with discrete torsion – a complete classification.* Nucl. Phys. **B411**, 97-121 (1994)
[59] Loke, T.: *Operator algebras and conformal field theory of the discrete series representations of* $\mathrm{Diff}(S^1)$. Dissertation Cambridge (1994)
[60] Longo, R.: *Solution of the factorial Stone-Weierstrass conjecture.* Invent. Math. **76**, 145-155 (1984)
[61] Longo, R: *Index of subfactors and statistics of quantum fields II.* Commun. Math. Phys. **130**, 285–309 (1990)
[62] Longo, R., Rehren, K.-H.: *Nets of subfactors.* Rev. Math. Phys. **7**, 567-597 (1995)
[63] Mack, G., Schomerus, V.: *Conformal field algebras with quantum symmetry from the theory of superselection sectors.* Commun. Math. Phys. **134**, 139-196 (1990)
[64] McKay, J.: *Graphs, singularities and finite groups.* Proc. Symp. in Pure Math. AMS **37**, 183-186 (1980)
[65] Moore, G., Seiberg, N.: *Polynomial equations for rational conformal field theories.* Phys. Lett.**B212**, 451-460 (1988)
[66] Moore, G., Seiberg, N.: *Naturality in conformal field theory.* Nucl. Phys. **B313**, 16-40 (1989)
[67] Müger, M.: *On charged fields with group symmetry and degeneracies of Verlinde's matrix S.* Ann. Inst. H. Poincaré (Phys. Théor.) **71**, 359-394 (1999)

[68] Ocneanu, A.: *Paths on Coxeter diagrams: From Platonic solids and singularities to minimal models and subfactors.* (Notes recorded by S. Goto) In: Rajarama Bhat, B.V. et al. (eds.), Lectures on operator theory, The Fields Institute Monographs, Providence, Rhode Island: AMS publications 2000, pp. 243-323

[69] Popa, S.: *Classification of subfactors of finite depth of the hyperfinite type III_1 factor.* Comptes Rendus de l'Académie des Sciences, Série I, Mathématiques, **318**, 1003-1008 (1994)

[70] Petkova, V.B., Zuber, J.-B: *From CFT to graphs.* Nucl. Phys. **B463**, 161-193 (1996)

[71] Petkova, V.B., Zuber, J.-B.: *Conformal field theory and graphs.* In: Proceedings Goslar 1996 "Group 21", hep-th/9701103

[72] Rehren, K.-H.: *Braid group statistics and their superselection rules.* In: Kastler, D. (ed.): The algebraic theory of superselection sectors. Palermo 1989, Singapore: World Scientific 1990, pp. 333-355

[73] Rehren, K.-H.: *Space-time fields and exchange fields.* Commun. Math. Phys. **132**, 461-483 (1990)

[74] Rehren, K.-H.: *Chiral observables and modular invariants.* Commun. Math. Phys. **208**, 689-712 (2000)

[75] Rehren, K.-H.: *Canonical tensor product subfactors.* Commun. Math. Phys. **211**, 395-406 (2000)

[76] Rehren, K.-H., Stanev, Y.S., Todorov, I.T.: *Characterizing invariants for local extensions of current algebras.* Commun. Math. Phys. **174**, 605-633 (1996)

[77] Schellekens A.N., Yankielowicz, S.: *Extended·chiral algebras and modular invariant partition functions.* Nucl. Phys. **B327**, 673-703 (1989)

[78] Schellekens A.N., Yankielowicz, S.: *Field identification fixed points in the coset construction.* Nucl. Phys. **B334**, 67-102 (1990)

[79] Takesaki, M.: *Tomita's theory of modular Hilbert algebras and its applications.* Heidelberg: Springer-Verlag 1970

[80] Toledano Laredo, V.: *Fusion of positive energy representations of $LSpin_{2n}$.* PhD Thesis, Cambridge 1997

[81] Verlinde, E.: *Fusion rules and modular transformations in 2D conformal field theory.* Nucl. Phys. **B300**, 360-376 (1988)

[82] Verstegen, D.: *New exceptional modular invariant partition functions for simple Kac-Moody algebras.* Nucl. Phys. **B346**, 349-386 (1990)

[83] Wassermann, A.: *Operator algebras and conformal field theory III: Fusion of positive energy representations of $LSU(N)$ using bounded operators.* Invent. Math. **133**, 467-538 (1998)

[84] Witten, E.: *Gauge theories and integrable lattice models.* Nucl. Phys. **B322**, 629-697 (1989)

[85] Xu, F.: *Orbifold construction in subfactors.* Commun. Math. Phys. **166**, 237-254 (1994)

[86] Xu, F.: *New braided endomorphisms from conformal inclusions.* Commun. Math. Phys. **192**, 347-403 (1998)

[87] Xu, F.: *3-Manifold invariants from cosets.* Preprint math.GT/9907077

SCHOOL OF MATHEMATICS, UNIVERSITY OF WALES CARDIFF, PO BOX 926, SENGHENNYDD ROAD, CARDIFF CF24 4YH, WALES, U.K.
E-mail address: BockenhauerJM@cardiff.ac.uk

SCHOOL OF MATHEMATICS, UNIVERSITY OF WALES CARDIFF, PO BOX 926, SENGHENNYDD ROAD, CARDIFF CF24 4YH, WALES, U.K.
E-mail address: EvansDE@cardiff.ac.uk

The classification of subgroups of quantum $SU(N)$

Adrian Ocneanu

1. Introduction

1.1. The classification and structure of subgroups of quantum $SU(2)$.
Drinfeld and Jimbo have shown that the simple Lie groups have quantum deformations. At roots of unity, the semisimple quotient of the quantum groups has only finitely many irreducibles, in a phenomenon called the Wess-Zumino-Witten cutoff. The irreducible representations of $SU(2)$ are labeled by the half-line $\{0, 1, 2, 3, \ldots\}$, and the quantum cutoff $SU(2)_l$ at the $l+2$-nd root of unity (here l is the level and $l + 2$ the Coxeter number) has irreducible representations labeled by the graph A_{l+1} with vertices $\{0, 1, 2, \ldots, l\}$; the level l is the highest degree of a representation which survives the cutoff. The graph D_k is obtained by folding in 2 the graph A_{2k-1} by a $\mathbf{Z}_2$ action which splits the irreducible in the middle into $\pm$ components. Thus the D_n series consists of orbifolds of the A_n series. There are also 3 exceptional graphs associated to $SU(2)$ namely E_6, E_7, E_8 at levels $10, 16$ and 28 respectively. Each of these graphs has eigenvalues among the eigenvalues of the graph A_n on the same level; these eigenvalues are labeled by the exponents of the graph.

In [1], we showed a new phenomenon: that in a way similar to the subgroups of the classical $SU(2)$ described by Felix Klein in his book "Das Ikosahedron", **the quantum cutoff $SU(2)_l$ has subgroups** as well. The irreducible representations of the Kleinian subgroups live, as observed by J. McKay, on the vertices of the affine ADE graphs. We have described subgroups of the quantum cutoffs $SU(2)_l$ for which the irreducibles live on the graphs A_n, D_{2n}, E_6 and E_8 respectively. The subfactors of Jones index < 4, which we had described earlier, correspond to these subgroups. In addition to subgroups there are also modules, for which the irreducibles can be tensored with irreducibles of $SU(2)_l$, but not tensored among themselves. The modules which are not subgroups correspond to the graphs D_{2n+1} and E_7. Each module is canonically obtained from a subgroup with an antiautomorphism, which we called the ambichiral twist; thus D_{2n+1} comes from A_{4n-1} and E_7 comes from an exceptional twist of the ambichiral part D_{10}^{even} of D_{10}. We also connected these subgroups and modules to the modular invariants for $SU(2)$ classified by Capelli, Itzykson and Zuber, and we showed that the modular invariants

---

1991 *Mathematics Subject Classification.* Primary 47LXX, 81RXX; Secondary 05C99, 46L37, 47N50, 47N55, 16W30, 20G42, 81R50, 81R60.

appeared in several different roles in the theory. A modular invariant for $SU(k)_l$ is a matrix M with entries $M_{i,j} \in \{0,1,2,\ldots\}$ labeled by pairs (i,j) of Young diagrams with $\leq k-1$ rows and $\leq l$ columns. The modular invariance requirement is that the matrix M be a self-intertwiner of a representation of the modular group $SL(2,\mathbf{Z})$ associated to $SU(k)_l$.

We have shown that the entries M_{ij} of the modular invariant corresponding to a graph G describe the following mathematical objects.

(i) the number of essential paths on a pair of chiral graphs with common vertices, corresponding to the common components of the irreducibles i and j of $SU(k)_l$ restricted to the subgroup. In an equivalent form, for $SU(2)$ this description is the following. There are 2 Hecke algebra connections for an ADE graph introduced by V. Jones, differing by a choice of $\sqrt{-1}$. Iterate the first one i times and the second one j times and decompose each into irreducible components. The number of common irreducible pairs is M_{ij}.

(ii) M_{ij} is the number of Kleinian invariants of degree (i,j). The original Kleinian invariants were polynomials left invariant by the action of a subgroup of $SU(2)$. Here there are 2 copies of the subgroup, complex conjugate to each other, and each invariant has a bidegree (i,j).

(iii) The fusion algebra of the connections between two copies of G can be block diagonalized and the numbers M_{ij} are the dimensions of the blocks.

1.2. Zuber's higher Coxeter graphs problem.

The theoretical physicists at the Centre d'Energie Atomique, Saclay, Paris, most notably Zuber and di Francesco have discovered what they called higher Coxeter graphs about 15 years ago.

The irreducible representations of $SU(3)$ (very important in physics, since they appear in the standard model) are labeled by a planar triangular lattice inside a Weyl chamber, and the level l cutoff $SU(3)_l$ has irreducibles labeled by the lattice points in an equilateral triangle, corresponding to Young diagrams with ≤ 2 rows and $\leq l$ columns; the edges correspond to tensoring with the generator of $\mathrm{Irr}\, SU(3)_l$. The triangular graph with these vertices will be called the A_l graph for $SU(3)$; it is the analog of the usual A_l graph for $SU(2)$.

One can obtain an orbifold series $A_l/3$, analogous to the D_n series for $SU(2)$. There are 2 more conjugate orbifold series, due to the fact that the graph A_2 from which $SU(3)$ itself is built has a symmetry called conjugation. Finally there are several exceptional graphs, among the list found by Zuber and collaborators empirically, with what they called computer aided flair. Zuber imposed spectral conditions analogous to the properties of the ADE graphs for $SU(2)$, such as the fact that their eigenvalues be among the eigenvalues of the A_l graph as described above. He also asked that the adjacency matrix of the graph be normal, due to the fact that with the Drinfeld-Jimbo braiding, the fusion algebra of $\mathrm{Irr}\, SU(3)_l$ is commutative.

A parallel list classifying $SU(3)$ modular invariants was due to T. Gannon, but the precise correspondence between it and the list of Zuber's graphs was unclear. Zuber stated about 10 years ago the problem of classifying the higher analogs of the ADE graphs and in an effort toward private support of research offered 1,2 and respectively 3 bottles of champagne for the classifications corresponding to $SU(3)$, $SU(4)$ and $SU(5)$ respectively.

2. The classification and structure of subgroups of quantum groups

2.1. The classification and structure of quantum subgroups of $SU(3)$.

The natural reformulation of Zuber's higher Coxeter graphs problem is the following. The quantum group $SU(k)_l$ has subgroups and modules. Each subgroup and module G gives raise to a family of graphs, having as common vertices $\operatorname{Irr} G$ and edges corresponding to tensoring with the $k-1$ generators of $\operatorname{Irr} SU(k)_l$. The proper way to look at the usual ADE graphs is to view them as the graphs of subgroups and modules of $SU(2)_l$.

As a result the higher Coxeter graphs problem is to be reformulated as the problem of classifying the subgroups and modules of $SU(k)_l$ (and more generally of any semisimple quantum Lie group at roots of unity), a version adopted by Zuber.

For the classification, we had to develop conditions which were possible to check on any given graph, and which insured that the graph corresponded to a subgroup of $SU(k)_l$. The problem here is that any computation which involves explicitly intertwiners of $SU(k)$ is completely out of reach.

The graph of a subgroup of $SU(3)$ or $SU(3)_l$ is made of triangles, corresponding to the fact that the generator $\sigma = \sigma_1 \in \operatorname{Irr} SU(3)$ satisfies $\sigma^{\otimes 3} \ni 1$. We have shown that in each triangle there is a complex number, which we call a **cell**, coming from the explicit composition of morphisms corresponding to the edges of the triangle. Together, the cells on a graph form what we call an **internal connection**, defined up to a gauge which comes from the choice of morphisms for edges. These cells satisfy quadratic and quartic equations of a cohomological nature, which are local, i.e. involve only cells in a small neighborhood of the cell. This is done in the same spirit in which checking that weights on the vertices of a graph form a Perron-Frobenius eigenvector for a given eigenvalue involves checking only neighboring vertices. One of the exceptional graphs in the empirical list proposed by Zuber fails this test and has to be eliminated, since according to our theory it is not related at a deep level with $SU(3)$. Later, Zuber and collaborators started to notice that its behavior is indeed aberrant from other points of view, which further justifies our reformulation of Zuber's problem.

The connection has been adopted rapidly by the physics community, where several papers on what are now called as "Ocneanu cells" have already appeared. The cells solve immediately and explicitly one of the major goals of the initial program of Zuber, the construction of new solutions of the QYB equation corresponding to each graph.

We then developed methods for an exhaustive description of all possible subgroups and modules of $SU(3)_l$ at all levels l. The starting point is the classification of modular invariants of $SU(3)$ by Gannon. We showed that given a modular invariant matrix M it is possible to construct a matrix

$$\mathcal{M}_{(i,j),(i',j')} = \sum_{i''} \sum_{j''} N_{i,i'}^{i''} N_{j,j'}^{j''} M_{i'',j''}$$

where $N_{i,i'}^{i''}$ are fusion numbers for $i, i', i'' \in \operatorname{Irr} SU(3)_l$. The matrix $\mathcal{M}_{(i,i'),(j,j')}$ decomposes as a product

$$\mathcal{M}_{(i,j),(i',j')} = \sum_x \widetilde{M}_{i,j}^x \widetilde{M}_{i',j'}^x$$

of matrices $\widetilde{M}^x$ indexed by a label x of a vertex of the graph of the subgroup. The matrices $\widetilde{M}^x = (M^x_{i,j})$, which we called the torus spectrum of the vertex x, generalize the modular matrix M and have natural numbers as entries; they describe for $SU(2)_q$ the dual asymptotic graph for the inclusion described by Goodman, Jones, and de la Harpe of the subalgebra generated by Jones projections into the algebra constructed from an ADE graph.

For $x = 1 \in \operatorname{Irr} G$ we have $\widetilde{M}^1 = M$. Let us now regard each $\widetilde{M}^x_{i',j'}$ as a vector and construct a matrix $\widetilde{\mathcal{M}} = (\widetilde{M}^x_{i',j'})_{(i,j),x}$ with nonnegative integer entries. We have to solve thus the equation $\mathcal{M} = \widetilde{M}\widetilde{M}^*$. A further reduction comes from the fact that the vertices x of the chiral left graph that we are looking for are obtained by solving the above equations with $j, j', j'' = 0$, i.e.

$$\sum_{i''} N^{i''}_{i,i'} M_{i'',0} = \mathcal{M}_{(i,0),(i',0)} = \sum_x \widetilde{M}^x_{i,0} \widetilde{M}^x_{i',0}$$

We called the above decomposition, in which the matrix $\mathcal{M}_{(i,0),(i',0)}$ is constructed from the first line $M_{i'',0}$ of the modular invariant M, and is then split into a product, the **chiral modular splitting**. This phenomenon has been, subsequent to our work, interpreted in conformal field theory by Petkova and Zuber.

If there is an upper triangular solution $\widetilde{M}$ with 1 on the diagonal, then the solution is unique; since $\widetilde{M}$ is easily computed from a given graph it becomes easy to check that a given graph is the unique solution corresponding to a given modular invariant M. This allows in the case of $SU(3)$ to check that there is precisely one graph of a subgroup (what the physicists call type I graph) for each first line of the modular invariants in Gannon's classification. There follows the classification of all ambichiral twists, and checking that there are, in the list of graphs that passed the necessary and sufficient cell test, enough graphs to account for each ambichiral twist. This completes the classification of the subgroups of $SU(3)_l$, which does not require machine help.

There are 4 series of orbifolds $A_l, A_l/3, (A_l)^c$ and $(A_l/3)^c = 3(A_l)^c$, where c is the conjugation coming from the symmetry of the Coxeter graph A_2 on which the Lie group $SU(3)$ is built; the subscript l denotes the level. Among these the graphs A_{3n} are flat, or type I, i.e. correspond to subgroups and the rest are modules, or type II. There are also 3 exceptional subgroups E_5, E_9 and E_{21} which come from conformal inclusions; E_5 and E_9 have a module-orbifold each, denoted by $E_5/3$ and $E_9/3$. There is also an exceptional twist $(A_9/3)^t$ of $A_9/3$ and its conjugate $(A_9/3)^{tc}$; these are analogous to the graph $E_7 = (D_{10})^t$ for $SU(2)$.

A remarkable fact is that, while the correspondence between the classical, Kleinian, subgroups of $SU(2)$ labeled by the affine ADE graphs and the quantum subgroups of $SU(2)_l$ indexed by the non-affine ADE graphs was bijective, the classical and quantum subgroups are quite far away from each other for $SU(3)$; about half of the quantum series and exceptionals have no classical correspondent, and several classical subgroups have no quantum correspondent. In fact going from $SU(3)$ to $SU(4)$ and $SU(5)$ increases the number of subgroups dramatically; in the quantum case, where subgroups of $SU(k)$ are not, in general, subgroups of $SU(k+1)$, the number of exceptional subgroups appears to decrease from $SU(3)$ to $SU(5)$.

2.2. The quantum subgroups of $SU(4)$: the non-conformal exceptional. We have almost completed the classification of quantum subgroups of

$SU(4)_l$ at all levels l. We have found 6 orbifold series, 3 exceptional subgroups and 3 exceptional modules. The nature of the problem was very different from $SU(3)$: no classification of modular invariants existed and no candidates for the subgroups existed either. The classification required new theoretical methods and very intensive machine computation. Among the surprises was the subgroup labeled E_8 (= the exceptional on level 8) of $SU(4)_8$. This is the first exceptional subgroup which does not come from a conformal inclusion, thus contradicting the running conjecture among physicists that the conformal inclusion method is exhaustive. The cell method was extended to $SU(4)$; the construction method is described in the next subsection. We do not have effective general methods to construct the modules, or type II graphs, which correspond to a type I graph although the methods that we do have were sufficient for $SU(4)$.

2.3. Rigidity, effective bounds and construction algorithms for subgroups of quantum groups.

We have described in a previous work (NSF proposal) our main rigidity results concerning subfactors and systems of bimodules.

For any fusion algebra there are finitely many 6j-symbols up to equivalence. This has as a consequence the fact that there are countably many finite depth subfactors of the hyperfinite II_1 factor up to conjugacy.

For any finite system of bimodules there are finitely many possibilities for a braiding.

The maximal atlas of any finite system of bimodules is finite. This latter result has as consequence that $SU(k)_l$ at any given rank k and level l has finitely many subgroups.

We can now show that for any given rank k the subgroups of $SU(k)_l$ at all levels l fit into a finite number of series which are orbifolds, and a finite number of exceptionals. We show that above a certain level there are no more exceptional subgroups. The maximal level of exceptionals is 28 for $SU(2)$ (the level of the E_8 subgroup), 21 for $SU(3)$, 8 for $SU(4)$ and probably 7 for $SU(5)$ (note the unexpected decrease in the maximal exceptional level with the rank in this range). The algorithms of T. Gannon allow the exhaustive computations of modular invariants up to very high levels (e.g. 10000 for $SU(4)$ in a few seconds on a current PC). However he found no exceptionals above very small levels. What his methods are lacking, except in the $SU(3)$ case, is a stopping point, and this is precisely what our methods provide.

We start from the modular splitting identity, which is nonlinear, and show that in the first row of the modular invariant the gap up to the first nonzero entry in the modular invariant is bounded by a universal bound. The only exception is for the case in which all nonzero entries in the first row of the modular invariant correspond to bimodules of index 1, or corners, of $SU(k)_l$; the latter case corresponds to the orbifold series analogous to D_n.

For the subgroup E_8 of $SU(2)$, the first of the modular invariants is $\chi_0 + \chi_{10} + \chi_{18} + \chi_{28}$, the gap to the first nontrivial entry is thus 10 and this gap manifests itself by the fact that the graph E_8 is identical to an A_n graph for the first 5=gap/2 vertices. Afterwards the 6th vertex of A_n splits into 2 vertices of E_8. We can show that the modular splitting identity gives a sharp bound on the gap, via the following **fundamental inequality**.

For a Young diagram λ denote by $|\lambda|$ its quantum dimension and by $\overline{\lambda}$ its conjugate. Then if the first row of the modular invariant has a nonzero coefficient for λ and if we denote by I the set of all the Young diagrams of level less than half the gap, we have

$$\sum_{\mu \in I} |\mu| N_{\lambda,\bar\lambda}^\mu \le |\lambda|$$

Note that $\sum_\mu |\mu| N_{\lambda,\bar\lambda}^\mu = |\lambda|^2$. Thus for a large gap, the left member is proportional to $|\lambda|^2$, and the only possibility is $|\lambda| = 1$, which produces the orbifold series.

On the other hand the first entry in the diagonal T matrix equal to 1 after the 0th entry arises on a level proportional to the level l of the cutoff, since the cutoff level appears in the denominator of the exponent. Thus the gap must be large if the level is large. Together with the previous absolute bound on the gap, this gives a bound on the maximal level l of exceptionals for $SU(k)_l$.

2.4. The internal mechanism of subfactors and QFT: the coefficients of the quantum symmetrizers. The theory of the irreducible representations of $SU(2)$ and its quantum cutoffs $SU(2)_l$ at the $l+2$-nd root of unity appears to be well understood; the 6j-symbols allow in principle the computation of the invariants of knots links and 3-manifolds. The irreducible representations σ_n of each degree n arise on symmetric tensors in this case homogeneous polynomials of degree n in 2 variables. The projection p_n from the space of all tensors of degree n corresponding to $(\sigma_1)^{\otimes n}$ to the symmetric tensors corresponding to the highest weight irreducible σ_n is the symmetrizer, which sums over all permutations of the n variables and then divides all by $n!$; in the quantum case there is an analogous formula, with braiding instead of permutations and with each term multiplied by a root of 1.

The trace τ_n of the symmetrizers $p_n = 1 - e_1 \vee \cdots \vee e_{n-1}$ for $SU(2)_l$, where e_i are the Jones projections satisfying $e_i e_{i\pm 1} e_i = [2]^{-2} e_i$ and $e_i e_j = e_j e_i$ if $|i-j| > 1$, has been shown by V.Jones to satisfy a recurrence relation $\tau_n = \tau_{n-1} - [2]^{-2}\tau_{n-2}$ and thus $\tau_n = [n+1]/[2]^n$, where $[k]$ is the quantum number k given by $[k] = (q^{k/2} - q^{-k/2})/(q^{1/2} - q^{-1/2}) = \sin(k\pi/(l+2))/\sin(\pi/(l+2))$ with $q = e^{2\pi i/(l+2)}$. This relation allowed him to prove that l must be an integer, i.e. the remarkable rigidity of the index $[2]^2 = 4\cos^2(\pi/(l+2))$, l integer, when the index is < 4.

Hans Wenzl has proved the recurrence relation

$$p_n = p_{n-1} - [2][n-1]/[n] p_{n-1} e_{n-1} p_{n-1}.$$

From this

$$p_1 = 1, \qquad p_2 = 1 - e_1,$$

and

$$p_3 = 1 - [2]^2/[3] e_1 - [2]^2/[3] e_2 + [2]^2/[3] e_1 e_2 + [2]^2/[3] e_2 e_1.$$

In general a repeated application of the Wenzl inductive formula allows one to express the symmetrizer p_n in terms of the linear basis consisting of monomials $(e_{i_1} e_{i_1+1} \ldots e_{j_1})(e_{i_2} e_{i_2+1} \ldots e_{j_2}) \ldots (e_{i_m} e_{i_m+1} \ldots e_{j_m})$ where $i_1 > i_2 > \cdots > i_m$ and $j_1 > j_2 > \cdots > j_m$. These monomials correspond, up to a normalization, to all planar diagrams without closed loops in a rectangle with n entrances and n exits. The complexity of the computation and the number of such monomials grows exponentially with n, however.

The problem, asked by Vaughan Jones, was to find **a closed formula for the coefficient of each monomial term in the expression of the symmetrizer**. The general opinion among mathematicians and physicists, who had been searching for such a formula for applications in quantum field theory, appeared to be that such a closed formula might not exist in general.

We have obtained the following closed formula for the coefficient ρ_B of a planar diagram without closed loops in a rectangle with n entrances and n exits. The formula is unexpected in the sense that it has no analog in the representation theory of $SU(2)_l$ among 6j-symbols, knot and 3-manifold invariants. It points out that there is a "micro level" structure, inside the symmetrizers, different from the "macro level" structure of intertwiners. It lead further to an unexpected discrete 1-dimensional QFT model with long distance interactions.

Let B be a planar box corresponding to a monomial in the e_i, with n entrances labeled from left to right $1, \ldots, n$ on the upper row and n exits labeled similarly on the lower row. The wires in B are divided into through wires connecting an entrance with an exit, cups connecting two entrances and caps connecting two exits. Through wires are grouped into maximal contiguous blocks as follows, from left to right. For $k = 1, \ldots, m$ there are sets of r'_k through wires called the k-th wall followed by the k-th room with p'_k cups and q'_k caps, where $r'_k > 0$ for $1 < k \leq m$ and $\max(p'_k, q'_k) > 0$ for $1 \leq k < m$. Cumulatively from the left we define $r_k = r'_1 + \cdots + r'_k$, $p_k = p'_1 + \cdots + p'_k$, $q_k = q'_1 + \cdots + q'_k$.

For a cap or cup with end points at positions k, l with $k < l$ we define the jump as $(l - k - 1)/2$ and we let J denote the set of strictly positive jumps of all cups and caps. Let $S = \{(s_1, \ldots, s_m) \in \mathbf{N}^m : 0 = s_1 \leq s_k \leq s_{k+1} \leq \min(p_k, q_k)$ for all $k = 1, \ldots, m-1\}$. Define the monomials

$$f_0(r) = \frac{1}{[r]!}$$

$$f_1(p, r; s) = \frac{[p-s]!^2 [p+r+s]!^2}{[2p+r]!}$$

$$f(p, q, r, r_1; s, s_1) = \frac{[p-s]![q-s]![p+r+s]![q+r+s]!}{[p-s_1]![q-s_1]![p+r_1+s_1]![q+r_1+s_1]!} \cdot \frac{[r_1 - r + s_1 - s - 1]![r_1 - 1 + s + s_1]![r_1 + 2s_1]}{[r_1 - r - 1]![r + s + s_1]![s_1 - s]!}$$

Let

$$\varepsilon_B = (-1)^{p_m + \sum_{k=1}^{m-1}(p_k - q_k)(r_{k+1} - r_k) + \sum_{j \in J} j}$$

and $\iota_B = \prod_{j \in J} [j+1]^{-1}$. We have the following expression for the weight of the planar box B:

$$\rho_B = \varepsilon_B \iota_B f_0(r_1) \sum_{(s_1, \ldots, s_m) \in S} \left(\prod_{k=1}^{m-1} f(p_k, q_k, r_k, r_{k+1}; s_k, s_{k+1}) \right) f_1(p_m, r_m; s_m).$$

The formula, which is a sum of quotients of factorials, is a discrete analog of a partition function in quantum field theory (QFT) and can be given the following physical interpretation. Consider n identical particles, say bosons, in a 1-dimensional model. Their partition function is symmetric and the symmetrizer acts by interchanging the particles and averaging over all such interchanges. However the process of interchanging particles is not physical. Instead, what is physical is the process of creation and annihilation. A pair of particles annihilate with the release of a photon followed by the creation of a new pair from that photon. Such a contiguous cup-cap pair corresponds to a Jones projection e_i. The problem then is to find in the symmetrizer the amplitude corresponding to each possible creation

SU(2)$_k$

Orbifold series

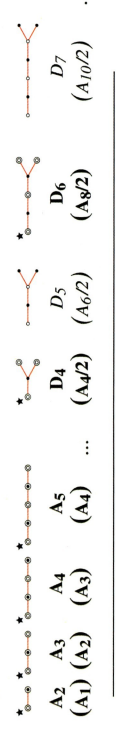

Exceptionals

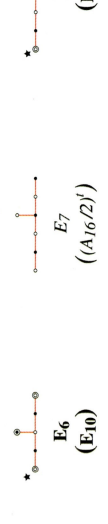

FIGURE 1. Classification of modules and subgroups of quantum $SU(2)$.

THE CLASSIFICATION OF SUBGROUPS OF QUANTUM $SU(N)$ 141

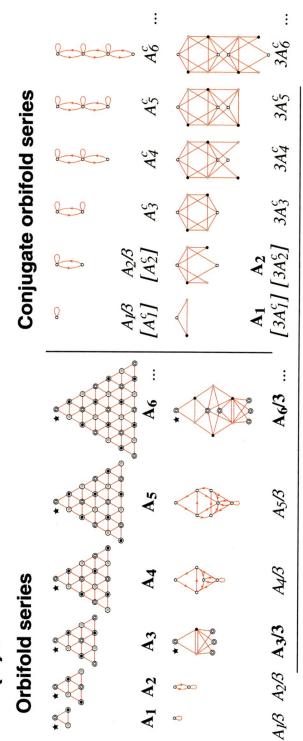
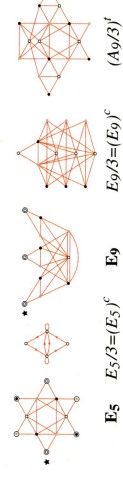

FIGURE 2. Classification of modules and subgroups of quantum $SU(3)$.

– annihilation mechanism (labeled by a planar diagram B or by a monomial in the e_i).

The formula given above for the coefficient ρ_B is additive, in the sense that all terms have the same sign. As a consequence, all basis elements have nonzero coefficients. The global sign is precisely the parity of the number of Jones projections in the monomial.

The above formula is directional, i.e. it gives a different expression when computed from right to left and when computed from left to right. The equality between the two expressions could lead to interesting and natural identities for basic (or $q-$)hypergeometric functions of higher type.

From the above formula we have developed a discrete analog of QFT, with -bra and -ket states which are half boxes and operators which are the middle part of a box, modulo symmetrizer relations, and we have found natural bases for each of these. The inner product is in this theory precisely the coefficient of the box computed above. It is remarkable, and unexpected, that the discrete operator product expansion of a product of two base elements has coefficients which are monomials. We have also extended the formula from the symmetrizer boxes as above, which are 2-gons, to general polygons.

We propose to develop the above theory in several natural directions. It should be possible to find global manifestations of this formula, i.e. to understand surface spaces, knot and 3-manifold invariants, as well as the 6j-symbols of $SU(2)_l$, in terms of the monomial coefficient expressions and the corresponding discrete form of QFT. In a different direction, there is a similar canonical basis for the projections corresponding to the projection onto the highest weight irreducible in a tensor product of generators of $\operatorname{Irr} SU(k)_l$, with planar diagrams consisting of hexagonal cells. The analogous coefficients of the highest weight projection of $SU(k)_l$ in the hexnet base are likely to have a very interesting mathematical structure.

2.5. The geometrization of quantum subgroups: a construction of weight lattices and roots from quantum subgroups.

Subfactors have been linked to many different branches of mathematics; however no geometrical structure of subfactors has been previously found. In particular links between subfactors connected to Coxeter ADE graphs on one side and the Lie algebras, Lie groups and quantum Lie groups built from the same Coxeter ADE graphs were missing. The ADE subfactors had their structure centered around the real valued Jones index, tensoring bimodules, composition of homomorphisms, and knot and 3-manifold invariants. The main structure of the ADE, i.e. simple unimodular, Lie groups is very geometrical with a weight lattice, roots, reflections and the Weyl group, and the deformation quantization.

We have found the natural link between the subgroups of quantum $SU(2)$ and the classical and quantum Lie groups, showing that the information for building a simple Lie group from copies of $SU(2)$ put together in a combinatorial way using the root lattice comes naturally from the fusion structure on representations of a quantum subgroup of $SU(2)$. The bridge between these two areas of research is a hitherto unobserved crystallographic property of homology theory: commuting squares and 6 term exact sequences are squares and hexagons in a lattice, i.e $A_1 \times A_1$ and A_2 root systems, and the 12 terms in the snake lemma form precisely an A_3 root system.

As a consequence, we have obtained a very elementary and natural construction of a canonical basis in the sense of Lusztig from an ADE graph. This elementary construction does not use in an over way quantum subgroups, although the mechanism behind it is based on the ADE subfactor and quantum subgroup.

The fact which shows that this link is indeed of a very general nature which transcends the quiver methods of Ringel and Lusztig is that the process works for our newly discovered quantum subgroups of $SU(3)$. More generally for any subgroup of a quantum semisimple Lie group at a root of unity we can associate in a natural way a finite dimensional weight lattice and a finite set of unimodular roots in it. The problem of associating roots and weights to the generalized Coxeter graphs had been set 15 years ago by Zuber and collaborators, although none of the attempts to make the vertices of such a graph into analogs of simple roots worked.

In our approach, even for the classical Lie groups, we obtain from the ADE graph all the roots at once, with no simple roots distinguished. It is precisely this fact that makes the process work in a very general framework.

For an ADE graph G with Coxeter number N consider the Cartesian product graded over $\mathbf{Z}_2$ of G (on the horizontal) with $\mathbf{Z}_{2N}$ (on the vertical), i.e. divide the vertices of G into even and odd vertices and retain from $\text{Vert}G \times \mathbf{Z}_{2N}$ only the pairs with the same parity. Equivalently, construct a Bratelli diagram on the whole ADE graph G and identify the bottom to the top after $2N$ levels, making the product graph $\mathcal{G}$ lie on the surface of a cylindrical band.

The graph $\mathcal{G}$ has now $|G|\,N$ vertices, exactly the number of roots (according to a theorem of Kostant) in the root system associated to the ADE graph G. We shall in fact **define the roots as the vertices of** $\mathcal{G}$. We now define the inner product between roots as follows. Denote by $\text{Path}^{(n)}_{(x,i),(y,j)}\mathcal{G}$ the (downward moving) paths between the vertices $(x,i),(y,j) \in \text{Vert}\mathcal{G} \subset \text{Vert}G \times \mathbf{Z}_{2N}$, of length $n \in \{0,\ldots,2N-1\}, n \equiv j-i \mod 2N$. Let $\text{HPath}^{(n)}_{(x,i),(y,j)}\mathcal{G}$ denote the Hilbert space with orthonormal basis $\text{Path}^{(n)}_{(x,i),(y,j)}\mathcal{G}$. Let μ denote the Perron Frobenius eigenvector of G, extended by $\mu((x,i)) = \mu(x)$ to $\mathcal{G}$ and let $\beta = 2\cos(\pi/N) = [2]$ be its eigenvalue. For $k < n$ denote by $c_k : \text{HPath}^{(n)}\mathcal{G} \to \text{HPath}^{(n-2)}\mathcal{G}$ the contraction operator, defined for a path $\xi = (\xi(1),\ldots,\xi(n))$ by

$$c_k(\xi) = \text{coef}\,\delta_{\xi(k),\xi(k+1)^{-1}}(\xi(1),\ldots,\widehat{\xi(k)},\widehat{\xi(k+1)},\ldots,\xi(n)),$$

in which $\text{coef} = \mu(s(\xi(k))^{-1/2}\mu(r(\xi(k))^{1/2}$, i.e. c_k cancels the edges $\xi(k),\xi(k+1)$ if they are inverse to each other, and is 0 otherwise. Then $\beta^{-1/2}c_k$ is a coisometry and $e_k = \beta^{-1}c_k^*c_k$ is a Jones projection. Define the **essential path subspace**

$$\text{EssPath}^{(n)}\mathcal{G} = \{\rho \in \text{HPath}^{(n)}\mathcal{G} : c_k(\rho) = 0 \text{ for any } k = 1,\ldots,n-1\}$$

consisting of the "non-repetitive" linear combinations of paths.

If we view the vertices of G as irreducibles of a subgroup or module S of $\text{Irr}\,SU(2)_{N-2}$ then paths of length n correspond to repeated tensor products n times with the generator σ_1 of $\text{Irr}\,SU(2)_{N-2}$ while essential paths correspond to tensor products with the irreducible $\sigma_n \subset \sigma_1^{\otimes n}$. The maximal length of essential paths is thus the level $l = N - 2$, since $\sigma_{N-1} = 0 \in \text{Irr}\,SU(2)_{N-2}$.

Let $N^{(y,j)}_{n,(x,i)} = \dim \text{EssPath}^{(n)}_{(x,i),(y,j)}\mathcal{G}$. Then $N^{(y,j)}_{n,(x,i)} = \dim \text{Hom}[\sigma_n \otimes x, y]$ is the fusion number corresponding to the module structure of the vertices of G over $\text{Irr}\,SU(2)_{N-2}$. Recall that $\sigma_{N-1} \in \text{Irr}\,SU(2)_{N-2}$ is killed by the cutoff. We shall

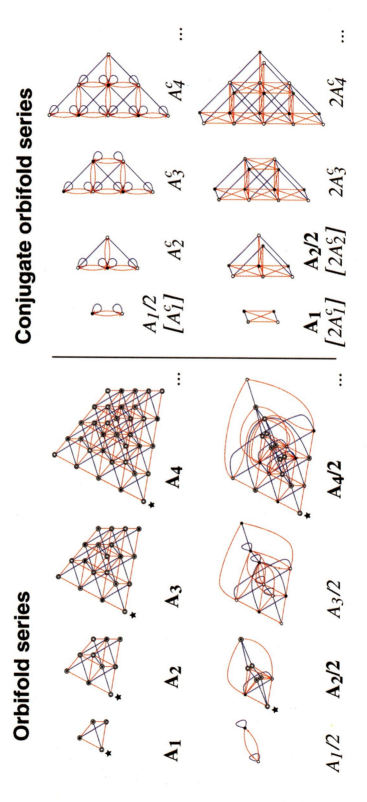

FIGURE 3

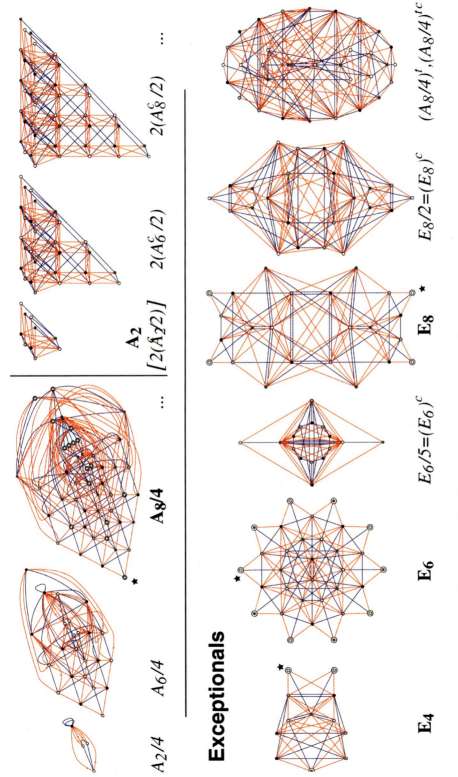

FIGURE 4. Classification of modules and subgroups of quantum $SU(4)$.

continue counting the irreducibles of $SU(2)_{N-2}$ by reflection around the killed level $N-1$, i.e. we take $\sigma_{N-1} = 0, \sigma_N = -\sigma_{N-2}, \sigma_{N+1} = -\sigma_{N-3}$, etc. Accordingly we let $N^{(y,j)}_{N-1+k,(x,i)} = -N^{(y,j)}_{N-1-k,(x,i)}$ and we further extend the fusion number $N^{(y,j)}_{n,(x,i)}$ to arbitrary $n \in \mathbf{Z}$ by replacing n with $n \mod 2N$. We have then $N^{(y,j)}_{-1+k,(x,i)} = -N^{(y,j)}_{-1-k,(x,i)}$.

We define the **inner product between roots** $(x,i), (y,j) \in \text{Vert}\mathcal{G}$ by

$$\langle (x,i), (y,j) \rangle = N^{(y,j)}_{j-i,(x,i)} + N^{(y,j)}_{i-j,(x,i)} = N^{(y,j)}_{j-i,(x,i)} - N^{(y,j)}_{j-i-2,(x,i)}$$

The above inner product extends by linearity to an inner product on the linear span of the roots; the nondegenerate quotient is positive definite and has dimension $|G|$. Although the fusion numbers $N^{(y,j)}_{j-i,(x,i)}$ can be as big as 6 for the graph E_8, the inner product $\langle (x,i), (y,j) \rangle$ always takes values in $\{-2, -1, 0, 1, 2\}$.

We have defined this way all the roots at once, entirely in terms of fusion on the subgroup or module S of $SU(2)_{N-2}$ (or elementarily in terms of essential paths on the graph.) In this setting the vertical shift by 2 is a Coxeter element of the Weyl group acting on roots.

A similarly simple construction gives the weight lattice as follows. For scalar valued functions on the vertices of the graph G define the Laplacian Δ_G as the sum of neighbors, i.e. the matrix of Δ_G is the adjacency matrix of G. Similarly $\Delta_{\mathbf{Z}_{2N}}$ is the Laplacian on $\mathbf{Z}_{2N}$. Call a $\mathbf{Z}$-valued function f on $\text{Vert}\mathcal{G} \subset \text{Vert}G \times \mathbf{Z}_{2N}$ **harmonic** if $\Delta_G(f) = \Delta_{\mathbf{Z}_{2N}}(f)$, i.e. if the sum of neighbors taken vertically equals the sum of neighbors taken horizontally. For instance for any fixed $(x,i) \in \text{Vert}\mathcal{G}$ the function $(y,j) \mapsto N^{(y,j)}_{j-i,(x,i)}$ is harmonic. The **weights are the integer valued harmonic functions** on $\text{Vert}\mathcal{G}$.

The remarkable fact about the above formulae is that they extend naturally to any subgroup or module S of any quantum Lie group $\mathbf{G}_l$ at a root of unity, such as the ones we have classified fro $SU(3)_l$ and $SU(4)_l$. Instead of having $\mathcal{G} \subset G \times_{\mathbf{Z}_2} \mathbf{Z} = G \times_{\mathbf{Z}_2} P_{SU(2)}$ where $P_{SU(2)}$ is the weight lattice of $SU(2)$ we let $\mathcal{G} \subset G \times_Z P_{\mathbf{G}}$ where the Cartesian product is graded over the center Z of $\mathbf{G}$.

The formula for the inner product between roots becomes, in the spirit of the Weyl character formula,

$$\langle (x,i), (y,j) \rangle = \sum_{w \in W} \varepsilon(w) N^{(y,j)}_{j-i+w\rho-\rho,(x,i)}$$

where W is the Weyl group and ρ the Weyl vector of the Lie group $\mathbf{G}$. All roots have square length $|W|$, i.e. the square length 2 for the classical unimodular lattice vectors comes from the fact that the Weyl group of $SU(2)$ has 2 elements.

This inner product is periodic, i.e. the points of $G \times_Z P_{\mathbf{G}}$ are naturally quotiented by a sublattice into a finite set of roots lying on a torus.

Thus for example from the first exceptional subgroup of $SU(3)_5$ one obtains 256 roots of square length 6 in a weight lattice of dimension 24.

The first nontrivial new root lattice corresponds to $SU(3)_1$, a graph with 3 points analogous to the graph $A_2 = SU(2)_1$ for $SU(2)$. The construction above yields from $SU(3)_1$ 16 roots of square length 6 in a lattice of dimension 6. A number theoretical formula for the theta function of this lattice has helped us identify it, using the online database of Conway, Sloane and Nebe as the lattice D_6^+, the union between D_6 and a translated copy of D_6. This lattice had not appeared before in representation theory (as opposed to D_8^+ which is the lattice E_8). The lattice D_6^+ exhibits in its new role a hexagonal symmetry not noticed before. This is a coincidence of small numbers (or rather small lattices); all the other lattices produced by our method from subgroups appear to be new.

For dimensions ≤ 8 the best sphere packing is provided by the classical root lattices, which from our point of view correspond to exceptional subgroups of quantum $SU(2)$. In higher dimensions, where good candidates are missing, the exceptional subgroups of quantum $SU(k)$, $k \geq 3$ could be candidates for optimal packing, a direction which we intend to investigate.

2.6. Homology and crystallography. In the previous paragraph we have described the way in which the roots and weight lattices of a simple Lie group arise out of fusion numbers for the irreducibles of a subgroup. It is in fact possible to go beyond counting and construct a natural basis of the universal enveloping algebra of a quantum simple Lie group.

Essential paths have a natural product, obtained by projecting onto the essential paths the concatenation of two paths. One extends this product to $V \otimes \text{EssPath}_{(x,i),(y,j)}^{(n)} \mathcal{G} \otimes \overline{W} = \text{EssPath}_{(x,i),(y,j)}^{(n)} \mathcal{G} \otimes \text{Hom}[V,W]$ where V, W are finite dimensional multiplicity vector spaces attached respectively to $(x,i), (y,j)$. We can now take kernels of maps; due to the presence of the essential path factor kernels have kernels again, and in 6 steps any exact sequence moves around the cylinder $2N$ levels and returns to the starting point.

Unimodular root systems are characterized by the fact that any 2 dimensional section is a square or a hexagon, and for us the squares correspond to commuting squares and the hexagons correspond to 6 term exact sequences. This points to the fact that in general homology theory appears to have a crystallographic aspect, manifested by the ubiquitous presence of 6-term exact sequences, but not of 5 or 7 terms exact sequences. The snake lemma, which connects 12 terms by means of 4 exact sequences with 6 terms each, is precisely the root system of type A_3, i.e. if we join the top to the bottom of the snake, the 4 exact sequences become the 4 hexagons joining the mid-edges of a cube. These observations appear to be new and deserve further investigation. This crystallographic aspect is essential for us as a bridge between an ADE subgroup of $SU(2)_l$ and the corresponding ADE Lie group.

An element of the canonical basis of the off diagonal universal enveloping algebra $\mathbf{U}^+ \cup \mathbf{U}^-$ is a formal exponential e^f where f is a function $f : \text{Vert}\mathcal{G} \to \mathbf{N}$. The product $e^f \circ e^g$ is defined as follows. Define vector spaces $F(x,i)$ of dimension $f((x,i))$ at every vertex (x,i) of $\mathcal{G}$, define similarly spaces $G(x,i)$ from g. Define

$$\text{Hom}[F,G] = \bigoplus_{(x,i),(y,j)} F(x,i) \otimes \text{EssPath}_{(x,i),(y,j)}^{(n)} \mathcal{G} \otimes \overline{G(y,j)} \ .$$

Construct now extensions H of G by F, i.e. find spaces $H(x,i)$ of dimension $h((x,i))$ at every vertex (x,i) of $\mathcal{G}$, and maps $\alpha \in \text{Hom}[F,H]$ and $\beta \in \text{Hom}[H,G]$ which form an exact sequence. Count the number of such extensions when the vector spaces are taken over a field with q elements and obtain (after a suitable normalization) a number $c_{f,g}^h(q)$ which is a polynomial in q. Then the product is

$$e^f \circ e^g = \sum_h c_{f,g}^h(q) \, e^h$$

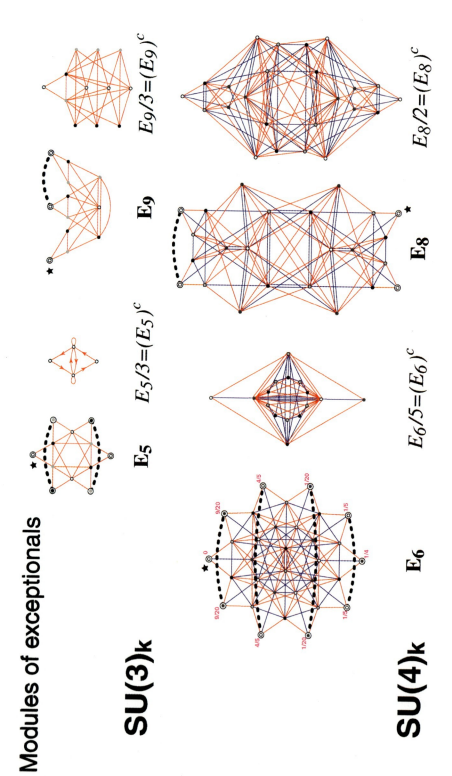

FIGURE 5. Modules of exceptionals.

Exceptionally twisted modules of orbifolds

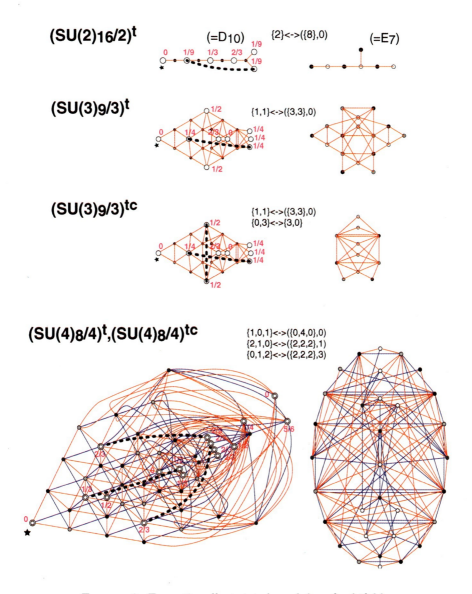

FIGURE 6. Exceptionally twisted modules of orbifolds.

This way we obtain directly the quantum deformation of the simple Lie groups. The snake lemma then shows that this product is associative. Another similar term takes care of the diagonal part $\mathbf{U}^0$.

3. Figures

The figures 1-12 describe our classification of the modules and subgroups of quantum $SU(N)$, $N = 2, 3, 4$ at the k-th roots of 1, for which the Young diagrams of the irreducible representations are constrained to less than N (=rank) rows and k (=level) columns. The graphs show the fusion with the generators of $SU(N)_k$, in a picture analogous to the McKay fusion graphs for the finite subgroups of $SU(2)$. They answer problems of Zuber concerning the classification of higher analogs of the Coxeter ADE graphs. In general we have shown that any WZW theory has a finite number of exceptionals.

Subgroups and modules. The modules of $SU(N)_k$ have a natural fusion with the irreducibles of $SU(N)_k$. Among modules, the subgroups have a naturally defined self-fusion; subgroups are titled bold. Each subgroup has a unit, which is starred. The subscript in a graph name denotes the level, except for $SU(2)_k$, for which the level notation is given in parentheses under the traditional name. Small level duplicates in the series are bracketed. Modules are interpreted in topological quantum field theories (TQFT) as new types of vertices, and also as boundary extensions of the theory. In the operator algebras bimodule picture they are new types of algebras and in 2-dimensional conformal field theory (CFT) they are boundary extensions.

Vertices. The vertices of the graphs are irreducible representations. The shades of gray from white to black indicate the grading. In CFT, vertices are primary operators, while in operator algebras the vertices are irreducible bimodules; in TQFT they label 1-dimensional edges.

Graphs. The red graph is the graph of tensoring with the standard irreducible σ_1 of $SU(N)_k$. The blue graph for $SU(4)_k$ is the graph of tensoring with $\sigma_2 = \sigma_1 \wedge \sigma_1$. Where necessary the edges of the graphs are oriented explicitly.

Self-connections. A necessary and sufficient condition for the existence of a module is given by a self-connection, which consists of a system of complex numbers called cells, satisfying certain local equations of cohomological nature. Cells reside in the triangles of the graphs and describe the composition of the intertwiners represented by edges. Cells are chosen up to a unitary gauge coming from the choice of intertwiners for the edges. The graphs and the self-connection determine a module completely. For $SU(2)_k$ the cells reside in the 2-gons of the graphs and the existence of the self-connection excludes the tadpoles. In the classical case, the existence of the self-connection on an $SU(N)$ graph is necessary and sufficient for the existence of a subgroup of $SU(N)$ with the corresponding irreducibles. The cells provide the data for a Hecke algebra representation yielding a solvable statistical mechanical plaques model for each graph.

Chirality. To each module M corresponds a subgroup G^+ called the chiral positive (or geometrically flat) part of M, a closed subsystem S of G^+ called the ambichiral set, and an automorphism θ called the ambichiral twist on S. The quantum automorphisms of the module M are a product of the flat part G^+ of M with the conjugate G^- of G^+, fibered over the ambichirals $S = G^+ \cap G^-$ with the twist θ. Among the ambichiral automorphisms of $SU(N)_k$, $N > 2$, the conjugation c replaces the generators σ_i of $SU(N)_k$ by their conjugates σ_{N-i}.

Exceptional twists. For each N there is an exceptional ambichiral twist, denoted by t, on an orbifold of $SU(N)_k$: on $SU(2)_{16}$ it gives $(SU(2)_{16})^t = E_7$; on $SU(3)_9$ it gives $(SU(3)_9)^t$, and on $SU(4)_8$ it gives $(SU(4)_8)^t$.

Orbifold series and exceptionals. $SU(N)_k$ is denoted by A_k. Its series subgroups are, for $SU(2)_k$: $A_k/2$ for $k = 0 \mod 4$; for $SU(3)_k$: $A_k/3$ for $k = 0 \mod 3$ and for $SU(4)_k$: $A_k/2$ for $k = 0 \mod 2$, $A_k/4$ for $k = 0 \mod 8$. The exceptional subgroups are denoted by E_k.

For $SU(2)_k$ the series $A_k/2 = D_{k/2+2}$ modules exist only for $k = 0 \mod 2$, and for $SU(4)_k$ the series $A_k/4$ modules and their conjugates $(A_k/4)^c = 2(A^c/2)_k$ exist only for $k = 0, 2, 6 \mod 8$.

The modules of $SU(N)_k$ arise from (i.e., have the chiral part G^+ equal to) the following subgroups: for $SU(3)_k$ and $k = 0 \mod 3$, $3A^c{}_k$ from $A_k/3$; for $SU(4)_k$ and $k = 0 \mod 8$, $2(A^c/2)_k$ from $A_k/4$; for $k = 0 \mod 2$, $2A^c{}_k$ from $A_k/2$; all other series modules from A_k. The notation of exceptional modules indicates the chiral part.

Conformal inclusions. All exceptional subgroups of $SU(2)_k$ and $SU(3)_k$ as well as some small orbifolds arise from conformal inclusions. It was conjectured that all exceptional subgroups and modular invariants come from conformal inclusions. For $SU(4)_k$ the exceptional $E8$ is not a conformal inclusion. The following subgroups arise from conformal inclusions: for $SU(2)_k$, D_4 from $SU(2)_4$ in $SU(3)_1$, E_6 from $SU(2)_{10}$ in $SO(5)_1$, and E_8 from $SU(2)_{28}$ in $(G_2)_1$; for $SU(3)_k$, $A_3/3$ from $SU(3)_3$ in $SO(8)_1$, E_5 from $SU(3)_5$ in $SU(6)_1$, E_9 from $SU(3)_9$ in $(E_6)_1$, and E_{21} from $SU(3)_{21}$ in $(E_7)_1$; for $SU(4)_k$, $2A^c{}_2 = A_2/2$ from $SU(4)_2$ in $SU(6)_1$, E_4 from $SU(4)_4$ in $SO(15)_1$, and E_6 from $SU(4)_6$ in $SU(10)_1$.

Modular invariants. Each module produces a modular invariant, which is a positive integer valued matrix, intertwiner of the modular group representation on the affine characters of $SU(N)_k$. The modular invariant is a manifestation of the fact that the corresponding TQFT and CFT are defined on the torus. The modular invariants for $SU(2)_k$ were classified by Capelli, Itzykson, and Zuber. The graphs for $SU(2)_k$ arose in our classification of small index subfactors. The $SU(3)_k$ modular invariants were classified by Gannon. The graphs for $SU(3)_k$ are a subset of the list proposed empirically by di Francesco and Zuber; the precise connection between graphs and modular invariants was an open problem. The natural interpretation of the graphs is part of our maximal atlas theory, which is a 3-dimensional TQFT analog of the Morita equivalence theory for 2-dimensional TQFT. The $SU(3)_k$ classification showed that different graphs can share the same modular invariant, and the exceptional E_8 of the $SU(4)_k$ classification showed that not all exceptional graphs come from conformal inclusions.

4. New directions of research

We have described in a very simple and natural way the construction of the simple Lie groups and of their quantum deformations from the quantum subgroups of $SU(2)$, and have developed a general theory of subgroups of quantum groups to a level which allowed the first classification results for $SU(3)$ and $SU(4)$, as well as general bounds on the level of exceptionals.

A first direction that should be investigated in this context is the construction of the irreducible representations of quantum groups.

The main question which follows is the **existence of higher analogs of the simple Lie groups**, constructed from the above lattices. The classical and quantum simple Lie groups, constructed by the above procedure from $SU(2)_l$ have a binary composition law related to the fact that the main building block of the classical root lattices, the hexagon $SU(2)_1$, is 2 dimensional. The higher analogs of the simple Lie groups obtained this way would likely model $SU(k)_1$ and the corresponding composition laws would be natural many-to-one laws. We have recently obtained experimental evidence for this mechanism.

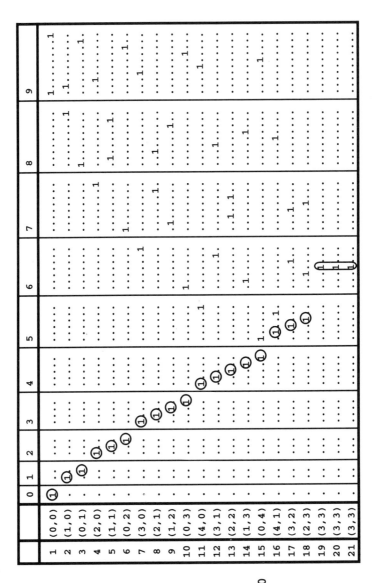

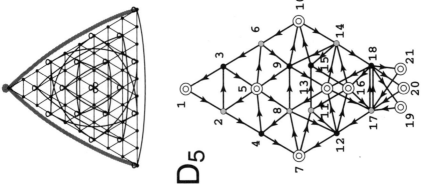

FIGURE 7. Modular ladder for the D_5 graph.

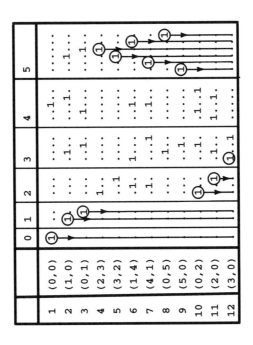
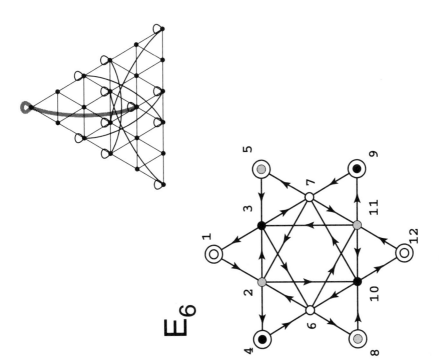

FIGURE 8. Modular ladder for the E_6 graph.

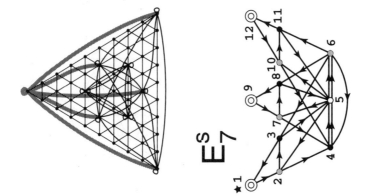

FIGURE 9. Modular ladder for the E_7^s graph.

THE CLASSIFICATION OF SUBGROUPS OF QUANTUM $SU(N)$ 155

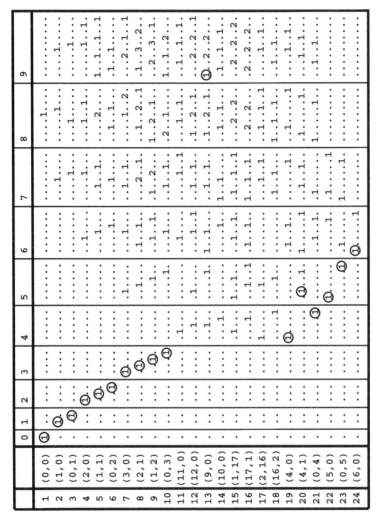

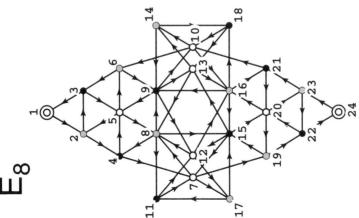

FIGURE 10. Modular ladder for the E_8 graph.

		10	11	12	13	14	15	16
1	(0,0)	....	.1.........	.....1......	....	....	....	....
2	(1,0)	1........1	.1....1..1	.1......1.1	1	....	....1.1.....	.1..1..1.1
3	(0,1)	.1........1	1...1.1...	.1..1.1..1	....	....1.1.....	....1.1.....	.1..1.1.1
4	(2,0)	.1..2......1	1..1.1.1...	.1..1.1..1.1	......1.....	....2..1....	.1..1.2..2.	.1..1.1.1.1
5	(1,1)	.1..2.1....	.2..1.1..2	1..1.1.2.1.1	....1..2.2.1	......1.3..1	.1..2.2.1.1	.1..2.2.2.1
6	(0,2)	1......2.1	..1.1.1.1.1	.1..1.1.1.1	....1..2.1.1	......1.2..	....2..2.1.1	.1..1.1.1.1
7	(3,0)	.1..2.1...	.1..3.1.1.1	1..2.2.2...	....1..2.2.1	..2..1.3..3.	....1.2.2.2.	.1..2.2.2.1
8	(2,1)	.1..3.2..2	..2..3.2.1	.2..3..3.2.	..1..2.4.3.1	..2..3.4..3.	..1..2.4.3..1	.1..2.2.2.1
9	(1,2)	.2..2..3..1	1..2.3..2.	.2..3.3.3.2	.1..3..4.2.1	..3..4.3..2.	..1..3.4.2.1	.1..3..3.3.2
10	(0,3)	.1..2..1...	.1..1.3..1	..2..2.2..1	.1..2.1.2...	....3.3.1..2	..2..2.2.1	.1..2.1.1
11	(11,0)	..2..1..1	..2.1.1...	.1..2..1...	.1..1.2..2.	.1..1.2..2.1	..1..1.2..2	.1..2.1.1
12	(12,0)	..2..2..2	..1..3.2.1	..2..3..2.	..2..3.3..2	..1..2.3.3..1	..1..2.3.3.2	.1..2.3..2.1
13	(9,0)	..2..2..2	..1..2.3.1	..2..3..2.1	..2..3.3..2	..1..3..3.2.1	..1..2.3.3.2.1	.1..2.3..2.1
14	(10,0)	..1..2 Ⓞ	..1..1.2..1	..1..2..2.1	..2..2.1.1	..1..2.3..2.1	..2..3..3.1	.1..1.2..1
15	(1,17)	.1..3..2.	..3..3.3.	..2..4..3.1	..2..3..3.3	..1..2..3.4..2	..2..3..3.3..1	.1..2..4.3..2.1
16	(17,1)	..2..3..1	..2..3..3	.1..3..4..2.	....3..3.3.2	..2..4..3.2.1	.1..3..3.3.2	1..2..3..4.2.1
17	(2,16)	..2..1..1	..1..2..1	.1..1.1..1	.1..1.2..2.	..1..1.2..2.	.1..1..2.2.1	.1..1.2..1.1
18	(16,2)	.1..1..2	..1..2..1	..1..1.2..1	.1..2..1.1.2	..2..2..1.1	..1..2..2.1	..1..1.2..1
19	(4,0)	...1..1..	..1..2..1	..2..1..1.	.1..1..1.2..	..1..1..1.1	.1..1..1.1.1	..2..1..1.1
20	(4,1)	.1..2..1	..2..2..	..2..2..2	..1..2..2.1	.1..2..2..1.1	.1..2..2.1.1	.1..2..2.2.1
21	(0,4)	..1..1..	..1..2..1	..1..1.2	..2..1..1.1	..1..1..1.1	..1..1..1.1	..1..1..1.1
22	(5,0)	..1..1..	..1..1..	..1..1..	...1..1..	..1..1..1.1	..1..1..1	..1..1.....1
23	(0,5)	..1..1..	..1..1..	..1..1..	..1..1..	..1..1..	.1..1..	1..1.....1
24	(6,0)	....	..1..1..	....	....	....1.......	...1.......	1..1.....1

FIGURE 11

FIGURE 12. Modular ladder for the E_8 graph.

The importance of such multi-nary laws is the following. Spaces of homomorphisms provide natural and universal models for quantum field theoretic Hilbert spaces, such as the ones which appear in conformal field theory. The main property is the **tensoriality**. The sections of ordinary vector bundles over a disjoint union of base sets are the direct sum of the sections over each set. In Quantum Field Theory, the direct sums must be replaced by tensor products. Tensoring over a common algebra is the gluing mechanism for overlapping base sets.

The model of a tensorial functor is $\text{Hom}[\alpha \otimes \beta, \gamma]$ for $\alpha, \beta, \gamma \in \text{Irr}\, G$ e.g. with G a group or quantum group, naturally associated to a triangle with edges labeled α, β, γ. This yields naturally 2-dimensional models, e.g. 2-dimensional conformal field theory (CFT). String theory then maps such 2-surfaces into higher dimensional spaces.

In order to make such models of QFT physical, which is a crucial problem in theoretical physics, a stronger approach would be to construct natural mathematical objects with tensorial behavior, which are attached, as functors, to the 3 or preferably 4 dimensional manifolds in general relativity. This problem is the **dimension barrier** in QFT. In our view, the impact of the last century's quantum mechanics has been strong enough to draw attention upon noncommutative mathematics, such as operator algebra, but the study of the higher dimensional tensorial objects required by quantum field theory is only beginning.

To break the dimension barrier, one must find for instance natural examples of $\text{Hom}[\alpha \otimes \beta \otimes \gamma, \delta]$, where now $\alpha, \beta, \gamma, \delta$ live on the faces of a tetrahedron and exhibit a tetrahedron symmetry and behavior.

This is in our view part of the potential of the higher analogs of the simple Lie groups for which we have now roots and weight lattices, built as described in these notes. The next step after the understanding of their structure would be the construction of higher dimensional QFT models, invariants for higher dimensional PL manifolds, etc.

5. Conclusions

It is our opinion that **the study of the group-like invariants of finite depth subfactors and of quantum subgroups is now in a situation similar to the study of simple Lie algebras a century ago**. While general rigidity results are now known, the discrete structure behind their classification is only beginning to appear in a few concrete cases. The further study of these structures, with strong motivation coming from the physics of quantum field theory (via topological quantum field theory), quantum gravity (from the spin models originated by Regge and Penrose to current spin foam models) and quantum chromodynamics (with the apparition of quantum $SU(3)$ in Connes's model based on noncommutative geometry) is likely to be an important direction of research in the coming century.

The inner structure of quantum symmetrizers, i.e. the closed form for coefficients of the Wenzl projectors, shows that unexpected new forms of discrete QFT appear inside mathematical objects which were considered well understood, and deserves further study.

A topic of particular interest would be the discovery of the higher analogs of the classical and quantum Lie groups, since these structures could impact constructive quantum field theory in a physical number of dimensions. This is also the most difficult of the topics to study, since the structures we are looking for, although very natural, are very different from any existing mathematical objects, but we have begun to obtain encouraging experimental data on it.

References

[1] A. Ocneanu, "*Paths on Coxeter Diagrams: From Platonic Solids and Singularities to Minimal Models and Subfactors*", AMS Fields Institute Monographs no. **13**, (1999), eds. B. V. Rajarama Bhat, George A. Elliott, Peter A. Fillmore, vol. "Lectures on Operator Theory".

DEPARTMENT OF MATHEMATICS, PENN STATE UNIVERSITY, MATHEMATICS DEPARTMENT, UNIVERSITY PARK, STATE COLLEGE, PA 16802, U.S.A.

E-mail address: adrian@math.psu.edu

Uses of Quantum Spaces

O. Ogievetsky

MAME

Contents

1. Introduction
2. Lie bialgebras
2.1. Deformation of the coproduct
2.1.1. Discrete groups
2.2. Lie algebras with an invariant scalar product
2.3. Belavin-Drinfeld triples
2.3.1. Maximal triples
3. Quantum spaces
3.1. GL type
3.2. Technics of checking the Poincaré series
3.2.1. Differential calculus and Poincaré series
3.3. Geometry of 3-dimensional quantum spaces
3.3.1. Gröbner base for $\mathcal{E}$
3.4. $sl_q(2)$ at roots of unity
3.4.1. Preliminaries
3.4.2. Formatted matrix algebras over graded rings
3.4.3. Matrix structure
3.4.4. Reduced function algebra
3.4.5. Centre
4. $\hat{R}$-matrices
4.1. Skew-invertibility
4.1.1. Generalities on Hopf algebras
4.1.2. Matrix picture
4.2. Ice $\hat{R}$-matrices
4.3. Construction of orthogonal and symplectic $\hat{R}$-matrices
5. Real forms
5.1. General linear quantum groups
5.2. Orthogonal and symplectic quantum groups
References

1991 *Mathematics Subject Classification.* 16W30, 16W35, 16W50, 81R50, 17B37, 20G42.

1. Introduction

Quite often, a group appears as a set of symmetries of some object - a set equipped with geometrical, algebraic or combinatorial data. The theory of quantum groups enlarges the notion of symmetry; a quantum group (often) describes "generalized symmetries" of an object. In the case of a linear (orthogonal, symplectic) quantum group, this object is a linear (orthogonal, symplectic) quantum space - an algebra with certain quadratic relations. A study of these underlying objects, the quantum spaces, helps to understand the structure of the quantum groups. In these lectures I will illustrate the role of the quantum spaces on two examples: non-perturbative effects in the theory of Yang-Baxter operators and real forms of quantum groups.

To talk about non-perturbative effects, one should explain first, what means "perturbative" or "deformational". This is the subject of the subsection 2.1. The initial data for a quantum deformation of a Lie algebra $\mathcal{L}$ is conveniently encoded in terms of another Lie algebra $D(\mathcal{L})$, the Drinfeld double of $\mathcal{L}$. The Lie algebra $D(\mathcal{L})$ has an invariant scalar product and I have included a subsection 2.2 on the structure of Lie algebras with an invariant scalar product.

For a semi-simple Lie algebra $\mathcal{L}$, the most important deformations are those which are called quasitriangular. They are classified by Belavin-Drinfeld triples. The subsection 2.3 contains some information about the combinatorics of the Belavin-Drinfeld triples.

In section 3, after a geometrical interpretation of the quantum deformations of Lie groups, we introduce an algebra of functions on a quantum group; a definition of GL-type quantum groups and quantum spaces is given in subsection 3.1. In subsection 3.2 we explain how to use a differential calculus on a GL-type quantum space for calculating the Poincaré series.

Subsection 3.3 is devoted to 3-dimensional quantum spaces. We exhibit an unexpected appearance of Yang-Baxter operators and give an example of a non-perturbative Yang-Baxter operator. We prove the Poincaré-Birkhoff-Witt theorem for the quantum space defined by this Yang-Baxter operator.

Subsection 3.4 deals with effects specific to quantum groups at roots of unity. We introduce a terminology of formatted matrix algebras over local graded rings, which is useful in the study of non semi-simple algebras. We describe the matrix structure of the reduced quantum enveloping algebra and the reduced function algebra for $sl_q(2)$.

Subsection 4.1 contains a summary of the theory of quasi-triangular Hopf algebras. In subsection 4.2 we classify Yang-Baxter matrices, which can have non-zero entries only at places where the pair of lower indices is a permutation of the pair of the upper ones. Subsection 4.3 gives a construction of the Yang-Baxter matrices for orthogonal and symplectic groups from the Yang-Baxter matrices for GL.

In section 5 we describe a method of classification of real forms of quantum groups. The method is based on the study of the corresponding quantum spaces.

Throughout the text, a sum over repeated indices is assumed. If $X = \{X_j^i\}$ and $Y = \{Y_j^i\}$ are two operators, the indices are summed as $(XY)_j^i = X_k^i Y_j^k$ in their product.

2. Lie bialgebras

A Hopf algebra H is a collection of data $\{H, m, \Delta, S, \epsilon\}$, where H is a vector space over a ground field k; $m : H \otimes H \to H$ a multiplication; $\Delta : H \to H \otimes H$ a comultiplication; $\epsilon : H \to k$ is a counit and $S : H \to H$ an antipode. For a precise formulation of various relations between these maps see *e.g.* [1]. Let me just remind that for a Hopf algebra H one knows how to build tensor products of representations and it is given universally by Δ; the counit gives rise to a trivial representation; the antipode is needed to build contragredient representations.

The classical examples of Hopf algebras are group algebras $k[G]$ of finite groups G and universal enveloping algebras $\mathcal{U}(\mathcal{L})$ of Lie algebras $\mathcal{L}$.

2.1. Deformation of the coproduct.

Let $\mathcal{L}$ be a Lie algebra over $\mathbb{C}$ and $\mathcal{U}$ its universal enveloping algebra. Denote by $\{X_i\}$ a basis of $\mathcal{L}$. The classical coproduct $\Delta_0 : \mathcal{U} \to \mathcal{U} \otimes \mathcal{U}$ is given on generators X_i by $\Delta_0 X_i = X_i \otimes 1 + 1 \otimes X_i$. The map Δ_0 is a coassociative homomorphism (coassociativity means $(\Delta_0 \otimes \mathrm{I})\Delta_0 = (\mathrm{I} \otimes \Delta_0)\Delta_0$ for the maps $\mathcal{U} \to \mathcal{U} \otimes \mathcal{U} \otimes \mathcal{U}$; here I is the identity map). In this subsection we shall study deformations of the coproduct Δ_0. A deformation of Δ_0 is, by definition, a coassociative homomorphism $\Delta : \mathcal{U} \to \mathcal{U} \otimes \mathcal{U}$,

$$\Delta(a) = \Delta_0(a) + \alpha \phi_1(a) + \alpha^2 \phi_2(a) + \ldots . \quad (2.1.1)$$

The right hand side is a formal power series in the parameter α, which is called a deformation parameter. The coefficients $\phi_k(a)$ are elements of $\mathcal{U} \otimes \mathcal{U}$.

Our task is to understand which deformations are "essential", in the sense that they cannot be removed by some redefinition of generators. Here is the answer modulo α^2.

Theorem 1. *Any deformation of Δ_0, after a change of generators, takes a form (in the first order in α)*

$$\Delta X_i = \Delta_0 X_i + \alpha \mu_i^{jk} X_j \otimes X_k. \quad (2.1.2)$$

The antisymmetric tensor μ_i^{jk} ($\mu_i^{jk} = -\mu_i^{kj}$) is a 1-cocycle with values in $\Lambda^2 \mathcal{L}$, $\mu \in Z^1(\mathcal{L}, \Lambda^2 \mathcal{L})$; explicitly:

$$N_{[ij]}^{[ab]} = \Gamma_{ij}^s \mu_s^{ab}, \quad (2.1.3)$$

where $N_{ij}^{ab} = \Gamma_{i\nu}^a \mu_j^{\nu b}$ and $[ab]$ means antisymmetrization in indices a and b, $t^{[ab]} = t^{ab} - t^{ba}$ for a tensor t^{ab}. Here Γ_{ij}^k are the structure constants of the Lie algebra $\mathcal{L}$, $[X_i, X_j] = \Gamma_{ij}^k X_k$.

Proof. Assume that Δ is a deformation of the classical coproduct Δ_0. On the generators X_i we have

$$\Delta(X_i) = \Delta_0(X_i) + \alpha \phi_i + \ldots \quad (2.1.4)$$

with some $\phi_i \in \mathcal{U} \otimes \mathcal{U}$, where dots denote higher powers in α.

The coassociativity, in order α^1, is equivalent to a following equation on ϕ_i in $\mathcal{U}^{\otimes 3}$

$$\phi_i \otimes 1 + (\Delta_0 \otimes \mathrm{I})\phi_i = 1 \otimes \phi_i + (\mathrm{I} \otimes \Delta_0)\phi_i, \quad (2.1.5)$$

I is the identity operator. The algebra $\mathcal{U}^{\otimes 3}$ is the enveloping algebra of $\mathcal{L} \oplus \mathcal{L} \oplus \mathcal{L}$. Let X_i, Y_i and Z_i be the generators of the first, second and third copies of $\mathcal{L}$, respectively. Then the equation (2.1.5) can be rewritten as

(2.1.6) $$\phi_i(X,Y) + \phi_i(X+Y,Z) = \phi_i(Y,Z) + \phi_i(X, Y+Z) \ .$$

The statement that Δ is a homomorphism reads, in terms of ϕ_i, as

(2.1.7) $$[X_i + Y_i, \phi_j] - [X_j + Y_j, \phi_i] = \Gamma^k_{ij} \phi_k$$

(the algebra $\mathcal{U} \otimes \mathcal{U}$ is the enveloping algebra of $\mathcal{L} \oplus \mathcal{L}$; X_i and Y_i are the generators of the first and second copies of $\mathcal{L}$).

Let $\sigma : \mathcal{U} \otimes \mathcal{U} \to \mathcal{U} \otimes \mathcal{U}$ be the flip, $\sigma(x \otimes y) = y \otimes x$. Decompose ϕ_i into symmetric and antisymmetric parts with respect to σ,

(2.1.8) $$\phi_i = s_i + a_i$$

with $\sigma(s_i) = s_i$ and $\sigma(a_i) = -a_i$.

Proposition 2. *If ϕ_i satisfies (2.1.5) and (2.1.7) then both s_i and a_i satisfy (2.1.5) and (2.1.7).*

Proof. We have $\Delta'(X_i) = \Delta_0(X_i) + \alpha \phi'_i + \ldots$, where $\phi'_i = \sigma(\phi_i) = s_i - a_i$.

If Δ is a coproduct then $\Delta' = \sigma \circ \Delta$ is a coproduct as well, so ϕ'_i satisfies (2.1.7),

(2.1.9) $$[X_i + Y_i, \phi'_j] + [X_j + Y_j, \phi'_i] = \Gamma^k_{ij} \phi'_k \ ,$$

and (2.1.5),

(2.1.10) $$\phi'_i \otimes 1 + (\Delta_0 \otimes \mathrm{I})\phi'_i = 1 \otimes \phi'_i + (\ \mathrm{I} \otimes \Delta_0)\phi'_i \ .$$

Take the sum and difference of (2.1.5) and (2.1.10) (respectively, (2.1.7) and (2.1.9)) to finish the proof. $\square$

In particular, each part (symmetric or antisymmetric) of ϕ_i alone defines a coproduct in order α^1.

Clearly, a redefinition of generators can change only the symmetric part of ϕ_i. We start by analyzing this case (the case of symmetric ϕ_i).

Proposition 3. *Assume that Δ is symmetric in order α^1, $\phi'_i = \phi_i$. Then the α^1 terms can be removed by a redefinition of generators.*

Proof. $\mathcal{U}$ is the algebra of polynomials in the generators X_i. It is filtered by the degree of polynomials, $F_k \mathcal{U}$ are polynomials of degree $\leq k$. The associated graded term $F_k \mathcal{U} / F_{k-1} \mathcal{U}$ is isomorphic to $S^k \mathcal{L}$, the symmetric power of $\mathcal{L}$. Any element $u \in \mathcal{U}$ has a well-defined "highest symbol": if $u \in F_k \mathcal{U} \setminus F_{k-1} \mathcal{U}$ (\ is the set-theoretic complement) then its highest symbol is the image of u in $S^k \mathcal{L}$. Denote by x_i the basis of commuting variables corresponding to generators X_i. The highest symbol is a homogeneous polynomial in a set of commuting variables x_i.

The algebra $\mathcal{U} \otimes \mathcal{U}$ is the enveloping algebra of $\mathcal{L} \oplus \mathcal{L}$, the highest symbols are homogeneous polynomials in two sets of variables, x_i and y_i.

Let f_i be the symbol of ϕ_i. Then f_i is a polynomial in two sets of variables, $f_i = f_i(x, y)$. The symmetry condition implies that $f_i(x, y) = f_i(y, x)$.

The coassociativity implies, in order α^1, an equation

(2.1.11) $$f(x+y, z) + f(x, y) = f(x, y+z) + f(y, z)$$

for each f_i.

Lemma 4. Let $f(x,y)$ be a homogeneous polynomial, symmetric with respect to the flip $x \leftrightarrow y$. The polynomial f satisfies (2.1.11) if and only if there exists a homogeneous polynomial $g(x)$ (a polynomial in only one set of variables x_i) such that

(2.1.12) $$f(x,y) = g(x+y) - g(x) - g(y) .$$

Proof. It is straightforward to see that $f(x,y) = g(x+y) - g(x) - g(y)$ satisfies (2.1.11).

Assume now that f satisfies (2.1.11). Let M be a total degree of f. If $M = 0$ then $f = c$ is a constant and it is enough to take $g = -c$. Assume that $M > 0$.

Applying $\frac{\partial}{\partial x_i}$ to (2.1.11) and evaluating at $x = 0$, we obtain an equation (after replacing $y \to x$ and $z \to y$)

(2.1.13) $$\partial_1^i f|_{x,y} = \partial_1^i f|_{0,x+y} - \partial_1^i f|_{0,x} ,$$

where ∂_1^i are the partial derivatives in the first set of variables.

Applying $\frac{\partial}{\partial z_i}$ to (2.1.11) and evaluating at $z = 0$, we obtain an equation

(2.1.14) $$\partial_2^i f|_{x,y} = \partial_2^i f|_{x+y,0} - \partial_2^i f|_{y,0} ,$$

where ∂_2^i are the partial derivatives in the second set of variables.

Since f is homogeneous of degree M, we have $(x_i \partial_1^i + y_i \partial_2^i) f = Mf$, which, together with (2.1.13) and (2.1.14), gives

(2.1.15) $$Mf = x_i \partial_1^i f|_{0,x+y} + y_i \partial_2^i f|_{x+y,0} - x_i \partial_1^i f|_{0,x} - y_i \partial_2^i f|_{y,0} .$$

The symmetry of f, $f(x,y) = f(y,x)$ implies that $\partial_i^1 f|_{x,y} = \partial_i^2 f|_{y,x}$. Therefore we can rewrite (2.1.15) in the form (2.1.11) with $g(x) = \frac{1}{M} x^i \partial_i^1 f|_{0,x}$. The proof of the Lemma 4 is finished. $\square$

We proved that for each i there exists $g_i(x)$ such that

(2.1.16) $$f_i(x,y) = g_i(x+y) - g_i(x) - g_i(y) .$$

Let $g_i^\vee(X)$ be an element whose highest symbol is $g_i(x)$. The combination $g_i^\vee(X+Y) - g_i^\vee(X) - g_i^\vee(Y)$ satisfies the equation (2.1.6). Therefore, an element $\phi_i(X,Y) - g_i^\vee(X+Y) + g_i^\vee(X) + g_i^\vee(Y)$, which has the filtration degree smaller than the degree of $\phi_i(X,Y)$, satisfies (2.1.6) as well, and we can apply the Lemma 4 again.

Repeating this process a needed number of times, we shall finally build a set of elements $\gamma_i \in \mathcal{U}$ such that

(2.1.17) $$\phi_i(X,Y) = \gamma_i(X+Y) - \gamma_i(X) - \gamma_i(Y) .$$

Let $X_i^\gamma = X_i - \alpha \gamma_i(X)$. It is straightforward to see that in the order α^1 the coproduct for the generators X_i^γ is classical, $\Delta(X_i^\gamma) = X_i^\gamma \otimes 1 + 1 \otimes X_i^\gamma$.

It is left to show that one can choose γ_i in such a way that the generators X_i^γ satisfy the same Lie algebraic relations as the original generator X_i. It will be so if and only if $[X_i, \gamma_j] - [X_j, \gamma_i] - \Gamma_{ij}^k \gamma_k = 0$ for all i and j.

Since Δ is a homomorphism, it follows immediately that elements $\gamma_{ij} = [X_i, \gamma_j] - [X_j, \gamma_i] - \Gamma_{ij}^k \gamma_k$ satisfy relations

(2.1.18) $$\gamma_{ij}(X+Y) = \gamma_{ij}(X) + \gamma_{ij}(Y) .$$

Equation (2.1.18) implies that the functions γ_{ij} are linear, $\gamma_{ij}(X) = \gamma_{ij}^k X_k$.

We shall need a short digression into the general theory of universal enveloping algebras (see, *e.g.* [**2**]).

An element $u \in \mathcal{U}$ can be uniquely decomposed into a sum

(2.1.19) $$u = \mathrm{symb}^0(u) + \mathrm{symb}^1(u) + \cdots + \mathrm{symb}^d(u) \; ,$$

where d is the filtration degree of u and $\mathrm{symb}^A(u) = c^{i_1\cdots i_A} X_{i_1} \ldots X_{i_A}$ for some completely symmetric tensor $c^{i_1\cdots i_A}$. Elements of the form $c^{i_1\cdots i_A} X_{i_1} \ldots X_{i_A}$ with a completely symmetric $c^{i_1\cdots i_A}$ form a subspace $\mathcal{U}^A \subset \mathcal{U}$ and the above decomposition of u implies that $\mathcal{U}$ is a direct sum of $\mathcal{U}^A$, $\mathcal{U} = \oplus_{A=0}^\infty \mathcal{U}^A$. Each $\mathcal{U}^A$ is a $\mathcal{L}$-module (that is, commutators of generators X_i with $\mathrm{symb}^A(u)$ are again in $\mathcal{U}^A$); in other words,

(2.1.20) $$\mathrm{symb}^A([X_i, u]) = [X_i, \mathrm{symb}^A(u)] \; .$$

The module $\mathcal{U}^A$ is isomorphic to the symmetric power $S^A \mathcal{L}$.

Let $\gamma_j = \sum \mathrm{symb}^A(\gamma_j)$ be a decomposition of the form (2.1.19) of the element γ_j. We have seen that γ_{ij} is in $\mathcal{U}^1$ for each i and j. It follows then from (2.1.20) that $\gamma_{ij} = [X_i, \mathrm{symb}^1(\gamma_j)] - [X_j, \mathrm{symb}^1(\gamma_i)] - \Gamma_{ij}^k \mathrm{symb}^1(\gamma_k)$. Therefore, $[X_i, \tilde{\gamma}_j] - [X_j, \tilde{\gamma}_i] - \Gamma_{ij}^k \tilde{\gamma}_k = 0$ for all i and j, where $\tilde{\gamma}_i = \gamma_i - \mathrm{symb}^1(\gamma_i)$. Therefore, the elements $\tilde{X}_i^\gamma = X_i^\gamma - \alpha \tilde{\gamma}_i(X)$ satisfy the same Lie algebraic relations as the original generators X_i, $[\tilde{X}_i, \tilde{X}_j] = \Gamma_{ij}^k \tilde{X}_k$.

Moreover, since for an element $g^1 \in \mathcal{U}^1$, the combination $g^1(X+Y) - g^1(X) - g^1(Y)$ vanishes, the elements $\tilde{\gamma}_i = \gamma_i - \mathrm{symb}^1(\gamma_i)$ still verify (2.1.16). Therefore, as before, the coproduct for the elements $\tilde{X}_i$ is classical.

Thus, the elements $\tilde{X}_i$ provide the needed redefinition of the generators X_i. The proof of the Proposition 3 is finished. $\square$

Using, if necessary, the redefinition of the Proposition 3, we get rid of the symmetric part of ϕ_i. Assume therefore that ϕ_i is antisymmetric. Again, let f_i be the highest symbol of ϕ_i. The symmetry condition is now $f_i(x,y) = -f_i(y,x)$. As before, the coassociativity implies, in order α^1, the equation (2.1.11) for each i.

Proposition 5. *Let $f(x,y)$ be a homogeneous polynomial, antisymmetric with respect to the flip $x \leftrightarrow y$. Assume that the polynomial f satisfies (2.1.11). Then*

(2.1.21) $$f(x,y) = \nu^{jk} x_j y_k$$

for some antisymmetric tensor ν, $\nu^{jk} = -\nu^{kj}$.

Proof. The derivatives of f satisfy equations (2.1.13) and (2.1.14). There is one more equation which we didn't need for the Lemma 4. It is obtained by applying $\frac{\partial}{\partial y_i}$ to (2.1.11) and evaluating at $y = 0$ (we change variables, $z \to y$)

(2.1.22) $$\partial_1^i f|_{x,y} + \partial_2^i f|_{x,0} = \partial_2^i f|_{x,y} + \partial_1^i f|_{0,y} \; .$$

The antisymmetry of f, $f(x,y) = -f(y,x)$, implies $\partial_i^1 f|_{x,y} = -\partial_i^2 f|_{y,x}$. Substituting $\partial_1^i f|_{x,y}$ and $\partial_2^i f|_{x,y}$ from (2.1.13) and (2.1.14) into (2.1.22) and using the antisymmetry, we find

(2.1.23) $$\partial_1^i f|_{0,x+y} = \partial_1^i f|_{0,x} + \partial_1^i f|_{0,y} \; .$$

Thus, $\partial_1^i f|_{0,x}$ is a linear function. Substituting (2.1.23) into (2.1.13), we find $\partial_1^i f|_{0,x+y} = \partial_1^i f|_{0,y}$. Thus, $\partial_1^i f|_{x,y}$ is a linear function which depends on the second set of variables only. In other words, $\partial_1^i f|_{x,y} = \nu_1^{ij} y_j$.

Similarly, $\partial_2^i f|_{x,y}$ is a linear function which depends on the first set of variables only, $\partial_2^i f|_{x,y} = \nu_2^{ij} x_j$.

The antisymmetry, $\partial_i^1 f|_{x,y} = -\partial_i^2 f|_{y,x}$, implies that $\nu_1^{ij} = -\nu_2^{ij}$. Let $\nu^{ij} = \nu_1^{ij}$. Then

(2.1.24) $$\partial_1^i f|_{x,y} = \nu^{ij} y_j \quad \text{and} \quad \partial_2^i f|_{x,y} = -\nu^{ij} x_j .$$

Since the derivatives of f are homogeneous of degree 1, the function f itself is homogeneous of degree 2. So $2f = (x_i \partial_1^i + y_i \partial_2^i) f$. Substituting expressions (2.1.24), we find

(2.1.25) $$2f = x_i \nu^{ij} y_j - y_i \nu^{ij} x_j = \nu^{[ij]} x_i y_j ,$$

where the square brackets mean antisymmetrization, $\nu^{[ij]} = \nu^{ij} - \nu^{ji}$ and the assertion of the Proposition 5 follows. □

After the Propositions 2, 3 and 5 it is only left to check a condition that Δ is a homomorphism in the first order in α. A straightforward calculation gives the cocycle condition (2.1.3). The proof of the Theorem 1 is finished. □

Remark. It is not necessary to assume that the ground field is $\mathbb{C}$. The Theorem 1 holds for an arbitrary field of characteristic 0 (and it is not true if the characteristic is different from 0).

Repeating the proof of the Proposition 3 consecutively in powers of α, one obtains a version of the Milnor-Moore theorem (for its general formulation see, e. g. [**3**]):

Corollary 6. A formal (i.e. given by a formal power series in α) cocommutative deformation of the coproduct on a universal enveloping algebra is always trivial, that is, it can be removed by a formal redefinition of generators.

In the rest of this subsection we explain what happens in the next order in α, in the α^2-terms.

It turns out that the consistency in the α^2 terms imposes new conditions on μ_i^{jk}.

Assume that we can extend the deformation (2.1.2) to α^2-terms,

(2.1.26) $$\Delta X_i = \Delta_0 X_i + \alpha \mu_i^{jk} X_j \otimes X_k + \alpha^2 \psi_i ,$$

and Δ is coassociative up to α^3. Coassociativity implies

(2.1.27) $$\psi_i \otimes 1 + (\Delta_0 \otimes \text{id})\psi_i + \mu_i^{abc} X_a \otimes X_b \otimes X_c$$
$$= 1 \otimes \psi_i + (\text{id} \otimes \Delta_0)\psi_i + \mu_i^{cba} X_a \otimes X_b \otimes X_c ,$$

with a notation: $\mu_i^{abc} = \mu_i^{jc} \mu_j^{ab}$ for a tensor μ_i^{jk}.

To cancel the μ_i^{abc}-terms, one has (in order to get trilinear in X expressions) to choose ψ_i in the form

(2.1.28) $$\psi_i = A_i^{abc} X_a X_b \otimes X_c + B_i^{abc} X_c \otimes X_a X_b .$$

with tensors A_i^{abc} and B_i^{abc} symmetric in indices $\{a,b\}$. In the next formulas, the lower index is omitted.

Coassociativity (2.1.27) gives

(2.1.29) $$2(B^{bca} - A^{abc}) = \mu^{abc} + \mu^{bca} .$$

Exchange b and c in (2.1.29) and subtract from (2.1.29), taking into account the symmetry of B^{abc} in $\{a,b\}$:

(2.1.30) $$2(A^{acb} - A^{abc}) = \mu^{bca} + J^{abc} ,$$

where

(2.1.31) $$J^{abc} = \mu^{abc} + \mu^{bca} + \mu^{cab} .$$

The tensor J^{abc} is totally antisymmetric. Under the exchange $a \leftrightarrow b$, eqn. (2.1.30) becomes

(2.1.32) $$2(A^{bca} - A^{bac}) = \mu^{acb} + J^{bac} .$$

Under the exchange $a \leftrightarrow c$, eqn. (2.1.30) becomes

(2.1.33) $$2(A^{cab} - A^{cba}) = \mu^{bac} + J^{cba} .$$

The combination (2.1.30) - (2.1.32) - (2.1.33) gives (due to the symmetry of A^{abc} in $\{a,b\}$)

(2.1.34) $$0 = 4J^{abc} .$$

This is the Jacobi identity (2.1.40) for μ.

Now from 6 permutations of $\{a,b,c\}$ one gets only two equations

(2.1.35) $$2(A^{acb} - A^{abc}) = \mu^{bca} ,$$

(2.1.36) $$2(A^{bca} - A^{bac}) = \mu^{acb} .$$

The tensor A^{abc}, being already symmetric in the first two indices, can have two types of symmetry, corresponding to Young diagrams ⊟ and ☐☐☐ .

The totally symmetric part (the diagram ☐☐☐) of A cannot be defined by (2.1.35)-(2.1.36), it is arbitrary. The part, corresponding to the diagram ⊟ satisfies

(2.1.37) $$A^{abc} + A^{bca} + A^{cab} = 0 .$$

Together, eqs. (2.1.35), (2.1.36) and (2.1.37) can be easily solved and we conclude (taking into account the totally symmetric part) that the general solution for A is

(2.1.38) $$A^{abc} = \frac{1}{6}(\mu^{cba} + \mu^{cab}) + \chi^{abc}$$

with totally symmetric χ^{abc}.

From (2.1.29) it follows then that

(2.1.39) $$B^{bca} = \frac{1}{6}(\mu^{abc} + \mu^{acb}) + \chi^{abc} .$$

In particular, $A^{abc} = B^{abc}$.

The totally symmetric part χ^{abc} can be removed by a redefinition (solve for g the equation (2.1.12) with $f(x,y) = \chi^{abc}(x_a x_b y_c + x_c y_a y_b)$).

We conclude that the coassociative extension of Δ to the α^2-terms is possible only if the Jacobi identity for μ is satisfied,

(2.1.40) $$\mu_i^{abc} + \text{cycle } in\ (a,b,c) = 0\ .$$

This extension has the form

(2.1.41)
$$\Delta X_i = \Delta_0 X_i + \alpha \mu_i^{jk} X_j \otimes X_k + \frac{\alpha^2}{6}(\mu_i^{cba} + \mu_i^{cab})(X_a X_b \otimes X_c + X_c \otimes X_a X_b)\ .$$

In general, Δ does not preserve the original commutation relations $[X_i, X_j] = \Gamma_{ij}^k X_k$, the multiplication structure of $\mathcal{U}$ has also to be deformed in the order α^2.

Exercise. The map Δ preserves the following relations [4]

(2.1.42) $$[X_i, X_j] = \Gamma_{ij}^k X_k + \frac{\alpha^2}{18} \mu_i^{sa} \mu_j^{tb} \Gamma_{st}^c X_{(a} X_b X_{c)}\ ,$$

with round brackets in $X_{(a} X_b X_{c)}$ denoting the symmetrization, $X_{(i_1} X_{i_2} X_{i_3)} = \sum X_{i_{\sigma(1)}} X_{i_{\sigma(2)}} X_{i_{\sigma(3)}}$, the sum is over all permutations $\sigma \in S_3$.

Given the multiplication (2.1.42) and the comultiplication (2.1.41), one needs to know the counit and the antipode to complete the Hopf algebra structure.

Exercise. The counit ϵ stays undeformed,

(2.1.43) $$\epsilon(X_i) \equiv 0 \mod \alpha^3\ ;$$

for the antipode mod α^3 one has

(2.1.44) $$S(X_i) = -X_i + \frac{1}{2}\alpha M_i^a X_a + \frac{1}{4}\alpha^2 \mu_i^{bl} M_l^a \Gamma_{ab}^c X_c\ ,$$

where $M_a^v = -\mu_a^{vc}\Gamma_{ck}^k - \mu_k^{kj}\Gamma_{ja}^v \equiv \mu_a^{st}\Gamma_{st}^v$. (Note that S stays linear in generators.)

It is known today that in higher orders in α no further restriction on μ appears; in other words, if μ_i^{jk} satisfies the Jacobi and cocycle conditions, there exist, as formal power series in α, the multiplication, which begins as (2.1.42), and the comultiplication, which begins as (2.1.41) (and the counit and antipode). Moreover, there exists such deformation that each term in the formal power series (for the multiplication, comultiplication and the antipode) is expressible in terms of the tensors μ_i^{jk} and Γ_{ij}^k only.

2.1.1. *Discrete groups.* The situation with discrete groups is different. It is an easy exercise to analyze the formal deformations of the coproduct for the group algebras of discrete groups. In contrast to the case of universal enveloping algebras, the result is trivial.

Let G be a discrete group, $U = \mathbb{C}[G]$ its group algebra over complex numbers.

Theorem 7. U does not admit a nontrivial deformation of the standard coproduct.

Proof. Assume that there is a first order deformation of a coproduct,

(2.1.45) $$\Delta g = g \otimes g + \alpha \sum_{k,l} C_g^{k,l} k \otimes l\ ,$$

where $\alpha^2 = 0$.

(*i*) The coassociativity condition in the first order in α gives

(2.1.46) $$\sum_{k,l} C_g^{k,l}(k\otimes l\otimes g + k\otimes k\otimes l - g\otimes k\otimes l - k\otimes l\otimes l) = 0 \ .$$

Collecting terms $a\otimes b\otimes c$ with fixed a and c, $a\neq g$ and $c\neq g$, one finds

(2.1.47) $$C_g^{a,c}a = C_g^{a,c}c \ .$$

This holds for all a and c different from g. Therefore, only $C_g^{g,g}$, $C_g^{k,g}$, $C_g^{g,k}$ and $C_g^{k,k}$ might differ from 0.

Now the condition (2.1.46) becomes

$$\sum_{k\neq g} \{C_g^{g,k}(g\otimes k\otimes g - g\otimes k\otimes k) + C_g^{k,g}(k\otimes k\otimes g - g\otimes k\otimes g)$$

(2.1.48) $$+ C_g^{k,k}(k\otimes k\otimes g - g\otimes k\otimes k)\} = 0 \ .$$

This implies that there is a set of constants B_g^k for $k\neq g$, s. t.

(2.1.49) $$C_g^{g,k} = C_g^{k,g} = B_g^k \ , \ C_g^{k,k} = -B_g^k \ , \ k\neq g \ .$$

This solves the coassociativity condition. Thus, we have

(2.1.50) $$\Delta g = (1+\alpha c_g)g\otimes g + \alpha \sum_{k\neq g} B_g^k(g\otimes k + k\otimes g - k\otimes k) \ .$$

(*ii*) The condition that Δ is a homomorphism implies (in the first order in α):

(2.1.51) $$B_h^{g^{-1}k} + B_g^{kh^{-1}} = B_{gh}^k \text{ and } c_{gh} = c_g + c_h \ .$$

Let

(2.1.52) $$g' = (1+\alpha c_g)g + \alpha \sum_{k\neq g} B_g^k k \ .$$

A direct calculation shows that (2.1.51) is exactly the condition saying that $g\to g'$ is an algebra homomorphism, $g'h' = (gh)'$.

Again a straightforward calculation shows that

(2.1.53) $$\Delta g' = g'\otimes g' \ .$$

Therefore, given a deformation of the standard coproduct, we can explicitly construct an isomorphism with the original bialgebra. The proof is finished. □

In the same way as the Corollary 6 followed from the Proposition 3, we obtain the information about the formal deformations in this case.

Corollary 8. At formal level, all deformations of the coproduct for the group algebras of discrete groups are trivial.

2.2. Lie algebras with an invariant scalar product. We have seen in the previous subsection that the essential role in the theory of deformations of the coproduct on universal enveloping algebras is played by a tensor μ_i^{jk}. All the conditions on the tensor μ are expressed in terms of the Lie algebra itself, without any reference to the deformation theory. The relevant classical notion is a "Lie bialgebra".

Definition. A Lie bialgebra is a Lie algebra $\mathcal{L}$ equipped with a map $\delta: \mathcal{L} \to \Lambda^2 \mathcal{L}$, $\delta X_i = \mu_i^{jk} X_j \otimes X_k$, where the tensor μ (antisymmetric in the upper indices) satisfies the Jacobi identity and belongs to $Z^1(\mathcal{L}, \Lambda^2 \mathcal{L})$.

Both Γ and μ satisfy the Jacobi identity. The condition $\mu \in Z^1$, written explicitly as (2.1.3), is symmetric in $\mu \leftrightarrow \Gamma$. So the notion of the Lie bialgebra is self-dual (like the notion of the Hopf algebra). In other words, if $\mathcal{L}$ is a Lie bialgebra then there is a Lie bialgebra structure on the dual space $\mathcal{L}^*$, the roles of Γ and μ being interchanged. There is an object which explicitly realizes this symmetry between μ and Γ. It turns out (*Exercise*: verify it) that all the data for a Lie bialgebra can be conveniently expressed as the Jacobi identity for a larger Lie algebra with generators X_i and X^i, satisfying

(2.2.1) $$[X_i, X_j] = \Gamma_{ij}^k X_k \quad, \quad \text{for generators of } \mathcal{L},$$

(2.2.2) $$[X^i, X^j] = \mu_k^{ij} X^k \quad, \quad \text{for generators of } \mathcal{L}^*,$$

(2.2.3) $$[X_i, X^j] = -\Gamma_{ik}^j X^k + \mu_i^{jk} X_k.$$

This Lie algebra is called a Drinfeld double of the Lie bialgebra $\mathcal{L}$ and denoted $D\mathcal{L}$. As a vector space, $D\mathcal{L}$ is isomorphic to $\mathcal{L} \oplus \mathcal{L}^*$.

Definition. A scalar product $\langle x, y \rangle$ on a Lie algebra $\mathcal{L}$ (i.e. a nondegenerate symmetric pairing $\mathcal{L} \otimes \mathcal{L} \to k$) is called invariant if $\langle [x,y], z \rangle = \langle x, [y,z] \rangle$ for all $x, y, z \in \mathcal{L}$.

Example: The Killing form on a semi-simple Lie algebra is invariant.

The natural pairing between $\mathcal{L}$ and $\mathcal{L}^*$, given by

(2.2.4) $$\langle X^i, X^j \rangle = 0, \ \langle X_i, X_j \rangle = 0, \ \langle X_i, X^j \rangle = \delta_i^j$$

is an invariant scalar product on $D\mathcal{L}$. Moreover, the commutation relations between X_i and X^j can be reconstructed (with relations (2.2.1) and (2.2.2) being given) from the demand of invariance of the natural pairing. Indeed, let $[X_i, X^j] = A_{ik}^j X^k + B_i^{jk} X_k$. Then

$$-A_{ia}^j = \langle [X^j, X_i], X_a \rangle = \langle X^j, [X_i, X_a] \rangle = \langle X^j, \Gamma_{ia}^b X_b \rangle = \Gamma_{ia}^j$$

and similarly for B_i^{jk}.

Definition. A set of data $\{\mathfrak{g}, \mathcal{L}_1, \mathcal{L}_2\}$ where $\mathfrak{g}$ is a Lie algebra with an invariant scalar product, $\mathcal{L}_1$ and $\mathcal{L}_2$ are isotropic Lie subalgebras of dimension $= \frac{\dim \mathcal{L}}{2}$ and $\mathfrak{g} = \mathcal{L}_1 \oplus \mathcal{L}_2$ is called a Manin triple.

A Lie bialgebra $\mathcal{L}$ defines a Manin triple $\{D\mathcal{L}, \mathcal{L}, \mathcal{L}^*\}$. Conversely, a Manin triple $\{\mathfrak{g}, \mathcal{L}_1, \mathcal{L}_2\}$ defines a Lie bialgebra $\mathcal{L}_1$. From this perspective, the study of Lie bialgebras splits into two parts: Lie algebras with an invariant scalar product; their maximal isotropic subalgebras. I will shortly comment on the first part.

Denote by $\{\mathfrak{g}, \phi\}$ a Lie algebra $\mathfrak{g}$ with an invariant scalar product ϕ ($\phi(x,y) = \langle x, y \rangle$). The pair $\{\mathfrak{g}, \phi\}$ is called indecomposable if it cannot be represented as a direct sum $\{\mathfrak{g}_1, \phi_1\} \oplus \{\mathfrak{g}_2, \phi_2\}$.

Example of $\{\mathfrak{g}, \phi\}$. Let $\mathcal{M}$ be a Lie algebra with generators X_i. Let $\mathfrak{g} = \mathcal{M} \ltimes \mathcal{M}^*$ (the semi-direct product with respect to the coadjoint action). Then the scalar product

(2.2.5) $$\langle X^i, X^j \rangle = 0 \ , \ \langle X_i, X_j \rangle = \xi_{ij} \ , \ \langle X_i, X^j \rangle = \delta_i^j \ ,$$

where $\xi(X_i, X_j) = \xi_{ij}$ is an arbitrary bilinear symmetric form, is an invariant scalar product.

Generalization of this example. Let $\{W, \phi\}$ be a Lie algebra with an invariant scalar product. Suppose that a Lie algebra $\mathfrak{g}$ acts on W by derivations, $T_a[x,y] = [T_a x, y] + [x, T_a y]$, where T is the action, $T : \mathfrak{g} \otimes W \to W$, $a \otimes w \mapsto T_a(w)$; suppose that the operators T_a, $a \in \mathfrak{g}$, are antisymmetric with respect to the scalar product on W, $\phi(T_a x, y) = -\phi(x, T_a y)$ for all $x, y \in W$ and $a \in \mathfrak{g}$.

Exercise. Show that the map $\beta : \Lambda^2 W \to \mathfrak{g}^*$ defined by $\langle a, \beta(x,y) \rangle = \phi(T_a x, y)$, where $\langle \cdot, \cdot \rangle$ is the natural pairing between $\mathfrak{g}$ and $\mathfrak{g}^*$, is a 2-cocycle, $\beta \in Z^2(W, \mathfrak{g}^*)$ ($\mathfrak{g}^*$ is considered here as a trivial W-module).

As a 2-cocycle, β defines a central extension of W by $\mathfrak{g}^*$. In other words, the bracket

(2.2.6) $$[x, y] = [x, y]_W + \beta(x, y)$$

where $[x, y]_W$ is the commutator of x and y in the Lie algebra W, defines a Lie algebra structure on $W \oplus \mathfrak{g}^*$. Denote this Lie algebra by $\tilde{W}$.

Exercise. For $a \in \mathfrak{g}$ $x \in W$ and $f \in \mathfrak{g}^*$ let

(2.2.7) $$\tilde{T}_a(x + f) = T_a x + \mathrm{ad}_a^* f \ ,$$

where ad^* is the coadjoint action. Show that the formula (2.2.7) defines an action of $\mathfrak{g}$ on $\tilde{W}$.

We have therefore a Lie algebra structure on the space $A = \mathfrak{g} \oplus W \oplus \mathfrak{g}^*$: a semi-direct product $\mathfrak{g} \ltimes \tilde{W}$ with respect to the action (2.2.7).

Define a scalar product ϕ_A on A: the pairings between the generators of $\mathfrak{g}$ and $\mathfrak{g}^*$ are given by (2.2.5); the restriction of ϕ_A on W is ϕ; all the other pairings are 0.

Exercise. The scalar product ϕ_A is invariant.

The Lie algebra A with the scalar product ϕ_A is called the double extension of $\{W, \phi\}$ by S (and the action of S on W).

Theorem ([5]). *If a Lie algebra with an invariant scalar product is not simple or 1-dimensional then it is either decomposable or a double extension. Moreover, one can always choose $\mathfrak{g}$ to be simple or 1-dimensional.*

This theorem gives a way to construct higher-dimensional Lie algebras with an invariant scalar product from lower-dimensional ones. However, this is not a classification.

Example of a nontrivial double extension: $\mathfrak{g} = so(n)$, W is the n-dimensional fundamental representation of $\mathfrak{g}$; consider W as an abelian Lie algebra. The cocycle, giving a bracket on $W \oplus \mathfrak{g}^*$ is given by the natural map $\beta : W \wedge W \to \mathfrak{g}^*$, and $A = \mathfrak{g} \ltimes (V \oplus \mathfrak{g}^*)$.

Exercises.

In dimension 2 there is only one non abelian Lie algebra; choose a basis $\{x,y\}$ in such a way that the commutation relation is $[x,y] = y$. Denote this Lie algebra by L_2.

1. Show that any bialgebra structure on L_2 can be written (after possible redefinitions) in one of two forms:

(2.2.8) $$\delta x = 0 , \quad \delta y = x \wedge y$$

or

(2.2.9) $$\delta x = x \wedge y , \quad \delta y = 0 .$$

2. Show that for the bialgebra structure (2.2.8) the double is $\mathfrak{gl}_2$; for (2.2.9) the double is a semi-direct product $\mathbb{C} \ltimes \mathcal{N}$ of a one-dimensional Lie algebra (with a generator W) and the three-dimensional Heisenberg algebra $\mathcal{N}$ (with generators X, Y, Z and relations $[X,Y] = Z$, $[Z,X] = [Z,Y] = 0$); the action of W on $\mathcal{N}$ is given by $[W, X] = 2X$, $[W, Y] = -2Y$ and $[W, Z] = 0$.

3. Show that operations

(2.2.10) $$\Delta x = x \otimes 1 + 1 \otimes x , \quad \Delta y = y \otimes 1 + e^{\alpha x} \otimes y$$

and

(2.2.11) $$\Delta x = x \otimes (1 - 2\alpha y)^{-1} + 1 \otimes x , \quad \Delta y = y \otimes 1 + (1 - 2\alpha y) \otimes y$$

provide Hopf algebra structures on corresponding completions of $\mathcal{U}L_2$.

4. Show that the Hopf algebra structure, defined by (2.2.10) (respectively, (2.2.11)) is a quantization of the Lie bialgebra structure (2.2.8) (respectively, (2.2.9)). Note that the terms of order 1 in the deformation parameter α are not antisymmetric, but, as you remember, the symmetric part can be removed by redefinitions.

5. Let $L = \mathfrak{sl}_2 \oplus \mathfrak{sl}_2$. Show that any invariant scalar product on L has a form $\nu \oplus c\nu$, where ν is the Killing form on $\mathfrak{sl}_2$ and c is a constant.

6. Let $c = -1$. Show that the diagonal $\mathfrak{g}_1 = \mathfrak{sl}_2$ is isotropic. A subalgebra $\mathfrak{g}_2$ with a basis $\{(e_+, 0), (0, e_-), (h, -h)\}$ is a complementary isotropic subalgebra ($\{h, e_+, e_-\}$ is a standard basis in $\mathfrak{sl}_2$, $[h, e_\pm] = \pm 2e_\pm$, $[e_+, e_-] = h$). Thus, this Manin triple provides a Lie bialgebra structure on $\mathfrak{sl}_2$.

7. Classify all Manin triples on L.

8. Classify 3-, 4- and 5-dimensional Lie algebras with an invariant scalar product.

2.3. Belavin-Drinfeld triples. Let $\mathcal{L}$ be a simple Lie algebra over $\mathbb{C}$. In this case, every 2-cocycle is a coboundary (see any textbook on Lie algebras, e.g., [**6**]) so one can solve the cocycle condition for μ: $\mu = \partial \rho$ or, explicitly, $\mu_i^{jk} X_j \otimes X_k = [\Delta_0 X_i, \rho]$, with $\rho = \rho^{ab} X_a \otimes X_b$ an element of the wedge square of $\mathcal{L}$ (that is, $\rho^{ab} = -\rho^{ba}$).

Now the Jacobi identity for μ can be rewritten as a non-linear equation for the element ρ.

Notation: for an element $A \in \mathcal{U} \otimes \mathcal{U}$, $A = \sum_\alpha x_\alpha \otimes y_\alpha$ let $A_{12} = \sum_\alpha x_\alpha \otimes y_\alpha \otimes 1$, $A_{13} = \sum_\alpha x_\alpha \otimes 1 \otimes y_\alpha$ and $A_{23} = \sum_\alpha 1 \otimes x_\alpha \otimes y_\alpha$; the elements A_{12}, A_{13} and A_{23} are from $\mathcal{U} \otimes \mathcal{U} \otimes \mathcal{U}$:

Exercise. Show that the Jacobi identity for μ can be rewritten in terms of ρ as

(2.3.1)
$$[[\rho_{12}, \rho_{13}] + [\rho_{12}, \rho_{23}] + [\rho_{13}, \rho_{23}], X_i \otimes 1 \otimes 1 + 1 \otimes X_i \otimes 1 + 1 \otimes 1 \otimes X_i] = 0$$

for all i.

The element $[\rho_{12}, \rho_{13}] + [\rho_{12}, \rho_{23}] + [\rho_{13}, \rho_{23}]$ belongs to the third wedge power of $\mathcal{L}$, i.e., it has a form $A^{ijk} X_i \otimes X_j \otimes X_k$ with totally antisymmetric A^{ijk}. The space of invariant elements in $\Lambda^3 \mathcal{L}$, for the simple $\mathcal{L}$, is known (see, e.g., [**6**]) to be one-dimensional; it is generated by an element $\gamma = \Gamma^{ijk} X_i \otimes X_j \otimes X_k$, where $\Gamma^{ijk} = \Gamma^k_{ab} B^{ai} B^{bj}$, $B_{ij} = \langle X_i, X_j \rangle$ for the Killing form $\langle \cdot, \cdot \rangle$ and B^{ij} is inverse to B_{ij}, $B^{ij} B_{jk} = \delta^i_k$ (δ^i_k is the Kronecker delta). We conclude that the Jacobi identity for $\mu = \partial \rho$ is satisfied iff $[\rho_{12}, \rho_{13}] + [\rho_{12}, \rho_{23}] + [\rho_{13}, \rho_{23}]$ is proportional to γ.

Let $C = B^{ij} X_i \otimes X_j$.

Exercise. Show that $[C_{12}, C_{13}] + [C_{12}, C_{23}] + [C_{13}, C_{23}]$ is proportional to γ.

Therefore we can find a combination $r = \rho + \text{const} \cdot C$ for which

(2.3.2)
$$[r_{12}, r_{13}] + [r_{12}, r_{23}] + [r_{13}, r_{23}] = 0 \ .$$

Note that we still have $\mu_i^{jk} X_j \otimes X_k = [\Delta X_i, r]$ since B commutes with $\Delta_0 X_i$ for all i. The equation (2.3.2) is called the classical Yang-Baxter equation (cYBe). We explained that for a simple Lie algebra $\mathcal{L}$ the problem of finding the Lie bialgebra structures on $\mathcal{L}$ reduces to cYBe for r which satisfies: $r + r'$ is proportional to C, $r + r' = xC$ with $x \in \mathbb{C}$ (r' is the flip of r, $r' = r^{ij} X_j \otimes X_i$ for $r' = r^{ij} X_i \otimes X_j$). If $x \neq 0$ one can set $x = 1$ by rescaling r.

The Yang-Baxter equation (which reduces to the cYBe in the classical limit) is

(2.3.3)
$$\mathcal{R}_{12} \mathcal{R}_{13} \mathcal{R}_{23} = \mathcal{R}_{23} \mathcal{R}_{13} \mathcal{R}_{12} \ .$$

Solutions of the cYBe for which $x \neq 0$ are the most interesting - their quantizations find lots of applications in statistical models, knot theory, representation theory etc.

Exercise. In the situation of the exercise 6 from the previous subsection, show that the corresponding coproduct on $\mathfrak{sl}_2$ arises from an r-matrix, $r = \frac{1}{4} h \otimes h + e_- \otimes e_+$. Verify the cYBe for this r.

We shall now explain how the solutions of cYBe with $r + r' = C$ are classified in terms of so called Belavin-Drinfeld triples. A procedure of quantizing these solutions is known today [**7, 8**]. It is however interesting to enumerate the Belavin-Drinfeld triples, which is a combinatorial question; in the end of this subsection we shall discuss and partly answer it.

Classification of solutions.

Fix a Cartan subalgebra $\mathfrak{h}$. Let R be the set of roots, $R = R_+ \cup R_-$, and Γ the set of simple positive roots.

Definition. A Belavin-Drinfeld triple $(\Gamma_1, \Gamma_2, \tau)$ consists of the following data: Γ_1 and Γ_2 are subsets in Γ and $\tau : \Gamma_1 \to \Gamma_2$ is a one-to-one mapping which satisfies properties:

(i) τ preserves the scalar product, that is, $\langle \tau(\alpha), \tau(\beta) \rangle = \langle \alpha, \beta \rangle$ for all α and β from Γ_1.

(ii) τ is "nilpotent". It means the following. Assume that $\tau(\alpha)$, which is an element from Γ_2 is still in Γ_1. Then $\tau^2(\alpha)$ is defined. If again $\tau^2(\alpha) \in \Gamma_1$ then there is $\tau^3(\alpha)$. Nilpotency means that the sequence must terminate, that is, for some $k \in \mathbb{N}$, an element $\tau^k(\alpha)$ is not any more in Γ_1 for any $\alpha \in \Gamma_1$.

Given a Belavin-Drinfeld triple, consider a system of equations for a tensor $r_0 \in \mathfrak{h} \otimes \mathfrak{h}$,

(2.3.4)
$$r_0 + r_0' = t_0,$$
$$(\tau(\alpha) \otimes \mathrm{id} + \mathrm{id} \otimes \alpha)(r_0) = 0 \quad \text{for all } \alpha \in \Gamma_1.$$

Here t_0 is the "Cartan part" of t: for a basis H_μ of $\mathfrak{h}$ let $B^o_{\mu\nu} = \langle H_\mu, H_\nu \rangle$; then $t_0 = B^{o\,\mu\nu} H_\mu H_\nu$ where $B^{o\,\mu\nu}$ is the inverse to $B_{\mu\nu}$, $B^{o\,\mu\nu} B_{\nu\rho} = \delta^\mu_\rho$.

The system (2.3.4) is compatible [9].

Recall that $\mathfrak{g} = \mathfrak{h} \oplus \bigoplus_{\alpha \in R} \mathfrak{g}_\alpha$, where $[h, x] = \alpha(h)x$ for $x \in \mathfrak{g}_\alpha$, $\dim \mathfrak{g}_\alpha = 1$.

Let $\mathcal{A}_i$ be a Lie subalgebra generated by e_α with $\alpha \in \Gamma_i$, $i = 1, 2$. Then $\mathcal{A}$ is the direct sum of those $\mathfrak{g}_\alpha$ for which the expansion of α in terms of simple roots contains simple roots from Γ_i only.

The map $\tau : \Gamma_1 \to \Gamma_2$ extends to an isomorphism $\tau : \mathcal{A}_1 \to \mathcal{A}_2$ (denoted also by τ), by the formula $e_\alpha \mapsto e_{\tau(\alpha)}$. It is an isomorphism because the only relations in $\mathcal{A}_i$ are Serre relations which depend on the scalar product $\langle ., . \rangle$ only and τ respects the scalar product.

For each $\alpha \in R$ choose e_α in such a way that

(i) $\langle e_{-\alpha}, e_\alpha \rangle = 1$,

(ii) $e_{\tau(\alpha)} = \tau(e_\alpha)$ whenever $\tau(e_\alpha)$ is defined ($e_\alpha \in \mathcal{A}_1$).

Define a partial order: $\alpha < \beta$ for $\alpha, \beta \in R$ means that there exists a natural k such that $\tau^k(\alpha) = \beta$.

Theorem [9]. Let

(2.3.5)
$$r = r_0 + \sum_{\alpha \in R_+} e_{-\alpha} \otimes e_\alpha + \sum_{\alpha, \beta \in R; \alpha < \beta} (e_{-\alpha} \otimes e_\beta - e_\beta \otimes e_{-\alpha}),$$

where r_0 is a solution of (2.3.4).

Then

(i) the tensor r satisfies $\mathrm{cYBE}(r) = 0$ and $r + r' = t$;

(ii) any solution of equations $\mathrm{cYBE}(r) = 0$ and $r + r' = t$, after a suitable change of the basis, is of the form (2.3.5).

The r corresponding to the trivial Belavin-Drinfeld triple (Γ_1 and Γ_2 are empty sets), with $r_0 = \frac{1}{2} t_0$, is called the "standard" r.

2.3.1. *Maximal triples.* I shall say several words about the combinatorics of Belavin-Drinfeld triples. The whole information about scalar products is contained in the Dynkin diagram for the algebra $\mathfrak{g}$. We shall consider the most interesting case of the Lie algebras of the type A (that is, Lie algebras $sl(n)$), for which the Dynkin diagram is

Fig. 1

Given a Belavin-Drinfeld triple, it is useful to draw a diagram, corresponding to it, like:

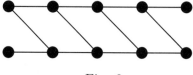

Fig. 2

The upper and lower rows are two copies of the Dynkin diagram A_5, the lines between the rows carry an information about the triple; the lines should be thought as going from the upper low to the lower one; the roots from Γ_1 are the roots at the upper row from which the lines start; the roots from the lower row are those at which the lines end; they are from Γ_2. The angles between the roots are determined by the number of edges connecting the corresponding vertices of the Dynkin diagram; it is therefore easy to understand, looking at the picture, whether the map τ preserves scalar products. To check the nilpotency one needs to draw more than two rows - depicting the powers of τ. For example, for the diagram on Fig. 2 one draws:

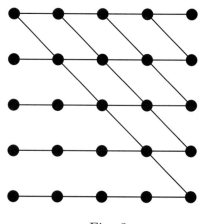

Fig. 3

The meaning of Fig. 3 is clear: the lines going from the first row to the third one represent τ^2, from the first row to the fourth one represent τ^3, etc.

There are two types of natural equivalences for triples:

(i) $T_\updownarrow$: $(\Gamma_1, \Gamma_2, \tau) \mapsto (\Gamma_2, \Gamma_1, \tau^{-1})$; this corresponds to a reflection of the picture of the triple in the horizontal mirror, $T_\updownarrow^2 = \text{id}$;

(ii) if a Dynkin diagram has a symmetry κ then $(\kappa(G_1), \kappa(G_2), \kappa\tau\kappa^{-1})$ is a triple and the equivalence is $T_\kappa : (G_1, G_2, \tau) \mapsto (\kappa(G_1), \kappa(G_2), \kappa\tau\kappa^{-1})$.

For A_n-diagram there is a symmetry: a reflection of Fig. 1 in the vertical mirror; let $T_\leftrightarrow$ be the corresponding equivalence; we have $T_\leftrightarrow^2 = \text{id}$.

If $\tilde{\Gamma}_1$ is any subset of Γ_1 then $(\tilde{\Gamma}_1, \tau(\tilde{\Gamma}_1), \tau|_{\tilde{\Gamma}_1})$ is clearly a triple. So it is interesting to look only for "maximal" triples, i. e. those to which one cannot add any more vertices.

For example, the only nontrivial triple for A_2 is

Fig. 4

Exercises. 1. Show that for A_3 there are, up to equivalences, two maximal triples:

Fig. 5 Fig. 6

2. Show that for A_4 there are, up to equivalences, four maximal triples:

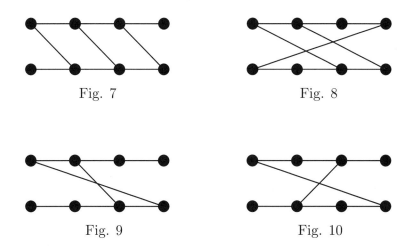

Fig. 7 Fig. 8

Fig. 9 Fig. 10

With the growth of rank it becomes more and more difficult to decide if a given triple is maximal. For example, the triple on Fig. 11

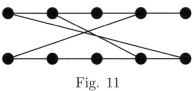

Fig. 11

is maximal, but one has to draw several rows (like on Fig. 3) to see that loops appear when one adds one more vertex.

If $\Gamma \setminus \Gamma_1$ consists of only one vertex (or $\#\Gamma_1 = \#\Gamma - 1$) then the triple is certainly maximal. We shall enumerate triples with $\#\Gamma_1 = \#\Gamma - 1$.

Proposition 9. For the Dynkin diagram A_l, the number of triples with $\#\Gamma_1 = l-1$ is $\frac{1}{2}\Phi(l+1)$, Φ is the Euler function, $\Phi(n) = \#\{j \in \{1,\ldots,n\} | j \text{ is coprime to } n\}$.

Proof.
(i) $\Phi(l+1)$ is the number of primitive roots of unity of order $(l+1)$.

We shall first associate a Belavin-Drinfeld triple to any primitive root of unity of order $(l+1)$. Let $\zeta = \exp(\frac{2\pi i}{l+1})$. Label the vertices of the Dynkin diagram A_l as shown on Fig. 12:

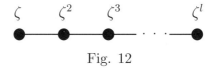

Fig. 12

If a and b are labels of two vertices then a is connected by an edge to b if and only if $a = \zeta^{\pm} b$.

Fix a primitive root q. Let $\Gamma_1 = \{q, q^2, \ldots, q^{l-1}\}$ and $\Gamma_2 = \{q^2, q^3, \ldots, q^l\}$ (more precisely, Γ_i, $i = 1, 2$, are the sets of vertices of A_l labeled by the corresponding roots of unity). Since q is primitive, each of the sets Γ_1 and Γ_2 contain $(l-1)$ distinct elements.

Let $\tau : \Gamma_1 \to \Gamma_2$ be the multiplication by q. Multiplying a label q^i by q, we obtain a sequence $q^i \to q^{i+1} \to \cdots \to q^l$, and the sequence terminates since $q^{l+1} = 1$ is not a label of any vertex. Thus, the map τ is nilpotent.

The condition of being neighbors, $q^i = \zeta^{\pm} q^j$ is stable under the multiplication by q, therefore τ preserves scalar products.

Thus, $(\Gamma_1, \Gamma_2, \tau)$ is a Belavin-Drinfeld triple. Call it $\mathcal{T}_q$.

Consider an arbitrary Belavin-Drinfeld triple $\mathcal{T} = (\Gamma_1, \Gamma_2, \tau)$ with $\#\Gamma_1 = l-1$. We shall prove that it coincides with one of $\mathcal{T}_q$'s.

(ii) Denote the vertex omitted from Γ_1 by q^{-1}. It divides the row of the diagram A_l (as on Fig. 1) into two segments I_1 and I_2:

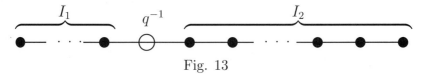

Fig. 13

We have $q^{-1} = \zeta^a$ for some a. Making, if necessary, a vertical reflection, we can, without loss of generality, assume that $\#I_1 \leq \#I_2$.

Let q' be a label of a vertex omitted from Γ_2. The lower row of the picture corresponding to $\mathcal{T}$ is also divided by q' into two segments J_1 and J_2 (J_1 to the left of q', J_2 to the right of it). The map τ preserves neighbors and it follows that either $\tau : I_1 \to J_1$, $I_2 \to J_2$ or $\tau : I_1 \to J_2$, $I_2 \to J_1$. The former case is excluded since otherwise $I_1 = J_1$, $I_2 = J_2$ and restrictions of τ on the sets I_i, $i = 1, 2$, are permutation of these sets and therefore τ cannot be nilpotent.

Thus, $\tau : I_1 \to J_2$, $I_2 \to J_1$ and $q' = q$.

We cannot have $\#I_1 = \#I_2$ – then restrictions of τ^2 on I_i, $i = 1, 2$, would be permutations of these sets. Therefore, $\#I_1 < \#I_2$.

(*iii*) Consider the restriction of τ on the set I_2, $\tau : I_2 \to J_1$. There are two possibilities: τ preserves the order or reverses it. We shall prove that τ cannot reverse the order. Indeed, if τ reverses the order then τ maps q to q^{-1} (it is useful to draw a picture here). Then τ induces a permutation on the set $\Gamma \setminus (q \cup q^{-1})$ and cannot be nilpotent.

Let us collect obtained information about the triple $\mathcal{T}$ in a picture:

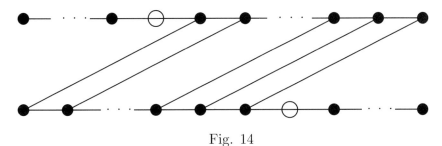

Fig. 14

(*iv*) For the restriction of τ on the set I_1 we have again two possibilities, the order is reversed or preserved. We shall prove that it is preserved. If the cardinality of I_1 is 0 or 1, there is nothing to prove, so, without a loss of generality we assume that $\#I_1 > 1$, in other words, $a \geq 3$. Thus we have $l \geq 6$ since $\#I_2 > \#I_1$.

Extend τ to a map $\tilde{\tau} : \Gamma \to \Gamma$ by $\tilde{\tau} : q^{-1} \mapsto q$ (and $\tilde{\tau} = \tau$ on Γ_1). The map $\tilde{\tau}$ is a permutation of Γ. Decompose $\tilde{\tau}$ into a product of cycles. Since q^{-1} maps to q, the decomposition contains a cycle $c = (\ldots q^{-1} q \ldots)$. If there are other cycles, $\tilde{\tau} = c \cdot c_1 \cdot c_2 \ldots$ then the product $c_1 \cdot c_2 \ldots$ is a permutation of some set S. This permutation is the restriction of τ on S, thus τ cannot be nilpotent. We conclude that $\tilde{\tau}$ is a cycle.

Explicitly, the action of τ is

(2.3.6)
$$\begin{cases} \zeta^{a+i} \mapsto \zeta^i, & i = 1, \ldots, n-a, \\ \zeta^i \mapsto \zeta^{-i}, & i = 1, \ldots, a. \end{cases}$$

We shall follow a sequence $\tilde{\tau}^n(q^{-1})$. First, $q^{-1} = \zeta^a$ maps to $q^{-1} = \zeta^{l+1-a}$. Then it goes back, $\zeta^{l+1-a} \mapsto \zeta^{l+1-2a} \mapsto \cdots \mapsto \zeta^{l+1-ka}$, where $l + 1 - ka \leq a$ but $l + 1 - (k-1)a > a$ or $l + 1 \leq (k+1)a$ but $l + 1 > ka$. This requires k steps (i.e. this is the result of the action of $\tilde{\tau}^k$ on q^{-1}). At the next step, ζ^{l+1-ka} maps to ζ^{ka} and then again goes back, $\zeta^{ka} \mapsto \zeta^{(k-1)a} \mapsto \cdots \mapsto \zeta^a$. This takes k more steps. Thus, $\tilde{\tau}^{2k}(q^{-1}) = q^{-1}$.

But $3k \leq ak < l + 1$. Therefore, $2k < \frac{2l+2}{3} < l$ because l is at least 6.

Therefore, the permutation $\tilde{\tau}^{2k}$ has a fixed point and $2k < l$. Thus $\tilde{\tau}$ cannot be a cycle.

We are left with only one possibility: the restriction of τ on I_1 preserves the order.

(*v*) The map τ preserves the order on I_2 by (*iii*) and on I_1 by (*iv*). Written explicitly, it means that the map τ is the multiplication by q. It will not be nilpotent if q is not primitive, therefore, the triple $\mathcal{T}$ coincides with one of $\mathcal{T}_q$'s.

(vi) The group, generated by flips $T_\updownarrow$ and $T_\leftrightarrow$, is $\mathbb{Z}_2 \times \mathbb{Z}_2$. But on triples $\mathcal{T}_q$ the operations $T_\updownarrow$ and $T_\leftrightarrow$ coincide: each of them is $\mathcal{T}_q \mapsto \mathcal{T}_{q^{-1}}$ (in general, it is not so, even for maximal triples: consider the triple corresponding to Fig. 11). Therefore, only $\mathbb{Z}_2$ acts on the space of triples $\mathcal{T}_q$. This action does not have fixed points and we conclude that the number of triples, under equivalences, is $\frac{1}{2}\Phi(l+1)$ as stated. The proof of the Theorem 9 is finished. □

3. Quantum spaces

We shall briefly, without going into details, give geometrical motivations which lead to the notion of quantum spaces.

Let G be a Lie group, $\mathfrak{g}$ its Lie algebra. In the section 2, Lie bialgebras appeared in the study of deformations of the coproduct on the universal enveloping algebra $\mathcal{U}\mathfrak{g}$. Geometrically, Lie algebra $\mathfrak{g}$ is the Lie algebra of left-invariant vector fields on G. The universal enveloping algebra of the Lie algebra $\mathfrak{g}$ can therefore be realized as the algebra of left invariant differential operators on G. Up to topological and functional analytic considerations (convergence, etc.), a function on $\mathbb{R}^n$ can be reconstructed, as a Taylor series, from the knowledge of its derivatives at the origin. For a Lie group G, the knowledge of derivatives of a function f at the origin is replaced by the knowledge of values on f of all left invariant differential operators at the unity of G. The elements of $\mathcal{U}\mathfrak{g}$ are linear functionals on the space $\mathcal{F}G$ of functions on G, so, up to topological considerations, the spaces $\mathcal{U}\mathfrak{g}$ and $\mathcal{F}G$ are dual to each other, the pairing between $X \in \mathcal{U}\mathfrak{g}$ and $f \in \mathcal{F}G$ is given by $\langle X, f \rangle = X(f)|_e$, where $e \in G$ is the unity element. It follows then that the coproduct on $\mathcal{U}\mathfrak{g}$ corresponds to the product on $\mathcal{F}G$ - the usual product of functions. Thus, deformations of the coproduct on $\mathcal{U}\mathfrak{g}$ correspond to deformations of the commutative algebra $\mathcal{F}G$ of functions. Infinitesimal deformations from Section 1 correspond to particular Poisson brackets on G – Poisson brackets which are compatible with the group structure. One says that Poisson brackets are compatible with the group structure if the multiplication $m : G \times G \to G$ is a Poisson map. In other words: define, for a given function f on G, a function $f\tilde{\,}$ on $G \times G$ by the rule $f\tilde{\,}(x,y) = f(x \cdot y)$; the compatibility of the Poisson brackets $\{.,.\}$ means $\{f, g\}\tilde{\,} = \{f\tilde{\,}, g\tilde{\,}\}$, where the Poisson brackets on $G \times G$ are Poisson brackets of the direct product of two Poisson manifolds. Groups with compatible Poisson brackets are called Poisson-Lie groups.

In the other direction, it is not difficult to check that if G carries compatible Poisson brackets then its Lie algebra $\mathfrak{g}$ gets a Lie bialgebra structure.

Compatible Poisson brackets are of a very special form. We shall illustrate it on an example of a matrix group G (a subgroup of the group of invertible matrices). Let a^i_j be matrix elements. Assume that $\{a^i_k, a^m_s\} = \Phi^{im}_{ks}(a)$ are compatible Poisson brackets. Then $\{a^i_j b^j_k, a^m_n b^n_s\} = \Phi^{im}_{ks}(ab)$ (this is the equality $\{f\tilde{\,}, g\tilde{\,}\} = \{f, g\}\tilde{\,}$ for $f = a^i_k$ and $g = a^m_s$). On the other hand, $\{a^i_j b^j_k, a^m_n b^n_s\} = \{a^i_j, a^m_n\} b^j_k b^n_s + \{b^j_k, b^n_s\} a^i_j a^m_n = \Phi^{im}_{jn}(a) b^j_k b^n_s + \Phi^{jn}_{ks}(b) a^i_j a^m_n$ because $G \times G$ is equipped with the Poisson structure of the direct product. Therefore Φ must be homogeneous of degree 2 (this reflects the fact that μ_i in the deformation of the coproduct on $\mathcal{U}\mathfrak{g}$ belongs to $\mathfrak{g} \otimes \mathfrak{g}$). Thus, the Poisson brackets are quadratic. To quantize constant or linear Poisson brackets, one simply replaces the Poisson brackets by the commutator. However, it is not obvious how to quantize quadratic Poisson brackets - we cannot replace Poisson brackets by the commutator because we don't know how to order consistently the quadratic right hand side.

Exercises. 1. Show that formulas
(3.0.1)
$$\{a,b\} = ab \,,\ \{a,c\} = ac \,,\ \{b,d\} = bd \,,$$
$$\{c,d\} = cd \,,\ \{b,c\} = 0 \,,\ \{a,d\} = 2bc$$
are Poisson brackets for four variables a, b, c and d.

2. Show that if a, b, c and d are matrix elements of a matrix $A = \begin{pmatrix} a & b \\ c & d \end{pmatrix}$ then the Poisson brackets (3.0.1) provide $GL(2)$ with a Poisson-Lie structure.

3. Show that the Poisson brackets (3.0.1) of the determinant of A, $\det A$ with all matrix elements vanish.

The main interest about the most commonly appearing groups, like GL, SO, Sp, ... - is that they arise as groups of symmetry (of a vector space, of a vector space with a bilinear form, ...). The Poisson-Lie groups one can interpret in this way too. One says that a Poisson-Lie group G acts on a Poisson manifold $\mathcal{M}$ in a Poisson way if

(i) G acts on M;

(ii) the action $G \times \mathcal{M} \to \mathcal{M}$ is a Poisson map, where $G \times \mathcal{M}$ is equipped with a Poisson structure of the direct product.

Again, for a matrix group G acting on a vector space V^2, $x^i \mapsto a^i_j x^j$, a similar calculation shows that the Poisson brackets of coordinates, $\{x^i, x^j\}$ must be quadratic in x^i.

Exercise. For a two dimensional vector space with coordinates x^1 and x^2 let
(3.0.2)
$$\{x^1, x^2\} = x^1 x^2 \,.$$
Show that $GL(2)$ with the Poisson structure given by (3.0.1) acts in a Poisson way on this Poisson vector space.

It turns out that from the point of view of the theory of quantum groups, the appropriate way to quantize the Poisson brackets (3.0.2) is provided by the following commutation relations:
(3.0.3)
$$x^1 x^2 = q x^2 x^1 \,.$$
This is the first example of a quantum vector space. Denote this quantum vector space (that is, the algebra of polynomials in x^1 and x^2 subjected to the relation (3.0.3)) by V_q^2.

The linear group of transformations preserving the relation (3.0.3) is poor, it consists only of rescalings $x^i \mapsto c_i x^i$ with some constants c_i. It is the quantum group which is the right analogue of the symmetry group of the quantum vector space.

General picture.

Let $\mathcal{U}$ be a quasitriangular Hopf algebra with a universal R-matrix $\mathcal{R}$. Let $\mathcal{U}^*$ be a dual Hopf algebra. The pairing between $\mathcal{U}$ and $\mathcal{U}^*$ satisfies $\langle \Delta a, x \otimes y \rangle = \langle a, xy \rangle$ and $\langle \alpha \otimes b, \Delta x \rangle = \langle ab, x \rangle$.

We shall think of $\mathcal{U}$ as of analogue of the universal enveloping algebra of a semi-simple Lie algebra. Ideologically (in the spirit of the Peter-Weyl theorem) the dual Hopf algebra is generated by matrix elements of representations of $\mathcal{U}$.

Let ρ be a representation of $\mathcal{U}$ on a vector space V, ρ maps an element $x \in \mathcal{U}$ to a matrix $T^i_j(x)$.

The coproduct Δ on matrix elements T^i_j looks especially simple:

(3.0.4)
$$\langle \Delta T^i_j, x \otimes y \rangle = \langle T^i_j, xy \rangle = T^i_j(xy) = T^i_k(x) T^k_j(y)$$
$$= \langle T^i_k, x \rangle \langle T^k_j, y \rangle = \langle T^i_k \otimes T^k_j, x \otimes y \rangle$$

and therefore

(3.0.5)
$$\Delta T^i_j = T^i_k \otimes T^k_j .$$

The commutation relations between the matrix elements T^i_j are expressed in terms of a numerical R-matrix R, the image of the universal R-matrix $\mathcal{R} = \sum_i a_i \otimes b_i$ in the representation, $R = \sum_i \rho(a_i) \otimes \rho(b_i)$. Let $\Delta x = \sum_\alpha u_\alpha \otimes v_\alpha$.

(3.0.6)
$$\langle T_1 T_2, x \rangle = \langle T_1 \otimes T_2, \sum u_\alpha \otimes v_\alpha \rangle = \sum T_1(u_\alpha) T_2(v_\alpha) .$$

By the definition of $\mathcal{R}$ we have $\sum u_\alpha \otimes v_\alpha = \mathcal{R}^{-1} \sum v_\alpha \otimes u_\alpha \mathcal{R}$. Therefore

(3.0.7)
$$\langle T_1 \otimes T_2, \sum u_\alpha \otimes v_\alpha \rangle = \langle T_1 \otimes T_2, \mathcal{R}^{-1} \sum v_\alpha \otimes u_\alpha \mathcal{R} \rangle$$
$$= T_1 \otimes T_2 (\mathcal{R}^{-1} \sum v_\alpha \otimes u_\alpha \mathcal{R})$$
$$= T_1 \otimes T_2(\mathcal{R}^{-1}) \cdot T_1 \otimes T_2(\sum v_\alpha \otimes u_\alpha) \cdot T_1 \otimes T_2(\mathcal{R})$$
$$= R^{-1} \sum T_1(v_\alpha) T_2(u_\alpha) R = R^{-1} \sum T_2(u_\alpha) T_1(v_\alpha) R .$$

The "$\cdot$" means matrix multiplication.

Arguments u_α and v_α are now in the same order as in (3.0.6). Therefore, $T_1 T_2 = R^{-1} T_2 T_1 R$ or

(3.0.8)
$$R T_1 T_2 = T_2 T_1 R .$$

Because of the form of this relation, this algebra is often called the "RTT"-algebra.

The algebra $\mathcal{U}^*$ was first written in the form (3.0.8) and (3.0.5) in [10].

We shall write the relations (3.0.8) in a different way. Let P be a permutation of factors in $V \otimes V$, $P(x \otimes y) = y \otimes x$. Let $\hat{R} = PR$. Then (3.0.8) is equivalent to

(3.0.9)
$$\hat{R} T_1 T_2 = T_1 T_2 \hat{R} .$$

A motivation to use $\hat{R}$ instead of R: the eigenvalues of $\hat{R}$ have a representation-theoretic meaning. A theorem, due to Drinfeld [11], says that there exists an element $\mathcal{F}$ such that $\mathcal{R} = \mathcal{F}_{21} q^t \mathcal{F}^{-1}$, where $q = \exp(\alpha)$ (α is the deformation parameter), t is the invariant tensor $B^{ij} X_i \otimes X_j$. Let $C = B^{ij} X_i X_j$. Now let $V \otimes V = \sum W_\gamma$ be a decomposition of the tensor square of the space V into irreducible representations. We have $\Delta_0 C = C \otimes 1 + 1 \otimes C + 2t$, where Δ_0 is the classical coproduct. Therefore, $t|_{W_\gamma} = \frac{1}{2} C|_{W_\gamma} - C|_V$. Denote this quantity by t_γ. We have $R = PFPq^{\rho \otimes \rho(t)} F^{-1}$, therefore, $\hat{R} = FPq^{\rho \otimes \rho(t)} F^{-1}$.

First of all, $V \otimes V$ decomposes into the symmetric and antisymmetric parts, $V \otimes V = S^2 V \oplus \Lambda^2 V$. The operator P takes the value $(+1)$ on $S^2 V$ and (-1) on $\Lambda^2 V$. Every W_γ is either is either in $S^2 V$ or in $\Lambda^2 V$ so the "sign" of W_γ, depending on whether W_γ is $S^2 V$ or $\Lambda^2 V$, sign$(W_\gamma) := P|_{W_\gamma}$ is well defined.

Since $q^{\rho\otimes\rho(t)}|_{W_\gamma} = q^{t_\gamma}$ we have

(3.0.10) $$\hat{R}|_{FW_\gamma} = FPq^{\rho\otimes\rho(t)}|_{W_\gamma} = q^{t_\gamma}\,\text{sign}(W_\gamma)\,.$$

Thus, the projector decomposition of $\hat{R}$ reflects the decomposition of the tensor product of representations.

Polynomials in the matrix elements T^i_j are "quantized" functions on the group. We also had the Poisson brackets on the coordinates x^i and $x^i \mapsto T^i_j x^j$ was a Poisson map. On the quantum level we have the $\hat{R}TT$ relations for T^i_j. Which relations can we impose on x's in such a way that the map $x^i \mapsto T^i_j x^j$ is an algebra homomorphism? These relations should also quantize the Poisson brackets for x's. Since the quantization of the Poisson brackets for T^i_j produced quadratic relations, we expect to have quadratic relations for the algebra of x's as well. Impose a set of quadratic relations for x's, $E^\alpha_{ij} x^i x^j = 0$, α labels relations. Then for Tx we have $E^\alpha_{ij} T^i_a x^a T^j_b x^b = E^\alpha_{ij} T^i_a T^j_b x^a x^b$, so we have to understand which tensors we can move through $T^i_a T^j_b$.

The defining relation (3.0.9) shows that we can move $\hat{R}$ and therefore any function of $\hat{R}$. As we have seen, $\hat{R} = \sum \nu_\gamma \Pi_\gamma$, where $\nu_\gamma = q^{t_\gamma}\,\text{sign}\,(W_\gamma)$ and Π_γ is the projector on the space FW_γ. So essentially, all functions of $\hat{R}$ are linear combinations of projectors.

Conclusion: covariant algebras are given by relations $(\Pi_\gamma)^{ij}_{kl} x^k x^l = 0$ for some γ (one or several). Denote by A^l_γ (l for "left") the quadratic algebra defined by one projector Π_γ. Equally well, there is a covariant "right" algebra A^r_γ (the algebra of covectors x_i), defined by $x_i x_j (\Pi_\gamma)^{ij}_{kl} = 0$, the covariance is $x_i \mapsto x_a T^a_i$.

An important fact is that the RTT-relations can be reconstructed from the requirement that all the algebras A^l_γ (or all the algebras A^r_γ) are covariant. Indeed, A^l_γ is covariant means

(3.0.11) $$(\Pi_\gamma)^{ij}_{kl} T^k_a T^l_b x^a x^b = 0\,,$$

therefore, $(\Pi_\gamma)^{ij}_{kl} T^k_a T^l_b$ must be proportional to Π_γ whose lower indices are a, b:

(3.0.12) $$(\Pi_\gamma)^{ij}_{kl} T^k_a T^l_b = \Sigma^{ij}_{uv} (\Pi_\gamma)^{uv}_{ab}\,.$$

Multiplying (3.0.12) by $(\Pi_\tau)^{ab}_{cd}$ with $\tau \neq \gamma$, we find

(3.0.13) $$\Pi_\gamma T_1 T_2 \Pi_\tau = 0$$

for all pairs $\{\tau, \gamma \,|\, \tau \neq \gamma\}$.

Lemma 10. The system (3.0.13) of relations for the matrix elements T^i_j is equivalent to the RTT-relations $\hat{R}T_1 T_2 = T_1 T_2 \hat{R}$.

Proof.

(i) (3.0.9) implies (3.0.13).

We have $\hat{R} = \sum \nu_\gamma \Pi_\gamma$; moreover, $\sum \Pi_g = 1$ is a decomposition of unity. Multiplying $\hat{R}T_1 T_2 = T_1 T_2 \hat{R}$ by Π_γ from the left and Π_τ from the right, we find $\nu_\gamma \Pi_\gamma T_1 T_2 \Pi_\tau = \nu_\tau \Pi_\gamma T_1 T_2 \Pi_\tau$. If $\gamma \neq \tau$ then $\nu_\gamma \neq \nu_\tau$ and it follows that the relation $\Pi_\gamma T_1 T_2 \Pi_\tau = 0$ is satisfied.

(ii) (3.0.13) implies (3.0.9).

We have $\hat{R}T_1T_2 = \hat{R}T_1T_2 \cdot 1 = \sum_{\gamma,\tau} \nu_\gamma \Pi_\gamma T_1 T_2 \Pi_\tau$. The last expression, due to the relations (3.0.13), can be rewritten as

(3.0.14) $$\sum_\gamma \nu_\gamma \Pi_\gamma T_1 T_2 \Pi_\gamma .$$

Similarly, $T_1T_2\hat{R} = 1 \cdot T_1T_2\hat{R} = \sum_{\gamma,\tau} \nu_\tau \Pi_\gamma T_1 T_2 \Pi_\tau$. Again, due to (3.0.13), this equals $\sum_\gamma \nu_\gamma \Pi_\gamma T_1 T_2 \Pi_\gamma$, which coincides with (3.0.14). Thus, (3.0.9) holds. □

If $\hat{R}$ appears in the process of a deformation then there is a candidate for an especially nice quantum space. Again, $V \otimes V = \bigoplus W_\gamma$ classically; denote the set of $\{\gamma\}$ by J, $J = J_+ \cup J_-$ where $J_\pm = \{\gamma | \text{sign} W_\gamma = \pm 1\}$. Then projectors $\Pi_\pm = \sum_{\gamma \in J_\pm} \Pi_\gamma$ have ranks $\text{rk} \Pi_\pm = \frac{N(N\pm 1)}{2}$. Therefore, a set $(\Pi_-)^{ij}_{kl} x^k x^l = 0$ contains $\frac{N(N-1)}{2}$ relations - exactly the number of relations which we have classically for commuting variables. The quantum space defined by relations $(\Pi_-)^{ij}_{kl} x^k x^l = 0$ is the only reasonable candidate for the quantization of $\mathbb{C}^N$ on which a group G acted in a Poisson way. Similarly, the quantum space defined by relations $(\Pi_+)^{ij}_{kl} x^k x^l = 0$ is the quantization of the algebra of odd (grassmanian) variables.

For $GL(N)$, in the decomposition $V \otimes V = S^2V \oplus \Lambda^2V$, the summands S^2V and Λ^2V are irreducible. It is natural therefore to call an $\hat{R}$, which contains only two projectors, an $\hat{R}$-matrix of GL type. One usually rescales $\hat{R}$ to have $\hat{R} = q\Pi_+ - q^{-1}\Pi_-$, where Π_+ and Π_- are projectors, which are called, due to their origin, the q-symmetrizer and the q-antisymmetrizer respectively (and we shall often denote Π_+ by S and Π_- by A).

To conclude:

- Geometrically and physically meaningful $\hat{R}$-matrices decompose into projectors

(3.0.15) $$\underbrace{\alpha_1 \Pi_1^+ + \ldots \alpha_k \Pi_k^+}_{S} + \underbrace{\beta_1 \Pi_1^- + \ldots \beta_l \Pi_l^-}_{A}$$

and we know which projectors constitute the q-symmetrizer S and the q-antisymmetrizer A, respectively (as shown by underbracing in (3.0.15)). The ranks of the projectors $S = \Pi_+ = \Pi_1^+ + \ldots \Pi_k^+$ and $A = \Pi_- = \Pi_1^- + \cdots + \Pi_l^-$ are classical, $\text{rk} \Pi_\pm = \frac{N(N\pm 1)}{2}$.

- Covariant algebras are quadratic algebras of the form $(\Pi_\gamma)^{ij}_{kl} x^k x^l = 0$ or $x_i x_j (\Pi_\gamma)^{ij}_{kl} = 0$ where Π_γ is one of the projectors in the decomposition (3.0.15).

- The algebra of functions on a quantum group is given by the relations (3.0.9) and these relations are equivalent to the condition that all the algebras A^l_γ are covariant; in other words, the RTT-algebra can be reconstructed from the A^l_γ algebras. The same holds if one replaces left algebras by the right ones.

3.1. GL type. For GL-type algebras we have only two projectors[1], $\hat{R} = qS - q^{-1}A$. In this case, to reconstruct the RTT-algebra, it is enough to require covariance of two algebras, A^l_+ and A^l_- defined by relations $S^{ij}_{kl} x^k x^l = 0$ and $A^{ij}_{kl} x^k x^l = 0$ respectively [12].

---

[1] When $\hat{R}$ has two eigenvalues, one says that $\hat{R}$ is of Hecke type. We require additionally that the ranks of the projectors are fixed $\text{rk} S = \frac{N(N+1)}{2}$, $\text{rk} A = \frac{N(N-1)}{2}$.

As we have seen, covariance of the algebra A_γ^l implies the condition $\Pi_\gamma T_1 T_2 \Pi_\tau = 0$ for all τ different from γ. Thus, in the GL case, the covariance of the algebra A_+^l implies

(3.1.1) $$ST_1 T_2 A = 0$$

while the covariance of A_-^l implies

(3.1.2) $$AT_1 T_2 S = 0 \ .$$

On the other hand, the covariance of the algebra A_+^r implies the same relation (3.1.2). Together, relations (3.1.1) and (3.1.2), are equivalent to the RTT-relations. This shows that in the GL case, one can interpret the RTT-relations in two ways: either as the condition of the covariance of the algebras A_+^l and A_-^l or as the condition of the covariance of the algebras A_+^l and A_+^r. We shall use the latter interpretation in the sequel.

The algebras $A_+^{l,r}$ are the left and right quantum spaces. If they are good deformations then the dimension $d(N,k)$ of the space of polynomials of degree k coincides with the dimension of the space of polynomials in N commuting variables,

(3.1.3) $$d(N,k) = \binom{N+k-1}{k} \ .$$

So, quantum spaces are quadratic algebras with correct Poincaré series.

As we shall see below, the behavior of Poincaré series is intimately related to the theory of quantum groups.

Definition. Given a set of tensors $\mathcal{E} = \{E^\alpha = E_{ij}^\alpha\}$, $i,j = 1,\ldots,n$, define an algebra $A_\mathcal{E}$ with generators x^i and relations

(3.1.4) $$E_{ij}^\alpha x^i x^j = 0 \quad \text{for all} \ \alpha \ .$$

Let $d(N,k)$ be the dimension of the space of polynomials of degree k in x^i. We say that $A_\mathcal{E}$ has a Poincaré-Birkhoff-Witt property (or that $A_\mathcal{E}$ is a PBW-algebra) if (3.1.3) is satisfied. In particular the range of the index α is $\{1,\ldots,\frac{N(N-1)}{2}\}$.

Relation $\Pi_{kl}^{ij} x^k x^l = 0$ (with Π a projector) is an example of (3.1.4) but in general the tensors E^α are not organized in a projector.

Requirement that x^i are covariant, that is, that the relations (3.1.4) are satisfied by $T_j^i x^j$ implies some relations between T_j^i's.

Assume that we are given two quantum spaces, $A_\mathcal{E}^l$ with generators x^i and relations $E_{ij}^\alpha x^i x^j = 0$ and $A_\mathcal{F}^r$ with generators x_i and relations $x_i x_j F_\beta^{ij} = 0$.

Definition. We say that the quantum spaces $A_\mathcal{E}^l$ and $A_\mathcal{F}^r$ are compatible if the covariance algebra of T_j^i has the PBW property (that is, its Poincaré series coincide with the Poincaré series for N^2 variables).

Next subsection is a digression on the Poincaré series; then we shall continue with a discussion of compatible quantum spaces in dimension 3.

3.2. Technics of checking the Poincaré series. Consider an algebra with generators x^i and relations (3.1.4). Sometimes it is useful to try to apply the diamond lemma [13, 14]. In its easiest form it says: assume that there is a basis $\{x^1,\ldots,x^N\}$ in which relations look: for $j > i$, $x^j x^i$ is a sum of monomials $x^a x^b$

with $a < b$ and all the monomials in the sum are lexicographically smaller or equal $x^j x^i$. Take these relations as instructions: replace $x^j x^i$ by the right hand side. Apply these instructions to $x^k x^j x^i$, where $k > j > i$, in two ways, starting from $x^k x^j$ or from $x^j x^i$. If both ways will eventually produce the same result, to which no more instructions can be applied then the ordered monomials $x_1^{\mu_1} \ldots x_N^{\mu_N}$ form a linear basis in the algebra, which implies that the algebra has the PBW property.

Note that whether one can apply this procedure depends on a choice of a basis in the algebra. Such a basis might not exist.

We shall now describe another way of checking the Poincaré series which one can apply in the GL case. It uses a differential calculus on quantum planes, developed in [15].

3.2.1. *Differential calculus and Poincaré series.* Assume that the quadratic relations are given by

$$(3.2.1) \qquad A^{ij}_{kl} x^k x^l = 0 \; ,$$

where A is the q-antisymmetrizer in the Hecke $\hat{R}$-matrix, $\hat{R} = qS - q^{-1}A$.

Let ξ^i be generators of the odd quantum space for $\hat{R}$, that is, the relations for ξ^i are

$$(3.2.2) \qquad S^{ij}_{kl} \xi^k \xi^l = 0 \; .$$

One can unify generators x^i and ξ^i into one quadratic algebra by requiring that

$$(3.2.3) \qquad x^i \xi^j = \hat{R}^{ij}_{kl} \xi^k x^l \; .$$

Exercise. Verify that relations (3.2.3) are compatible with (3.2.1) and (3.2.2). The compatibility here means the following. Let ϕ be a combination of quadratic expressions $S^{ij}_{kl} \xi^k \xi^l$. Then $\phi = 0$ in the algebra with generators ξ^i. Take an element $x^i \phi$ with some i and use (3.2.3) to move x to the right. We obtain an element of the form $\sum_j \psi_j x^j$, with some quadratic (in ξ's) elements ψ_j. Since ϕ was 0, we must have $\psi_j = 0$. In other words, for each j, the element ψ_j must be a combination of expressions $S^{ij}_{kl} \xi^k \xi^l$. In the same manner, there is a compatibility check when one moves ξ^i to the left through relations (3.2.1) for x's.

At the next step, one adds "q-derivatives" ∂_i in the generators x^i. An algebra of the derivatives ∂_i is the algebra with generators ∂_i and relations

$$(3.2.4) \qquad \partial_i \partial_j A^{ji}_{kl} = 0$$

(note the order of i, j).

To add ∂_i, one needs cross-commutation relations with the already existing generators x^i and ξ^i. These relations are:

$$(3.2.5) \qquad \partial_i x^j = 1 + q \hat{R}^{js}_{it} x^t \partial_s \; ,$$

$$(3.2.6) \qquad \partial_i \xi^j = (\hat{R}^{-1})^{js}_{it} \xi^t \partial_s \; .$$

Exercise. Verify that relations (3.2.5) and (3.2.6) are compatible with (3.2.1), (3.2.2) and (3.2.3) (in the above sense of compatibility).

Finally, one adds derivatives δ_i in ξ^i. An algebra of the derivatives δ_i is the algebra with generators δ_i and relations

(3.2.7) $$\delta_i \delta_j S^{ji}_{kl} = 0 \ .$$

The cross-commutation relations between δ_i and the generators x^i, ξ^i and ∂_i are:

(3.2.8) $$\delta_i x^j = \hat{R}^{js}_{it} x^t \delta_s \ ,$$

(3.2.9) $$\delta_i \xi^j = 1 - q \hat{R}^{js}_{it} \xi^t \delta_s \ ,$$

(3.2.10) $$\delta_i \partial_j = (\hat{R}^{-1})^{kl}_{ji} \partial_l \delta_k \ .$$

Exercise. Verify that relations (3.2.8), (3.2.9) and (3.2.10) are compatible with (3.2.1)- (3.2.7).

We shall need the following singlets, the Euler operators E_e and E_o and the differentials d and δ:

(3.2.11)
$$E_e = x^i \partial_i, \ E_o = \xi^i \delta_i \ ,$$
$$d = \xi^i \partial_i \ , \ \delta = x^i \delta_i \ .$$

Their relations with the generators of the algebra are (*Exercise*: verify the relations):

(3.2.12)
$$E_e x^i = x^i (1 + q^2 E_e) \ , \ \partial_i E_e = (1 + q^2 E_e) \partial_i \ ,$$
$$E_e \xi^i = \xi^i E_e \ , \ \delta_i E_o = E_o \delta_i \ ,$$

(3.2.13)
$$E_o x^i = x^i E_o \ , \ \partial_i E_o = E_o \partial_i \ ,$$
$$E_o \xi^i = \xi^i (1 + q^2 E_o) \ , \ \delta_i E_o = (1 + q^2 E_o) \delta_i \ ,$$

(3.2.14)
$$dx^i = \xi^i + q x^i d \ , \ \partial_i d = q d \partial_i \ ,$$
$$d\xi^i = -q \xi^i d \ , \ \delta_i d = (1 + (q^2 - 1) E_o) \partial_i - q d \delta_i \ ,$$

(3.2.15)
$$\delta x^i = q x^i \delta \ , \ \partial_i \delta = (1 + (q^2 - 1) E_e) \delta_i + q \delta \partial_i \ ,$$
$$\delta \xi^i = x^i - q \xi^i \delta \ , \ \delta_i \delta = -q^{-1} \delta \delta_i \ .$$

Using operators $N_e = 1 + (q^2 - 1) E_e$ and $N_o = 1 + (q^2 - 1) E_o$, appearing in the right hand sides of (3.2.14) and (3.2.15), one can rewrite (3.2.12) and (3.2.13) in a form

(3.2.16)
$$N_e x_{\langle 1|} = q^2 x_{\langle 1|} N_e \ , \ N_e \partial_{|1\rangle} = q^{-2} \partial_{|1\rangle} N_e \ ,$$
$$N_e \xi_{\langle 1|} = \xi_{\langle 1|} N_e \ , \ N_e \delta_{|1\rangle} = \delta_{|1\rangle} N_e \ ,$$

(3.2.17)
$$N_o x_{\langle 1|} = x_{\langle 1|} N_o \ , \ N_o \partial_{|1\rangle} = \partial_{|1\rangle} N_o \ ,$$
$$N_0 \xi_{\langle 1|} = q^2 \xi_{\langle 1|} N_o \ , \ N_o \delta_{|1\rangle} = q^{-2} \delta_{|1\rangle} N_o \ .$$

Exercise. Verify:

1. The Euler operators commute,

(3.2.18) $$E_e E_o = E_o E_e .$$

2. Commutation relations between the Euler operators and the differentials are

(3.2.19)
$$dE_e = (1 + q^2 E_e)d \quad \text{or} \quad N_e d = q^{-2} dN_e d ,$$
$$E_o d = d(1 + q^2 E_o) \quad \text{or} \quad N_o d = q^2 dN_o ,$$
$$E_e \delta = \delta(1 + q^2 E_e) \quad \text{or} \quad N_e \delta = q^2 \delta N_e ,$$
$$\delta E_o = (1 + q^2 E_o)\delta \quad \text{or} \quad N_o \delta = q^{-2} \delta N_o .$$

3. The differentials square to zero,

(3.2.20) $$d^2 = 0 , \quad \delta^2 = 0 .$$

4. For the anticommutator of d and δ we have

(3.2.21) $$\Lambda := d\delta + \delta d = E_e + E_o + (q^2 - 1)E_o E_e = \frac{1}{q^2 - 1}(N_o N_e - 1) .$$

The last exercise implies that

(3.2.22) $$\Lambda x^i = x^i(q^2 \Lambda + 1) , \quad \Lambda \xi^i = \xi^i(q^2 \Lambda + 1) .$$

Let $M_{a,b}$ be a space of polynomials in x and ξ, of degree a in x and of degree b in ξ. For $\phi \in M_{a,b}$ one finds, by induction,

(3.2.23) $$\Lambda \phi = \phi((a+b)_{q^2} + q^{2(a+b)} \Lambda) ,$$

where the q-number $(n)_q$ is defined by $(n)_q = \frac{1 - q^n}{1 - q} = 1 + q + q^2 + \cdots + q^{n-1}$.

Let $M_n = \oplus_{a+b=n} M_{a,b}$ and $M = \oplus_{n=0}^{\infty} M_n$. The space M is a $\mathbb{Z}_2$-graded vector space, the grading is given by the degree of a monomial in ξ's.

One can consider ∂_i and δ_i as operators acting on the space M. To this end, one introduces a vacuum Vac, which satisfies ∂_i Vac $= 0$ and δ_i Vac $= 0$. Let X be an expression in x^i, ξ^i, ∂_i and δ_i. To evaluate it on an element $\phi \in M$, take an element $X\phi$. Using the commutation relations, we move all ∂_i and δ_i to the right and evaluate on the vacuum. This gives an element of M which we denote $X(\phi)$. The consistency requires only that the vacuum is a representation of the algebra of ∂_i and δ_i which is clearly true.

For instance, we have $\Lambda(\phi) = (a+b)_{q^2} \phi$ for $\phi \in M_{a,b}$.

For each n, the space $M_n = \oplus_{a+b=n} M_{a,b}$ is finite dimensional. We have $\dim M_{a,b} = d_e(a) d_o(b)$, where $d_e(a)$ and $d_o(b)$ are dimensions of the spaces of polynomials in x of degree a and in ξ of degree b, respectively. The grading of $M_{a,b}$ is $(-1)^b$.

The space M_n is closed under the action of d and δ. Therefore, the supertrace of their anticommutator (of the operator Λ) vanishes, $\text{Str}\,\Lambda = 0$, which implies

(3.2.24) $$\sum_{a+b=n} d_e(a) d_o(b) (-1)^b (a+b)_{q^2} = 0$$

for each n.

One can write the set (labeled by n) of identities (3.2.24) in a compact form. Let t be an indeterminate. Denote by P_e and P_o the Poincaré series for even and odd variables, respectively; that is, P_e and P_o are the generating functions for the dimensions d_e and d_o, $P_e(t) = \sum_a d_e(a) t^a$ and $P_o(t) = \sum_a d_o(a) t^a$. We have

(3.2.25) $$P_e(t) P_o(-t) = \sum_n \sum_{a+b=n} d_e(a) d_o(b) (-1)^b .$$

Introduce a q-derivative in t. It satisfies, by definition, a relation $\partial_t t = 1 + q^2 t \partial_t$. By induction,

(3.2.26) $$\partial_t t^n = (n)_{q^2} t^{n-1} + q^{2n} t^n \partial_t .$$

As above, ∂_t becomes an operator after we define a vacuum - a one dimensional representation of the algebra of polynomials in ∂_t, Vac_t, by $\partial_t \text{Vac}_t = 0$. In particular, $\partial_t(t^n) = (n)_{q^2} t^{n-1}$.

The formula (3.2.26) shows that the action of ∂_t on the formal power series in t is well defined.

Now, (3.2.24) implies

(3.2.27) $$\partial_t(P_e(t) P_o(-t)) = 0 .$$

Note that the series P_e and P_o start with 1, $P_e(t) = 1 + \mathcal{O}(t)$, $P_o(t) = 1 + \mathcal{O}(t)$. Therefore,

(3.2.28) $$P_e(t) P_o(-t) = 1 + \mathcal{O}(t)$$

as well. Classically ($q = 1$), equations (3.2.27) and (3.2.28) imply that

(3.2.29) $$P_e(t) P_o(-t) = 1 .$$

For a generic q the same conclusion (3.2.29) holds. Here "generic" means that q is not a root of unity. However, if q^2 is a primitive root of unity of order l one can conclude only that $P_e(t) P_o(-t) = 1 + t^l F(t^l)$ for some power series F. By a different method, without using the differential operators, the formula (3.2.29) for generic q was obtained in [**16**].

The advantage of having a formula like (3.2.29) is that in the GL case the relations for the odd generators ξ are strong enough to force the space of polynomials in ξ to be finite-dimensional. Then $P_o(t)$ is a polynomial and instead of checking the infinite number of coefficients in $P_e(t)$ one has only finite number of checks for $P_o(t)$.

3.3. Geometry of 3-dimensional quantum spaces. In dimension 2, a quantum vector space is a quadratic algebra with two generators and one relation. This situation can be quickly analyzed [**17**] and we shall not stop at it here.

For a 3-dimensional quantum space we need 3 generators and 3 relations. Let

(3.3.1) $$E_{ij}^\alpha x^i x^j = 0 , \alpha = 1, 2, 3 ,$$

be the three relations. The number of independent monomials of degree k in dimension 3 is $\binom{k+2}{k}$, so we need to have $\binom{3+2}{2} = 10$ cubics.

In the free associative algebra with three generators there are 27 cubics. Thus we need 17 relations for cubics. How many relations can we deduce from (3.3.1)? We have 9 relations of the form $(E_{ij}^\alpha x^i x^j) x^k = 0$ and 9 relations of the form $x^k (E_{ij}^\alpha x^i x^j) = 0$, altogether 18 relations. Therefore they cannot be independent, there should 18-17=1 linear combination of them which vanishes. Therefore,

(3.3.2) $$e_{i\alpha} E_{jk}^\alpha = E_{ij}^\beta f_{\beta k}$$

for some tensors $e_{i\alpha}$ and $f_{\beta k}$. We shall assume that the tensors $e_{i\alpha}$ and $f_{\beta k}$ are nondegenerate.

Let $E_{ijk} = e_{i\alpha} E_{jk}^{\alpha}$ (all the indices of E_{ijk} are now of the same nature; before, α, β labeled relations, while i, j, k labeled variables). The equation (3.3.2) now becomes

$$(3.3.3) \qquad E_{ijk} = Q_k^l E_{lij} ,$$

where $Q_j^i = f_{\alpha j}(e^{-1})^{\alpha i}$ (e^{-1} is inverse to e,

$$(3.3.4) \qquad e_{\alpha i}(e^{-1})^{\alpha j} = \delta_i^j \quad \text{and} \quad e_{\alpha i}(e^{-1})^{\beta i} = \delta_\alpha^\beta ,$$

δ_i^j and δ_α^β are Kronecker delta's; the fact that two relations (3.3.4) hold is because in dimension 3, both indices i and α run from 1 to 3).

A direct inspection shows that classically (for commuting variables) E_{ijk} is the ϵ-tensor. The ϵ-tensor has a good behavior under all permutations of indices. The moral is that for the PBW-algebras, it is enough that the E-tensor behaves well only under cyclic permutations of indices - the effect of a cyclic permutation is a rotation in one index by an operator Q.

This simple behavior under cyclic permutations makes possible a classification of PBW-algebras in dimension 3: go to a basis in which Q has a normal form then solve the cyclicity equation (3.3.3) for the E-tensor and select nondegenerate solutions (which give exactly three relations for quadrics). The article [18] contains the result of the classification. The list of PBW-algebras is quite large; for us it is the beginning of the work: one has to classify compatible pairs of quantum spaces.

We have now two tensors, E_{ijk} and F^{ijk}. The analysis is quite lengthy - because one has to work with the Poincaré series of nine variables T_j^i. But the final result [19] is surprisingly simple.

It turns our that $E_{jmn}F^{mni} = x\delta_j^i$ where x is a number (in fact, this relation describes a little more narrow $SL(3)$-case, when the quantum group has a central determinant and one can define a corresponding special linear quantum group; for the general situation, see [19]).

Define $A_{kl}^{ij} = xE_{klm}F^{mij}$. Then the resulting equations say that A is a projector, $A^2 = A$ and

$$(3.3.5) \qquad (1+\kappa)\,\mathrm{tr}_3(A_{13}A_{23}Q_3^{-1}) = x^{-1}P_{12}Q_1^{-1} + 1 ,$$

where P_{12} is a permutation of spaces 1 and 2 and $\kappa = x\,\mathrm{tr}\,Q$.

Surprise: the equations for A imply that $\hat{R} = 1 + (1-q)A$ satisfies the Yang-Baxter equation, where q is a solution of $q^2 + (1-\kappa)q + 1 = 0$ (the other root defines $\hat{R}^{-1}$). Classically, $\kappa = 3$, $q = 1$ and $\hat{R} = P$.

This $\hat{R}$-matrix is of $GL(3)$-type and the relations for T_j^i ensuring that the compatible left and right spaces are covariant are nothing else but the RTT-relations.

In the beginning there was no demand to have a solution of the Yang-Baxter equations. The demands were to have PBW-algebras and the compatibility between them. So, unexpectedly, the study of the correct Poincaré series is a machine to produce $\hat{R}$-matrices and quantum groups.

In the list of $\hat{R}$-matrices found in this way in [**19**] there is an example which stands out for several reasons. The left quantum space is defined by relations

$$\zeta zx + \zeta^5 y^2 + xz = 0 ,$$

(3.3.6)
$$\zeta^2 z^2 + yx + \zeta^4 xy = 0 ,$$

$$zy + \zeta^7 yz + \zeta^8 x^2 = 0 .$$

Here ζ is a primitive 9-th root of unity; the operator Q has the form

(3.3.7)
$$Q = \begin{pmatrix} \zeta & & \\ & \zeta^4 & \\ & & \zeta^7 \end{pmatrix} .$$

The left quantum space (3.3.6) is compatible with an (isomorphic as an algebra) right quantum space; one can take $x = 1$. Thus we have a quantum group and an $\hat{R}$-matrix.

The $\hat{R}$-matrix is given by

(3.3.8)
$$\hat{R} = \zeta^6 + D\hat{\Sigma} ,$$

where

(3.3.9)
$$\hat{\Sigma} = \begin{pmatrix} 1 & & & & & & \zeta^4 & & \zeta^2 \\ & \zeta^8 & & 1 & & & & & \zeta \\ & & \zeta^4 & & \zeta^8 & & 1 & & \\ & 1 & & \zeta & & & & & \zeta^2 \\ & & \zeta^5 & & 1 & & \zeta & & \\ & \zeta^7 & & & & \zeta^2 & & 1 & \\ & & 1 & & \zeta^4 & & \zeta^5 & & \\ \zeta^5 & & & & & 1 & & \zeta^7 & \\ & \zeta^7 & & \zeta^8 & & & & & 1 \end{pmatrix} ,$$

and

(3.3.10)
$$D = \text{diag}(\sigma_2, \sigma_2, \sigma_1, \sigma_2, \sigma_3, \sigma_3, \sigma_1, \sigma_3, \sigma_1)$$

with

(3.3.11)
$$\begin{aligned} 3\sigma_1 &= (\zeta^4 + \zeta^{-4}) - (\zeta^2 + \zeta^{-2}) , \\ 3\sigma_2 &= (\zeta^1 + \zeta^{-1}) - (\zeta^4 + \zeta^{-4}) , \\ 3\sigma_3 &= (\zeta^2 + \zeta^{-2}) - (\zeta^1 + \zeta^{-1}) . \end{aligned}$$

We have $q = \zeta^3$.

For the standard Drinfeld-Jimbo deformation, the left quantum space is given by $x^i x^j = q x^j x^i$ for $i < j$. When q is a primitive root of unity of order l, then the left quantum space has a center generated by elements $(x^i)^l$. If one requires that the covariance algebra of T^i_j preserves relations $(x^i)^l = c_i$, one obtains additional relations for T^i_j. The quotient algebra of the algebra of T^i_j by these relations is called a "small quantum group" [**20**].

The center of the algebra (3.3.6) is a polynomial ring generated by three elements of degrees 3,6 and 9. The algebra (3.3.6) is finite-dimensional over its center, the dimension equals 162 ([**18**]). Therefore, the quantum group defined by the $\hat{R}$-matrix (3.3.8) has finite-dimensional quotients as well.

The algebra (3.3.6) does not admit ordering. In other words, in any basis, the defining relations are not ordering relations, ordering of $x^3 x^2 x^1$ will always produce loops. The algebra (3.3.6) is the first example of a PBW-algebra with this property. Therefore, the $\hat{R}$-matrix (3.3.8) is a very particular point in the moduli space of solutions of the Yang-Baxter equation in dimension 3. Another peculiarity is that the $\hat{R}$-matrix (3.3.8) is an isolated point in the space of solutions of the Yang-Baxter equation; it cannot be obtained as a deformation of any other solution; in particular, one cannot reach it starting from the classical solution (the permutation). In this sense, this $\hat{R}$-matrix is non-perturbative.

Call $\mathcal{E}$ the algebra (3.3.6). In the next subsection we prove some of its properties mentioned above, in particular, the PBW property.

3.3.1. *Gröbner base for $\mathcal{E}$*. For a homogeneous element f of a free associative algebra $\mathcal{A}$ with generators $\{x^1, \ldots, x^N\}$, let $\hat{f}$ be a "highest symbol" of f, the lexicographically highest word in f.

Let $\mathcal{B}$ be a quotient algebra of the algebra $\mathcal{A}$ by some homogeneous relations $\mathcal{S}_0 = \{r_1, \ldots, r_M\}$. Every relation r we write in the form $\hat{r}$ = terms, smaller than $\hat{r}$; we understand it as an instruction to replace $\hat{r}$ by the right hand side. Taking, if necessary, linear combinations of relations, we always assume that all $\hat{r}$ are different.

Let $\hat{\mathcal{S}}_0 = \{\hat{r}_1, \ldots, \hat{r}_M\}$.

A word can contain several entries of the form $\hat{r}_\alpha$ for some α. Comparing different ways of applying instructions to this word, we may obtain new instructions - relations, whose highest symbols do not belong to $\hat{\mathcal{S}}_0$. We add these relations to $\mathcal{S}_0$ and obtain a new set $\mathcal{S}_1$. Let again $\hat{\mathcal{S}}_1$ be the set of highest symbols.

Continuing the process, we shall build an (eventually infinite) set $\mathcal{S} = \cup_{i=0}^\infty \mathcal{S}_i$, which is called a Gröbner base for the algebra $\mathcal{A}$ (it depends on a choice of generators $\{x^1, \ldots, x^N\}$ and on a choice of an order). Let $\hat{\mathcal{S}}$ be the corresponding set of highest symbols.

Now the basis of $\mathcal{A}$, as a vector space, consists of "normal" words - words, which do not have subwords belonging to $\hat{\mathcal{S}}$. This gives sometimes a way to estimate the Poincaré series of the algebra. See, *e.g.*, [21] for further information about Gröbner bases.

For the algebra $\mathcal{E}$, written in generators $\{x, y, z\}$, as in (3.3.6), the Gröbner base seems to be infinite ant non analyzable.

There are several other nice sets of generators and one of them leads to a finite Gröbner base. Let

$$x = \gamma \zeta^2 (A + \zeta^3 B + \zeta^6 C) ,$$

(3.3.12)
$$y = -\tfrac{1}{1+\zeta}(A + B + C) ,$$

$$z = \gamma^2 (A + \zeta^6 B + \zeta^3 C) ,$$

where γ satisfies $\gamma^3 = \zeta \dfrac{1+\zeta^2}{1+\zeta}$.

In terms of new generators A, B and C the relations are

$$\zeta A^2 + \zeta^5 AB + B^2 = 0 ,$$

(3.3.13)
$$\zeta C^2 + \zeta^5 CA + A^2 = 0 ,$$

$$\zeta B^2 + \zeta^5 BC + C^2 = 0 .$$

Choose the order $A > B > C$. Then the set (3.3.13) of relations gives the following set of instructions

$$A^2 \rightsquigarrow -\zeta C^2 - \zeta^5 CA \ ,$$

(3.3.14)
$$B^2 \rightsquigarrow -\zeta^4 BC - \zeta^8 C^2 \ ,$$

$$AB \rightsquigarrow -C^2 + \zeta CA + \zeta^8 BC \ .$$

Possible overlaps are A^2B, AB^2, A^3 and B^3. This leads to new instructions

$$ACA \rightsquigarrow BC^2 + \zeta^2 CAC + \zeta^6 CBC + \zeta^8 C^2 A + \zeta^3 C^2 B - \zeta C^3 \ ,$$

(3.3.15)
$$BCB \rightsquigarrow -\zeta^4 C^2 B - \zeta^8 C^3 \ ,$$

$$AC^2 \rightsquigarrow -\zeta^4 BC^2 - \zeta^6 CAC - \zeta CBC - \zeta^3 C^2 A - \zeta^7 C^2 B \ .$$

One has new overlaps and they, in turn, lead to new instructions with highest symbols BC^3 and $ACBC$. We shall not give more details, but it turns out that now overlaps are all compatible, so the construction of the Gröbner base is completed and we have

(3.3.16) $\qquad \hat{\mathcal{S}} = \{A^2, B^2, AB, ACA, BCB, AC^2, BC^3, ACBC\} \ .$

For such $\hat{\mathcal{S}}$ it is possible to explicitly describe the normal form.

For a word $w = x_1 x_2 \ldots x_k$, let $\beta_j(w)$ be the beginning, of the length j (that is, first j symbols), of a word $www\ldots w$ (the word w repeated sufficiently many times). For example, $\beta_4(ACB) = ACBA$, $\beta_5(ACB) = ACBAC$.

Lemma. For $\hat{\mathcal{S}}$, as in (3.3.16), the normal words have a form

(3.3.17) $\qquad C^i \beta_j(BCC) \beta_k(ACB) \ .$

Corollary. The algebra $\mathcal{E}$ has the PBW property.

Proof. Normal words (3.3.17) are characterized by ordered triples of numbers $\{i, j, k\}$, as for monomials $x_1^i x_2^j x_3^k$. $\qquad \square$.

3.4. $sl_q(2)$ at roots of unity. The simplest example of a quantum space is the algebra V_q^2 with two generators x^1 and x^2 subjected to the relation (3.0.3). If one chooses for a left quantum space $(V^*)_q^2$ an algebra with two generators x_1 and x_2 and the same relation

(3.4.1) $\qquad x_1 x_2 = q x_2 x_1 \ ,$

then the quantum group (rather, bialgebra, for the moment we will not talk about invertibility of quantum matrices) which preserves the relations (3.0.3) and (3.4.1) is the standard $Mat_q(2)$, the matrix elements of $T = \begin{pmatrix} a & b \\ c & d \end{pmatrix}$ satisfy relations

(3.4.2)
$$ab = qba \ , \ ac = qca \ , \ bd = qdb \ , \ cd = qdc \ ,$$
$$bc = cb \ , \ ad = da + (q - q^{-1})bc$$

(this is a correct quantization of the Poisson brackets (3.0.1)).

If q is a primitive l-th root of unity, the bialgebra $Mat_q(2)$ has a finite dimensional quotient $\overline{Mat}_q(2)$, one adds

(3.4.3) $$a^l = d^l = 1, b^l = c^l = 0$$

to the relations (3.4.2).

The bialgebra $\overline{Mat}_q(2)$ has a symmetry interpretation as well. The elements x_1^l and x_2^l lie in the center of V_q^2. Let $V_{q\mu_1\mu_2}^2$ be a quotient of V_q^2 by relations $(x^1)^l = \mu_1$ and $(x^2)^l = \mu_2$ for some constants μ_1 and μ_2. If one requires that all the algebras $V_{q\mu_1\mu_2}^2$ are preserved by the coaction, one finds the extra relations (3.4.3); the same relations (3.4.3) one finds from a demand that all the left algebras $(V^*)_{q\mu_1\mu_2}^2$ are preserved.

In this subsection we shall illustrate, on this simple 2-dimensional example, some phenomena, pertinent to a situation when a non-commutative quantum space has a large center: loss of quasi-triangularity, loss of semi-simplicity, appearance of finite-dimensional Hopf quotients *etc.*

We shall give a description of the reduced universal enveloping algebra and of the reduced function algebra in terms of matrix algebras over local rings. This language seems to be quite appropriate to talk about such algebraic concepts as Ext-groups, a scheme of an algebra, its Cartan matrix *etc.* The material on the matrix structure is partly taken from [**22**].

Notation: n_q is a q-number, $n_q = \dfrac{q^n - q^{-n}}{q - q^{-1}}$ and $n_q! = 1_q 2_q \ldots n_q$ is a q-factorial. Let q be a l-th primitive root of unity, $l > 2$ (so $q^2 \neq 1$). Denote

(3.4.4) $$\tilde{l} = \begin{cases} l & , \quad l \equiv 1 \ (mod\ 2) \\ l/2 & , \quad l \equiv 0 \ (mod\ 2) \end{cases}$$

Thus, $q^{2n} = 1 \iff 2n \equiv 0 \ (mod\ l) \iff n \equiv 0 \ (mod\ \tilde{l})$; $n_q = 0 \iff n \equiv 0 \ (mod\ \tilde{l})$. Denote $\nu = l/\tilde{l}$ and $\tilde{q} = q^\nu$; $\tilde{q}^{\tilde{l}} = 1$, $\tilde{q}$ is a primitive $\tilde{l}$-th root of unity.

3.4.1. *Preliminaries.* The Hopf algebra which gives rise (as in the section 3) to the quantum space V_q^2 is an algebra $\mathcal{U} = \mathcal{U}_q(sl_2)$, generated by elements K, K^{-1}, E and F and relations

(3.4.5) $$KK^{-1} = K^{-1}K = 1, \quad KE = q^2 EK,$$
$$KF = q^{-2} FK, \quad [E, F] = \frac{K - K^{-1}}{q - q^{-1}};$$

the coproduct is defined on the generators by

(3.4.6) $\quad \Delta K = K \otimes K, \quad \Delta E = E \otimes K + 1 \otimes E, \quad \Delta F = F \otimes 1 + K^{-1} \otimes F;$

the counit ε and the antipode S are defined on the generators by

(3.4.7) $$\varepsilon(K) = 1, \quad \varepsilon(E) = 0, \quad \varepsilon(F) = 0,$$

(3.4.8) $$S(K) = K^{-1}, \quad S(E) = -EK^{-1}, \quad S(F) = -KF.$$

The algebra $\mathcal{U}_q(sl_2)$ has a central element, a q-deformed Casimir operator:

(3.4.9) $$C = qK + q^{-1}K^{-1} + (q - q^{-1})^2 FE.$$

If $q = \exp(\alpha)$ and $K = \exp(\alpha H)$, the combination $\dfrac{C-2}{(q-q^{-1})^2} - \dfrac{1}{4}$ tends to the standard Casimir operator $\dfrac{H^2}{4} + \dfrac{H}{2} + FE$ in the classical limit $\alpha \to 0$.

Consider a vector space $V(\sigma, j)$, $j \in \mathbf{Z}/2$, $j = 0, \frac{1}{2}, \ldots, \frac{\tilde{l}-1}{2}$ and $\sigma = \pm 1$ with a basis $\{e_j^m, \ m = j, j-1, \ldots, -j\}$. Denote by $K(\sigma, j)$, $E(\sigma, j)$ and $F(\sigma, j)$ the operators

(3.4.10)
$$K(\sigma, j)e_j^m = \sigma q^{2m} e_j^m,$$
$$E(\sigma, j)e_j^m = e_j^{m+1},$$
$$F(\sigma, j)e_j^m = \sigma (j+m)_q (j-m+1)_q e_j^{m-1}.$$

In these formulas, the right hand side should be replaced by 0 if $m \pm 1$ runs out of the allowed range.

The operators (3.4.10) realize standard representations of U_q. When q is not a root of unity, the representations $V(\sigma, j)$ exhaust the list of all irreducible representations.

The expression

(3.4.11)
$$\mathcal{R} = e^{\alpha \frac{H \otimes H}{2}} \sum_{m \geq 0} \frac{(q - q^{-1})^m}{m_q!} q^{\frac{m(m-1)}{2}} (E \otimes F)^m,$$

being understood informally, intertwines the coproduct with the opposite coproduct. However, because of the denominators, the expression (3.4.11) does not make sense when q is a root of unity. One may ask whether it is possible to redefine $\mathcal{R}$ at these values of q. The answer is negative. A standard argument goes as follows. If $\mathcal{R}$ existed, we would have an isomorphism $V \otimes W \simeq W \otimes V$ for any two representations of $\mathcal{U}$, for which $\mathcal{R}$ is defined ($\mathcal{R}$ would intertwine the tensor products).

Elements $x = E^{\tilde{l}}$, $y = F^{\tilde{l}}$ and $z = K^{\tilde{l}}$ are central; we have

(3.4.12) $\quad \Delta z = z \otimes z, \Delta x = x \otimes z + 1 \otimes x, \Delta y = y \otimes 1 + z^{-1} \otimes y.$

There is a family of representations $W_{\mu ab}$ of dimension $\tilde{l}$ (the index j runs from 0 to $\tilde{l} - 1$):

(3.4.13)
$$K : v_j \mapsto \mu q^{-2j} v_j,$$
$$F : v_j \mapsto v_{j+1} \ \text{for} \ j < \tilde{l} - 1,$$
$$E : v_j \mapsto \left(\frac{\mu q^{1-j} - \mu^{-1} q^{j-1}}{q - q^{-1}} j_q + ab\right) v_{j-1}, \ \text{for} \ j > 0,$$
$$F : v_{\tilde{l}-1} \mapsto b v_0, \ E : v_0 \mapsto a v_{\tilde{l}-1}.$$

The values of the parameters μ, a and b are not restricted (one only needs $\mu \neq 0$).

In the representation $W_{\mu ab}$, the value of the element y is b, the value of the element z is $\mu^{\tilde{l}}$.

Assume that $V \otimes W \simeq W \otimes V$. Then, applying the formula (3.4.12) for the coproduct of the element y, we find

(3.4.14) $\quad y_V + z_V^{-1} y_W = y_W + z_W^{-1} y_V,$

where y_V and z_V are the operators, corresponding to the elements y and z in the representation V (the same for W).

Take $V = W_{\mu ab}$ and $W = W_{\nu cd}$. Then (3.4.14) implies a relation between μ, ν, b and d, a contradiction.

1. Hopf ideals

Here we collect some information about Hopf ideals of a finite codimension in $\mathcal{U}$.

The Hopf subalgebra of Laurent polynomials in K coincides with the group algebra $\mathbf{C}[\mathbf{Z}]$ of the additive group $\mathbf{Z}$ of integers.

Lemma. Let I be a proper Hopf ideal in $\mathbf{C}[\mathbf{Z}]$. Then I is generated by $(K^j - 1)$ for some j.

Proof. Any ideal I in $\mathbf{C}[\mathbf{Z}]$ is a principal ideal, $I = (f)$, where $f(t) = t^j + a_{j-1}t^{j-1} + \cdots + a_0$ is a polynomial with $a_0 \neq 0$.

The element

$$(3.4.15) \qquad \Delta f(K) = K^j \otimes K^j + a_{j-1}K^{j-1} \otimes K^{j-1} + \cdots + a_0$$

equals

$$(a_{j-1}K^{j-1} + \cdots + a_0) \otimes (a_{j-1}K^{j-1} + \cdots + a_0) + a_{j-1}K^{j-1} \otimes K^{j-1} + \cdots + a_0$$

in the algebra $\mathbf{C}[\mathbf{Z}]/I \otimes \mathbf{C}[\mathbf{Z}]/I$. If I is a Hopf ideal then the element (3.4.15) must be zero. In particular, in the expression above, the coefficient in $1 \otimes K^i$ for $1 \leq i \leq j-1$ must vanish, which gives $a_0 a_i = 0$. Therefore $a_i = 0$. Vanishing of the coefficient in $1 \otimes 1$ gives $a_0(a_0 + 1) = 0$, therefore $a_0 = -1$. Thus, $f(K) = K^j - 1$, and it is straightforward to check that (f) is a Hopf ideal for such f. □

Consider a Hopf ideal I of a finite codimension. If $E \in I$ then $K^2 - 1 = (q - q^{-1})K[E, F] \in I$, therefore

$$(K^2 - 1)F = F(q^{-2}K^2 - 1) \equiv F(q^{-2} - 1)(mod\ I) \in I ,$$

so $F \in I$. Thus the factor-algebra is $\mathbf{C}[\mathbf{Z}_2]$.

Assume now that $E \notin I$. Let $\bar{\mathcal{U}}$ be the factor-algebra of $\mathcal{U}$ by I.
According to the Lemma, $(K^j - 1) \in I$ for some j. Therefore,

$$(K^j - 1)E = E(q^{2j}K^j - 1) \equiv E(q^{2j} - 1)(mod\ I) \in I ,$$

which implies $j = m\tilde{l}$, so $\bar{z}^m = 1$ in the factor-algebra $\bar{\mathcal{U}}$ ($\bar{z}$ is the image in $\bar{\mathcal{U}}$ of the central element $z = K^{\tilde{l}} \in \mathcal{U}$).

Lemma. The central elements $x = E^{\tilde{l}}$ and $y = F^{\tilde{l}}$ belong to I.

Proof. Let $\bar{x}$ be the image of x in $\bar{\mathcal{U}}$. Let $f(t) = t^i + b_{i-1}t^{i-1} + \cdots + b_0$ be a characteristic polynomial of $\bar{x}$ in $\bar{\mathcal{U}}$. We have

$$\Delta f(x) = (x_1 z_2 + x_2)^i + b_{i-1}(x_1 z_2 + x_2)^{i-1} + \cdots + b_0 ,$$

where $x_1 = x \otimes 1$, $x_2 = 1 \otimes x$ and $z_2 = 1 \otimes z$. Thus, in $\bar{\mathcal{U}} \otimes \bar{\mathcal{U}}$ one has

$$0 = \Delta f(\bar{x}) = -(b_{i-1}(\bar{x}_1 \bar{z}_2)^{i-1} + \cdots + b_0) - (b_{i-1}\bar{x}_2^{i-1} + \cdots + b_0)$$

$$(3.4.16) \qquad + \sum_{s=1}^{i-1} \binom{i}{s} (\bar{x}_1 \bar{z}_2)^s \bar{x}_2^{i-s} + b_{i-1}(\bar{x}_1 \bar{z}_2 + \bar{x}_2)^{i-1} + \cdots + b_0 .$$

where $\bar{x}_1 = \bar{x} \otimes 1$, $\bar{x}_2 = 1 \otimes \bar{x}$ and $\bar{z}_2 = 1 \otimes \bar{z}$. The coefficient in, for example, $(\bar{x}_1 \bar{x}_2^{i-1})$ is $\bar{z}_2 i$. Thus, $i = 0$, therefore, $\bar{x} = 0$ or $x \in I$. Similarly, $y \in I$. □

Denote by I_m the ideal

(3.4.17)
$$I_m = \{E^{\tilde{l}}, F^{\tilde{l}}, K^{m\tilde{l}} - 1\} .$$

We shall call it a congruence ideal, and the number m - level.

We have shown that each Hopf ideal of a finite codimension contains a congruence ideal I_m for some m. The minimal m for which it happens, we shall call the level of the ideal.

Denote $\mathcal{U}_q/I_m$ by $\bar{\mathcal{U}}_{q,m}$ and the images of the elements E, F and K by $\bar{E}$, $\bar{F}$ and $\bar{K}$.

We shall give a complete description of $\bar{\mathcal{U}}_{q,1}$ (and of $\bar{\mathcal{U}}_{q,2}$) as an algebra[2].

2. Equation for C

We shall find a polynomial $\chi(x)$ such that $\chi(\bar{C}) = 0$ ($\bar{C}$ is the image of the Casimir operator C (3.4.9)). Later we shall prove that χ is a minimal polynomial for $\bar{C}$.

One has

(3.4.19)
$$(q - q^{-1})^2 FE = C - (qK + q^{-1}K^{-1}) .$$

Lemma. The following relation holds in U_q:

(3.4.20)
$$(q - q^{-1})^{2i} F^i E^i = \prod_{a=0}^{i-1}(C - (q^{1+2a}K + q^{-1-2a}K^{-1})) .$$

Proof. For $i = 1$ this is (3.4.19). Induction in i:

(3.4.21)
$$(q - q^{-1})^{2i+2} F^{i+1} E^{i+1} = (q - q^{-1})^2 F \prod_{a=0}^{i-1}(C - (q^{1+2a}K + q^{-1-2a}K^{-1}))E$$
$$= (q - q^{-1})^2 FE \prod_{a=0}^{i-1}(C - (q^{1+2a+2}K + q^{-1-2a-2}K^{-1})) .$$

Use (3.4.19) to finish the proof. □

Corollary. In $\bar{\mathcal{U}}_{q,m}$ one has

(3.4.22)
$$\prod_{a=0}^{\tilde{l}-1}(\bar{C} - (q^{1+2a}\bar{K} + q^{-1-2a}\bar{K}^{-1})) = 0 .$$

Proof. For $i = \tilde{l}$, the lhs of (3.4.20) belongs to the ideal I_m. □

Denote by $p(x)$ a polynomial $p(x) = 1 + x + \ldots x^{\tilde{l}-1}$ and let

(3.4.23)
$$p_a = \frac{1}{\tilde{l}} p(\tilde{q}^a \bar{K}) , \quad a = 0, \ldots, \tilde{l} - 1 .$$

---

[2]We shall not use it, but it is known (see, e.g. [23]) that $\bar{\mathcal{U}}_{q,1}$ and $\bar{\mathcal{U}}_{q,2}$ are quasitriangular, say, for $\bar{\mathcal{U}}_{q,1}$ the universal $\mathcal{R}$-matrix is

(3.4.18)
$$\mathcal{R} = \frac{1}{\tilde{l}} \sum_{i,j=0}^{\tilde{l}} q^{-2ij} K^i \otimes K^j \sum_{s=0}^{\tilde{l}} \frac{(q-q^{-1})^s}{s_q!} q^{\frac{s(s-1)}{2}} \bar{E}^s \otimes \bar{F}^s .$$

Then

(3.4.24) $$\bar{K}p_a = \tilde{q}^{-a}p_a .$$

The elements p_a are the usual idempotents, decomposing 1:

(3.4.25) $$p_a p_b = \delta_{ab} p_b , \quad 1 = p_0 + \cdots + p_{\tilde{l}-1} .$$

Define a polynomial $\chi(x)$,

(3.4.26) $$\chi(x) = \prod_{a=0}^{\tilde{l}-1}(x - q^{1+2a} - q^{-1-2a}) .$$

More precisely,

(3.4.27) $$\chi(x) = \begin{cases} \prod_{a=0}^{l-1}(x - q^a - q^{-a}) , & \nu = 1 \\ \prod_{a=0}^{\tilde{l}-1}(x - q\tilde{q}^a - q^{-1}\tilde{q}^{-a}) , & \nu = 2 \end{cases}$$

Lemma. In $\bar{\mathcal{U}}_{q,1}$, one has

(3.4.28) $$\chi(\bar{C}) = 0 .$$

Proof. We have (using (3.4.24), the Corollary above and the fact that $\bar{C}$ and $\bar{K}$ commute)

(3.4.29)
$$0 = \prod_{a=0}^{\tilde{l}-1}(\bar{C} - (q^{1+2a}\bar{K} + q^{-1-2a}\bar{K}^{-1}))p_{-b}$$
$$= \prod_{a=0}^{\tilde{l}-1}(\bar{C} - (q^{1+2a}\tilde{q}^b + q^{-1-2a}\tilde{q}^{-b}))p_{-b} = \chi(\bar{C})p_{-b}$$

for all b. Summing over b and using (3.4.25) we conclude that $\chi(\bar{C}) = 0$. □

Remarks. 1. For odd l the eigenvalue in (3.4.27) corresponding to $a = 0$ is simple, the others have the multiplicity 2 - pairs $(a, l-a)$. For l even, $\tilde{l}$ odd: the eigenvalue corresponding to $a = (\tilde{l} - 1)/2$ is simple, the others have the multiplicity 2 - pairs $(a, \tilde{l} - 1 - a)$. For l even, $\tilde{l}$ even: all eigenvalues have the multiplicity 2 - pairs $(a, \tilde{l} - 1 - a)$.

2. If we knew that χ is a minimal polynomial, we could immediately state that there are indecomposable but not irreducible representations: the center of a semisimple algebra is semisimple.

3.4.2. *Formatted matrix algebras over graded rings.* Let Γ be a finite abelian group, $\hat{\Gamma}$ its dual.

Let A be a Γ-graded ring over $\mathbf{C}$, that is, $A = \oplus_{\chi \in \hat{\Gamma}} A_\chi$ and if $a \in A_\chi$, $b \in A_{\chi'}$ then $ab \in A_{\chi\chi'}$.

Let $\mathcal{I}$ be a set. A couple ϕ, consisting of the set $\mathcal{I}$ and a map $\mathcal{I} \to \hat{\Gamma}$ we shall call a "format".

Definition. A set of matrices $X = \{X^i_j\}$ with indices belonging to the set $\mathcal{I}$ and with entries in A will be called a matrix algebra of format ϕ over A (and denoted by $M_\phi(A)$) if $X^i_j \in A_{\phi(i)\phi(j)^{-1}}$.

Clearly, $M_\phi(A)$ is an algebra.

In our examples the ring A will satisfy two conditions:

C1 A is local, that is, $A/\operatorname{rad} A$ is isomorphic to $\mathbf{C}$.

C2 The part of A which corresponds to the trivial representation of Γ is $\mathbf{C}$ itself; in other words, $\operatorname{rad} A = \oplus_{\text{nontrivial } \chi} A_\chi$.

The simplest example is the algebra $M_n(\mathbf{C})$ of matrices of size $n \times n$. Here the group Γ is trivial. The algebra $M_n(\mathbf{C})$ has only one representation - a column of the matrix.

Similarly, for any algebra $M_\phi(A)$, the columns provide representation spaces. Now there might be several types of columns, corresponding to the chosen format ϕ.

An advantage of the introduced terminology is summarized in the following lemma, which generalizes the properties of $M_n(\mathbf{C})$.

Lemma. Assume that A satisfies conditions C1 and C2. Then the columns realize principal projective modules of $M_\phi(A)$. The set of principal projective modules is in 1-1 correspondence with types of columns (that is, with the image of ϕ). The classes of isomorphism of the quotients of the principal projective modules of each type are in 1-1 correspondence with the graded quotients of A.

Therefore, a knowledge that some algebra is isomorphic to $M_\phi(A)$ gives a complete information about the representation theory for this algebra; there is no need to study first irreducible representations, then their extensions *etc.*

Let Λ_2 be the Grassmann algebra in two variables ξ and η. It is graded by the parity; the group Γ is $\mathbf{Z}_2$. Essentially, the format is specified by two numbers, m and n; we shall write $\phi = m|n$. The algebra $M_{m|n}(\Lambda_2)$ is the algebra of matrices

$$(3.4.30) \qquad \left(\begin{array}{c|c} X & Y \\ \hline Z & W \end{array} \right),$$

the entries of the matrices X and W are even, the entries of the matrices Y and Z are odd elements of Λ_2.

Remark. Let B be an algebra. Suppose that we know that it belongs to the class of formatted algebras over graded rings, *i. e.* it can be represented as $M_\phi(A)$ for some choice of Γ, A, $\mathcal{I}$ and ϕ. One may ask, how intrinsic the ring A is, whether it is defined by the algebra B. It turns out that for different rings A and A', the formatted matrix algebras over them can be isomorphic. In this case we shall say that A and A' are GM-equivalent (GM stands for "Graded Matrices").

Example. Let A' be a ring over $\mathbf{C}$ generated by two elements θ_1 and θ_2, satisfying $\theta_1^2 = \theta_2^2 = 0$ and $\theta_1 \theta_2 = \theta_2 \theta_1$. The ring A' is graded by $\mathbf{Z}_2$, the grading is given by a degree in the variables θ_i, $\deg \theta_i = 1$, $i = 1, 2$. The format is again specified by two numbers, $\phi = m|n$. The algebras $M_{1|1}(\Lambda_2)$ and $M_{1|1}(A')$ are isomorphic, the isomorphism is given by

$$(3.4.31) \qquad \left(\begin{array}{c|c} a_1 + a_2\xi\eta & b_1\xi + b_2\eta \\ \hline c_1\xi + c_2\eta & d_1 + d_2\xi\eta \end{array} \right) \mapsto \left(\begin{array}{c|c} a_1 + a_2\theta_1\theta_2 & b_1\theta_1 - b_2\theta_2 \\ \hline c_1\theta_1 + c_2\theta_2 & d_1 - d_2\theta_1\theta_2 \end{array} \right),$$

where $a_i, b_i, c_i, d_i \in \mathbf{C}$, $i = 1, 2$.

Thus, the rings Λ_2 and A' are GM-equivalent.

3.4.3. *Matrix structure.* In [24], after a description of irreducible and some indecomposable representations of $\tilde{\mathcal{U}}_{q,2}$, a regular representation of $\tilde{\mathcal{U}}_{q,2}$ was decomposed into a direct sum of indecomposable representations. As a consequence, the algebra $\bar{\mathcal{U}}_{q,2}$ decomposes into a direct sum of ideals. It was noticed in [**24**] that each of these ideals is isomorphic to a subalgebra in the matrix algebra whose matrix elements belong to a Grassmann algebra in two variables.

We shall adopt an opposite point of view and start by establishing homomorphisms into the matrix algebras with Grassmanian entries. Then we shall prove that the reduced enveloping algebras are direct sums of formatted matrix algebras over the local ring Λ_2. As explained above, this immediately provides an entire information about all modules, in particular, the principal projective modules.

Some homomorphisms into matrix algebras, odd l

Here we shall consider the case when the number l is odd.

Let $i = 2j + 1$. Let $K(i)$, $E(i)$ and $F(i)$ be operators corresponding to $\sigma = 1$ in formulas (3.4.10) (pay attention to the order of the basis vectors: $m = j, j-1, \ldots, -j$; say, the matrix of the operator $E(i)$ is upper-triangular).

We shall also use a matrix $M(i)$, defined by

$$(3.4.32) \qquad M(i) e_j^m = e_j^{m-1}$$

on the same basis as in (3.4.10).

Let ξ and η be two Grassmann variables, $\xi^2 = \eta^2 = \xi\eta + \eta\xi = 0$.
Let

$$(3.4.33) \quad \mathcal{K}(i) = \begin{pmatrix} K(i) & \bullet \\ \hline \bullet & K(\frac{l}{2} - i - 1) \end{pmatrix},$$

$$(3.4.34) \quad \mathcal{E}(i) = \begin{pmatrix} E(i) & \begin{matrix} \vdots & \bullet \\ \xi & \cdots \end{matrix} \\ \hline \begin{matrix} \vdots & \bullet \\ \xi & \cdots \end{matrix} & E(\frac{l}{2} - i - 1) \end{pmatrix},$$

$$(3.4.35) \quad \mathcal{F}(i) = \begin{pmatrix} F(i) & \begin{matrix} \cdots & -\eta \\ \bullet & \vdots \end{matrix} \\ \hline \begin{matrix} \cdots & \eta \\ \bullet & \vdots \end{matrix} & F(\frac{l}{2} - i - 1) \end{pmatrix} + \xi\eta \begin{pmatrix} M(i) & \bullet \\ \hline \bullet & -M(\frac{l}{2} - i - 1) \end{pmatrix}.$$

Dots mean that the corresponding entries are zero.

The diagonal entries of the operator $\mathcal{K}(i)$ form a sequence $\{a_n\}$, $n \in \mathbf{Z}/l\mathbf{Z}$,

$$\{a_n\} = \{q^{2i}, q^{2(i-1)}, \ldots, q^{-2i}, q^{2(\frac{l}{2}-i-1)}, \ldots, q^{-2(\frac{l}{2}-i-1)}\}.$$

Since $q^l = 1$, we have $a_{n+1} = q^{-2} a_n$ for all $n \in \mathbf{Z}/l\mathbf{Z}$. The non-zero entries of the operators $\mathcal{E}(i)$ and $\mathcal{F}(i)$ are exactly on those places which are allowed by relations

$KE = q^2EK$ and $KF = q^{-2}FK$. Next, one finds

$$\mathcal{E}(i)\mathcal{F}(i) = \begin{pmatrix} E(i)F(i) & \\ \hline & E(\frac{l}{2}-i-1)F(\frac{l}{2}-i-1) \end{pmatrix} + \xi\eta\mathcal{P} \tag{3.4.36}$$

and

$$\mathcal{F}(i)\mathcal{E}(i) = \begin{pmatrix} F(i)E(i) & \\ \hline & (F\frac{l}{2}-i-1)E(\frac{l}{2}-i-1) \end{pmatrix} + \xi\eta\mathcal{P}, \tag{3.4.37}$$

where

$$\mathcal{P} = \begin{pmatrix} 1 & \\ \hline & -1 \end{pmatrix}. \tag{3.4.38}$$

We have $[E(k), F(k)] = \dfrac{K(k) - K(k)^{-1}}{q - q^{-1}}$ for all k, so

$$[\mathcal{E}(i), \mathcal{F}(i)] = \dfrac{\mathcal{K}(i) - \mathcal{K}(i)^{-1}}{q - q^{-1}}.$$

Therefore, the operators $\mathcal{E}(i)$, $\mathcal{F}(i)$ and $\mathcal{K}(i)$ provide a representation of the algebra $\mathcal{U}$.

It is easy to verify that $\mathcal{E}(i)^l = \mathcal{F}(i)^l = 0$ due to nilpotency of the Grassmann variables. The relation $\mathcal{K}(i)^l = 1$ is evident. Thus, the matrices $\mathcal{E}(i)$, $\mathcal{F}(i)$ and $\mathcal{K}(i)$ realize a representation of $\bar{\mathcal{U}}_{q,1}$.

Matrix structure of $\bar{\mathcal{U}}_{q,1}$, odd l

Formulas (3.4.33)-(3.4.35) provide homomorphisms $(j = 2i + 1)$

$$\rho_j : \bar{\mathcal{U}}_{q,1} \to M_{j|l-j}(\Lambda_2) \tag{3.4.39}$$

for $j = 1, \ldots, l-1$ and a homomorphism

$$\rho_0 : \bar{\mathcal{U}}_{q,1} \to M_l(\mathbf{C}), \tag{3.4.40}$$

corresponding to $j = \frac{l-1}{2}$.

All the eigenvalues of the operator $\mathcal{K}(i)$ are different, so the diagonal matrices $\text{diag}(0, \ldots, 0, 1, 0, \ldots, 0)$ are polynomials in $\mathcal{K}(i)$ (projectors on the eigenspaces of $\mathcal{K}(i)$) and belong to the image of ρ_i. Now, looking at the matrices for the operators $\mathcal{E}(i)$ and $\mathcal{F}(i)$, one concludes immediately that ρ_j is an epimorphism for all $j = 1, \ldots, l$.

For the Casimir element $\bar{C}$ one computes

$$\rho_j(\bar{C}) = (q^j + q^{-j})I + (q - q^{-1})^2 \xi\eta\mathcal{P}, \tag{3.4.41}$$

(I is the identity operator) for $j = 1, \ldots, l-1$ and $\rho_0(\bar{C}) = 2$.

Because of the $\xi\eta$-term in (3.4.41), $\chi(t) = \prod_{a=0}^{l-1}(x - q^a - q^{-a})$ is indeed the minimal polynomial for $\bar{C}$ in $\bar{\mathcal{U}}_{q,1}$.

Let $P_j \in \bar{\mathcal{U}}_{q,1}$, $j = 0, \ldots, [\frac{l}{2}]$ ($[z]$ is the integer part of z) be central idempotents corresponding to the eigenvalues $q^j + q^{-j}$ of the semi-simple part of $\bar{C}$; $P_0 + \cdots + P_{[\frac{l}{2}]} = 1$ is the central decomposition of unity. We have $\bar{\mathcal{U}}_{q,1} = \oplus_j P_j \bar{\mathcal{U}}_{q,1}$.

Let $Y_j = P_j \bar{\mathcal{U}}_{q,1}$, $j = 0, \ldots, [\frac{l}{2}]$ and let B_j be the matrix algebras, $B_0 = M_l(\mathbf{C})$ and $B_a = M_{a|l-a}(\Lambda_2)$, $a = 1, \ldots, [\frac{l}{2}]$. Then $\rho_j : \bar{\mathcal{U}}_{q,1} \to B_j$ vanishes on Y_k for $k \neq j$ because of the value of $\bar{C}$. Thus, we have a collection of epimorphisms $Y_j \to B_j$, so their direct sum is an epimorphism

$$(3.4.42) \qquad \rho : \bar{\mathcal{U}}_{q,1} \to \oplus_{j=0}^{[\frac{l}{2}]} B_j \ .$$

Let $B = \oplus_{j=0}^{[\frac{l}{2}]} B_j$. We have $\dim(B_0) = l^2$ and $\dim(B_a) = 2l^2$, $a = 1, \ldots, [\frac{l}{2}]$. Therefore, $\dim(B) = l^2 + \frac{l-1}{2} \cdot 2l^2 = l^3$.

On the other hand, relations (3.4.5) clearly allow an ordering: we can rewrite any expression as, say, a linear combination of monomials $\bar{K}^a \bar{F}^b \bar{E}^c$. Therefore, $\dim(\bar{\mathcal{U}}_{q,1}) \leq l^3$. But (3.4.42) is the epimorphism, so $\dim(\bar{\mathcal{U}}_{q,1}) = l^3$ and (3.4.42) is an isomorphism. We proved:

Proposition. For odd l, the algebra $\bar{\mathcal{U}}_{q,1}$ is isomorphic to a direct sum of formatted matrix algebras,

$$(3.4.43) \qquad \bar{\mathcal{U}}_{q,1} \simeq M_l(\mathbf{C}) \oplus \bigoplus_{a=1}^{[\frac{l}{2}]} M_{a|l-a}(\Lambda_2) \ .$$

As a byproduct, we saw that the monomials $\bar{K}^a \bar{F}^b \bar{E}^c$ are linearly independent. This is a version of the Poincaré-Birkhoff-Witt theorem for $\bar{\mathcal{U}}_{q,1}$: the monomials $\bar{K}^a \bar{F}^b \bar{E}^c$, with $a, b, c = 1, \ldots, l$ form a basis.

Exercise. Describe the matrix structure of $\bar{\mathcal{U}}_{q,2}$ (that is, $K^{2l} = 1$): replace the operators $K(i)$, $E(i)$ and $F(i)$ in formulas (3.4.33)-(3.4.35) by the operators corresponding to $\sigma = -1$ in (3.4.10), verify the defining relations for $\bar{\mathcal{U}}_{q,2}$ and show

$$(3.4.44) \qquad \bar{\mathcal{U}}_{q,2} \simeq M_l(\mathbf{C}) \oplus M_l(\mathbf{C}) \oplus \bigoplus_{a=1}^{l-1} M_{a|l-a}(\Lambda_2) \ .$$

Remark. The algebra $\bar{\mathcal{U}}_{q,1}$ (or $\bar{\mathcal{U}}_{q,2}$) is unimodular, that is, the left and the right integrals coincide, $\int = \int_L = \int_R$ (they are defined by $x \int_R = \varepsilon(x) \int_R$ and $\int_L x = \varepsilon(x) \int_L$). The location of the integral inside the matrix blocks is very natural. In the direct sum, describing the matrix structure of the algebra, there is exactly one block $M_{1|\bar{l}-1}(\Lambda_2)$, for which the 1×1 sub-block realizes the trivial representation (the same holds for even l). The integral is

$$(3.4.45) \qquad \int = \left(\begin{array}{c|c} \xi\eta & \cdot \\ \hline \cdot & \cdot \end{array} \right)$$

(so the evaluation on the integral might remind to someone a true integration over Grassmann variables).

Example: $l = 3$

For $q^3 = 1$ we have $(q-q^{-1})^2 = -3$ and $2_q = -1$. The Casimir element satisfies
$$\bar{C}^3 - 3\bar{C} - 2 = (\bar{C}+1)^2(\bar{C}-2) = 0 . \tag{3.4.46}$$

For the block M_3 (the value of the Casimir element is 2) we have:

$$\rho_0(\bar{K}) = \begin{pmatrix} q^2 & . & . \\ . & 1 & . \\ . & . & q^{-2} \end{pmatrix} , \tag{3.4.47}$$

$$\rho_0(\bar{E}) = \begin{pmatrix} . & 1 & . \\ . & . & 1 \\ . & . & . \end{pmatrix} , \tag{3.4.48}$$

$$\rho_0(\bar{F}) = \begin{pmatrix} . & . & . \\ -1 & . & . \\ . & -1 & . \end{pmatrix} . \tag{3.4.49}$$

Irreducible representations of dimensions 1 and 2 have the same value (-1) of the Casimir element, they can be glued indecomposably into a block $M_{1|2}(\Lambda_2)$:

$$\rho_1(\bar{K}) = \left(\begin{array}{c|cc} 1 & . & . \\ \hline . & q & . \\ . & . & q^{-1} \end{array}\right) , \tag{3.4.50}$$

$$\rho_1(\bar{E}) = \left(\begin{array}{c|cc} . & \xi & . \\ \hline . & . & 1 \\ \xi & . & . \end{array}\right) , \tag{3.4.51}$$

$$\rho_1(\bar{F}) = \left(\begin{array}{c|cc} . & . & -\eta \\ \hline \eta & . & . \\ . & 1-\xi\eta & . \end{array}\right) . \tag{3.4.52}$$

The algebra $\bar{\mathcal{U}}_{q,1}$ has two blocks, $\bar{\mathcal{U}}_{q,1} \simeq M(3) \oplus M_{1|2}(\Lambda_2)$.

Case when l is even, $\tilde{l}$ is odd

Now $l = 2\tilde{l}$ and $\tilde{l} = 2s+1$. We have $q^{2s+1} = -1$, so $q' = -q$ is a primitive $\tilde{l}$-th root of unity. A substitution
$$\bar{E}' = -\bar{E} , \quad \bar{K}' = \bar{K} , \quad \bar{F}' = \bar{F} , \quad q' = -q \tag{3.4.53}$$
establishes an isomorphism of the algebra $\bar{\mathcal{U}}_{q,1}$ and the algebra $\bar{\mathcal{U}}_{q',1}$ whose matrix structure we know already.

Matrix structure for even $\tilde{l}$

We have $l = 4s$, $\tilde{l} = 2s$ and $q^{2s} = -1$.

We shall describe simultaneously the matrix structures of $\bar{\mathcal{U}}_{q,1}$ ($K^{2s} = 1$) and $\bar{\mathcal{U}}_{q,2}$ ($K^{4s} = 1$).

Let $K(\sigma,j)$, $E(\sigma,j)$ and $F(\sigma,j)$ be the operators as in (3.4.10), $j = 0, \frac{1}{2}, \ldots, s-1, s-\frac{1}{2}$. Note that (3.4.10) gives a representation of $\bar{\mathcal{U}}_{q,1}$ only when j (the "spin") is integer.

Let $j' = s - 1 - j$ for $j = 0, \frac{1}{2}, \ldots, s-1$. Let $C(\sigma, j)$ be the value of the Casimir element in the representation $V(\sigma, j)$. We have $C(\sigma, j) = \sigma(q^{1+2j} + q^{-1-2j})$. Thus, $C(-1, j') = C(1, j)$.

On the representation $V(\sigma, s - \frac{1}{2})$, the Casimir element takes a value (-2σ).

Now the assignment

$$(3.4.54) \quad \bar{K} \mapsto \left(\begin{array}{c|c} K(1,j) & \bullet \\ \hline \bullet & K(-1,j') \end{array} \right),$$

$$(3.4.55) \quad \bar{E} \mapsto \left(\begin{array}{c|c} E(1,j) & \begin{matrix} \vdots & \bullet \\ \xi & \cdots \end{matrix} \\ \hline \begin{matrix} \vdots & \bullet \\ \xi & \cdots \end{matrix} & E(-1,j') \end{array} \right),$$

$$(3.4.56) \quad \bar{F} \mapsto \left(\begin{array}{c|c} F(1,j) & \begin{matrix} \cdots & -\eta \\ \bullet & \vdots \end{matrix} \\ \hline \begin{matrix} \cdots & \eta \\ \bullet & \vdots \end{matrix} & F(-1,j') \end{array} \right) + \xi\eta \left(\begin{array}{c|c} M(2j+1) & \bullet \\ \hline \bullet & -M(2j'+1) \end{array} \right)$$

(dots mean that the corresponding entries are zero) establishes homomorphisms of $\bar{\mathcal{U}}_{q,2}$ into graded matrix algebras over Λ_2.

There are also two homomorphisms

$$(3.4.57) \quad \bar{\mathcal{U}}_{q,2} \to M_{2s}(\mathbf{C}),$$

corresponding to the representations $V(\sigma, s - \frac{1}{2})$.

We have a collection $\mathcal{C}$ of homomorphisms (3.4.54)-(3.4.56) and (3.4.57). Parallelly to the case of odd l, one shows that these are epimorphisms and then, by counting dimensions, that the direct sum of the homomorphisms from $\mathcal{C}$ is an isomorphism. This proves:

Proposition. For even $\tilde{l} = 2s$, the algebra $\bar{\mathcal{U}}_{q,2}$ is isomorphic to a direct sum of formatted matrix algebras,

$$(3.4.58) \quad \bar{\mathcal{U}}_{q,2} \simeq M_{2s}(\mathbf{C}) \oplus M_{2s}(\mathbf{C}) \oplus \bigoplus_{a=1}^{2s-1} M_{a|2s-a}(\Lambda_2).$$

The algebra $\bar{\mathcal{U}}_{q,1}$ is a direct sum of those terms in (3.4.58) for which a is odd,

$$(3.4.59) \quad \bar{\mathcal{U}}_{q,1} \simeq \bigoplus_{b=1}^{s} M_{2b-1|2s-2b+1}(\Lambda_2).$$

As for odd l, the matrix description implies the Poincaré-Birkhoff-Witt theorem: The monomials $\bar{K}^a \bar{F}^b \bar{E}^c$, $a, b, c = 1, \ldots, l$ for $\bar{\mathcal{U}}_{q,1}$ ($a = 1, \ldots, 2l$ for $\bar{\mathcal{U}}_{q,2}$), are linearly independent and hence form a basis.

Remark. The appearance of the sign σ in the formulas (3.4.10) is related to the existence of the following involution ϕ in the case when $\tilde{l} = 2s$ is even:

(3.4.60) $$\phi : \bar{K} \mapsto -\bar{K} \, , \ \bar{E} \mapsto \bar{E} \, , \ \bar{F} \mapsto -\bar{F} \, .$$

The subalgebra $\bar{\mathcal{U}}_{q,2,\phi}$ of fixed points of the involution ϕ consists of polynomials in $\bar{K}^2$, $\bar{F}\bar{K}$ and $\bar{E}$.

To describe the matrix structure of the algebra $\bar{\mathcal{U}}_{q,2,\phi}$, let $Q_s(\Lambda_2)$ be an algebra of matrices

(3.4.61) $$\left(\begin{array}{c|c} A & B \\ \hline B & A \end{array} \right) ,$$

A and B are $s \times s$ matrices, the entries of A are even, the entries of B are odd elements of the ring Λ_2.

Then

(3.4.62) $$\bar{\mathcal{U}}_{q,2,\phi} \simeq M_{2s}(\mathbf{C}) \oplus \bigoplus_{a=1}^{s-1} M_{a|2s-a}(\Lambda_2) \oplus Q_s(\Lambda_2)$$

As for the algebra $\bar{\mathcal{U}}_{q,1,\phi}$, one keeps those terms in the direct sum (3.4.62) which correspond to an integer spin. Now the answer depends on the parity of s (the appearance of the algebra Q), that is, on the residue of l modulo 8.

Example: $l = 4$

The algebra $\bar{\mathcal{U}}_{q,1}$:

(3.4.63) $$\bar{K}\bar{E} = -\bar{E}\bar{K} \, , \ \bar{K}\bar{F} = -\bar{F}\bar{K} \, , \ [\bar{E},\bar{F}] = 0 \, ,$$

and

(3.4.64) $$\bar{K}^2 = 1 \, , \ \bar{E}^2 = \bar{F}^2 = 0 \, .$$

The Casimir operator is $\bar{C} = -4\bar{F}\bar{E}$; it satisfies $\bar{C}^2 = 0$.

The realization:

(3.4.65) $$\bar{K} \mapsto \left(\begin{array}{c|c} 1 & \\ \hline & -1 \end{array} \right) , \ \bar{E} \mapsto \left(\begin{array}{c|c} & \xi \\ \hline \xi & \end{array} \right) , \ \bar{F} \mapsto \left(\begin{array}{c|c} & -\eta \\ \hline \eta & \end{array} \right) .$$

This realization is faithful, the algebra $\bar{\mathcal{U}}_{q,1}$ has only one block, $\bar{\mathcal{U}}_{q,1} \simeq M_{1|1}(\Lambda_2)$.

We have $\bar{E}\bar{F}' + \bar{F}'\bar{E} = 0$, where $\bar{F}' = \bar{F}\bar{K}$, so the algebra $\bar{\mathcal{U}}_{q,1,\phi}$ is isomorphic to the ring Λ_2 itself.

3.4.4. Reduced function algebra. A reduced function algebra $\bar{\mathcal{F}}_q$ on $SL_q(2)$ at roots of unity is the algebra with generators a, b, c and d, subjected to relations (3.4.2), (3.4.3) and

(3.4.66) $$ad - qbc = 1 \, .$$

This last relation, together with $d^l = 1$, allows to express a in terms of d, b and c; the algebra $\bar{\mathcal{F}}_q$ is generated by d, b and c only.

The algebra $\bar{\mathcal{F}}_q$ also has a formatted matrix structure. Let ξ_1 and ξ_2 be two variables which satisfy

(3.4.67) $$\xi_1^l = \xi_2^l = 0 \, , \ \xi_1 \xi_2 = \xi_2 \xi_1 \, .$$

The algebra $\mathbf{C}[\xi_1, \xi_2]$ is graded by the degree in the variables ξ_i, $\deg \xi_1 = \deg \xi_2 = 1$. The group Γ is the cyclic group $\mathbf{Z}/l\mathbf{Z}$. The format ϕ is specified by a set of

l numbers, $\phi = n_0|\ldots|n_{l-1}$, the number n_j corresponds to the character $z \mapsto \exp(\frac{2\pi i}{l}j)$, where z is a given generator of Γ.

A map

$$b \mapsto \xi_1 \begin{pmatrix} \cdot & 1 & & & \\ \cdot & & 1 & & \\ & & \cdot & \cdot & \\ & & & \cdot & 1 \\ 1 & & & & \cdot \end{pmatrix}, \quad c \mapsto \xi_2 \begin{pmatrix} \cdot & 1 & & & \\ \cdot & & 1 & & \\ & & \cdot & \cdot & \\ & & & \cdot & 1 \\ 1 & & & & \cdot \end{pmatrix},$$

$$d \mapsto \begin{pmatrix} 1 & & & & \\ & q & & & \\ & & \cdot & & \\ & & & \cdot & \\ & & & & q^{l-1} \end{pmatrix}$$

establishes an isomorphism

(3.4.68) $$\bar{\mathcal{F}}_q \xrightarrow{\sim} M_{1|1\ldots|1}(\mathbf{C}[\xi_1, \xi_2])$$

(all the numbers in the format equal 1). As for reduced enveloping algebras, this isomorphism implies the Poincaré-Birkhoff-Witt theorem.

3.4.5. *Centre.* We conclude the subsection by several remarks concerning the centers of the algebras $\bar{\mathcal{U}}_{q,m}$.

1. The center of the formatted matrix algebra $M_{m|n}(\Lambda_2)$ consists of matrices

(3.4.69) $$\left(\begin{array}{c|c} \alpha + \beta\xi\eta & \cdot \\ \hline \cdot & \alpha + \gamma\xi\eta \end{array} \right),$$

with some constants α, β and γ. It is 3-dimensional.

There is a conjecture by Kaplansky: "A Hopf algebra of characteristic zero has no non-zero central idempotents" (the citation is according to [25]).

This conjecture is false, the algebras $\bar{\mathcal{U}}_{q,m}$ provide a counter-example.

2. We have seen (eq. (3.4.41)) that the image of the Casimir element is of the form

(3.4.70) $$\left(\begin{array}{c|c} \alpha + \beta\xi\eta & \cdot \\ \hline \cdot & \alpha - \beta\xi\eta \end{array} \right).$$

Therefore, the Casimir element does not generate the whole center.

For the algebra $\mathcal{U}$, when q is a primitive l-th root of unity, a theorem (see [26]) states that the center of $\mathcal{U}$ is generated by the elements $E^{\tilde{l}}$, $F^{\tilde{l}}$, $K^{\tilde{l}}$ and C (and that there is a polynomial relation between these elements, which is eq. (3.4.20) at $i = \tilde{l}$; for $i = \tilde{l}$, the r.h.s. of (3.4.20) depends only on C and $K^{\tilde{l}}$). The image (in $\bar{\mathcal{U}}_{q,1}$) of the algebra, generated by $E^{\tilde{l}}$, $F^{\tilde{l}}$, $K^{\tilde{l}}$ and C, is the algebra of polynomials in $\bar{C}$. As we saw above, it is a strict subalgebra in the center.

3. Let $C(K)$ be a centralizer of K in $\mathcal{U}$. One has $C(K) = \bigoplus_{i=0}^{\tilde{l}} A_i$ where A_i is spanned by elements $F^i K^a E^i$. The subspace $A_{>0} = \bigoplus_{i=1}^{\tilde{l}} A_i$ is an ideal in $C(K)$ and A_0 is a complementary subalgebra, $A = A_0 \oplus A_{>0}$.

We have a well-defined projection $\pi : C(K) \to A_0$.

Let $\mathcal{Z}$ be the center of $\mathcal{U}$, it is a subalgebra in $C(K)$. The restriction $\phi = \pi_{\mathcal{Z}}$ of the projection ϕ to the center $\mathcal{Z}$ is called a Harish-Chandra homomorphism. It is known to be injective when q is not a root of unity.

For the reduced algebras $\bar{\mathcal{U}}_{q,1}$ (or $\bar{\mathcal{U}}_{q,2}$) the Harish-Chandra homomorphism is defined in the same way. However, the injectivity is lost, because the center $\tilde{\mathcal{Z}}$ is not semi-simple while the algebra $\bar{A}_0$ is. One verifies that the kernel of the Harish-Chandra homomorphism coincides with the Rad$\mathcal{Z}$. It is natural to conjecture that this holds for quantum deformations for all semi-simple Lie algebras.

4. $\hat{R}$-matrices

The first subsection is a summary of some essential facts from the theory of quasi-triangular Hopf algebras and their representations.

The $\hat{R}$ matrix for the standard quantum group $GL_q(N)$ is [**27, 28**],

$$(4.0.1) \qquad \hat{R}^{ij}_{kl} = q^{\delta_{ij}} \delta^i_l \delta^j_k + (q - q^{-1})\Theta(l - k)\delta^i_k \delta^j_l ,$$

where $\Theta(i) = 1$ for $i > 0$ and $\Theta(i) = 0$ otherwise. The indices run from 1 to N.

The $\hat{R}$-matrix (4.0.1) belongs to a class of "ice" $\hat{R}$-matrices; the precise definition of the ice condition is in the second subsection. There we give a classification of ice $\hat{R}$-matrices. The main result is that they are all of GL type.

The final subsection establishes a way to build, starting from an arbitrary $\hat{R}$-matrix of GL-type, $\hat{R}$-matrices for orthogonal and symplectic quantum groups.

4.1. Skew-invertibility. The first part of this chapter is a short reminder on the general theory of quasi-triangular Hopf algebras, originating mostly from [**29**].

Then we discuss an important notion of "skew-invertibility" and explain how it arises in the context of the quasi-triangular Hopf algebras.

In the second part we derive, on a representation level, matrix analogues of some identities in Hopf algebras. These matrix identities will be needed for the discussion of the $\hat{R}$-matrices for orthogonal and symplectic quantum groups.

4.1.1. *Generalities on Hopf algebras.* Let A be a Hopf algebra.

We recall that

$$(4.1.1) \qquad m(S \otimes \text{id})\Delta(a) = \epsilon(a)\,1 ,$$

$$(4.1.2) \qquad m(\text{id} \otimes S)\Delta(a) = \epsilon(a)\,1 ,$$

$$(4.1.3) \qquad (\epsilon \otimes \text{id})\Delta = (\text{id} \otimes \epsilon)\Delta = \text{id} ,$$

where S is the antipode and ϵ is a counit.

We use a standard notation omitting a summation index, for example, instead of writing $\Delta(x) = \sum_i x^i_1 \otimes x^i_2$ we shall simply write $\Delta(x) = x_{(1)} \otimes x_{(2)}$.

1. The Hopf algebra A is called almost cocommutative if there exists an invertible element $\mathcal{R} \in A \otimes A$ such that

$$(4.1.4) \qquad \Delta'(x)\mathcal{R} = \mathcal{R}\Delta(x)$$

for any element $x \in A$. Here Δ' is the flipped coproduct, $\Delta'(x) = x_{(2)} \otimes x_{(1)}$ for $\Delta(x) = x_{(1)} \otimes x_{(2)}$.

We symbolically write $\mathcal{R} = a \otimes b$ instead of $\mathcal{R} = \sum_i a_i \otimes b_i$.

Let $\Delta^2(x) = x_{(1)} \otimes x_{(2)} \otimes x_{(3)}$ ($\Delta^2 = (\Delta \otimes \mathrm{id})\Delta = (\mathrm{id} \otimes \Delta)\Delta$). By (4.1.4), we have

(4.1.5) $\qquad x_{(2)}a \otimes x_{(1)}b \otimes x_{(3)} = ax_{(1)} \otimes bx_{(2)} \otimes x_{(3)}$,

(4.1.6) $\qquad x_{(1)} \otimes x_{(3)}a \otimes x_{(2)}b = x_{(1)} \otimes ax_{(2)} \otimes bx_{(3)}$.

Let $u = S(b)a$. Applying $\mathrm{id} \otimes S \otimes S^2$ to (4.1.5) and multiplying terms in the inverse order, one obtains

(4.1.7) $\qquad\qquad S^2(x)u = ux$.

Applying $\mathrm{id} \otimes S \otimes S^2$ to (4.1.6) and multiplying terms, one obtains

(4.1.8) $\qquad\qquad xS(u) = S(u)S^2(x)$.

Eqs. (4.1.7) and (4.1.8) hold for an arbitrary $x \in A$ so the element $S(u)u$ is central.

Exercises.

1. Take a flip of (4.1.4): $\Delta(x)\mathcal{R}_{21} = \mathcal{R}_{21}\Delta'(x)$, and derive, parallelly to (4.1.7) and (4.1.8), identities

(4.1.9) $\qquad\qquad xv = vS^2(x)$,

(4.1.10) $\qquad\qquad S^2(x)S(v) = S(v)x$,

where $v = aS(b)$.

2. This exercise is taken from [30].

Let A be a Hopf algebra (not necessarily almost cocommutative). Let T be an operator on $A \otimes A$ defined by

(4.1.11) $\qquad T(a \otimes b) = aS(b_{(1)})b_{(3)} \otimes b_{(2)}$.

Show that T satisfies the Yang-Baxter equation, $T_{12}T_{13}T_{23} = T_{23}T_{13}T_{12}$. Show that for a cocommutative A, the operator T reduces to the identity operator.

2. The Hopf algebra A is called quasi-triangular if

(4.1.12) $\qquad (\Delta \otimes \mathrm{id})\mathcal{R} = \mathcal{R}_{13}\mathcal{R}_{23}$,

(4.1.13) $\qquad (\mathrm{id} \otimes \Delta)\mathcal{R} = \mathcal{R}_{13}\mathcal{R}_{12}$.

Exercise. Show that any of these formulas, together with (4.1.4), implies the Yang-Baxter equation

(4.1.14) $\qquad\qquad \mathcal{R}_{12}\mathcal{R}_{13}\mathcal{R}_{23} = \mathcal{R}_{23}\mathcal{R}_{13}\mathcal{R}_{12}$.

Applying $\epsilon \otimes \mathrm{id} \otimes \mathrm{id}$ to the formula (4.1.12) gives $\mathcal{R} = (\epsilon \otimes \mathrm{id})(\mathcal{R})\mathcal{R}$ or, upon canceling by $\mathcal{R}$,

(4.1.15) $\qquad\qquad (\epsilon \otimes \mathrm{id})\mathcal{R} = 1$.

Similarly, an application of $\mathrm{id} \otimes \mathrm{id} \otimes \epsilon$ to (4.1.13) gives

(4.1.16) $\qquad\qquad (\mathrm{id} \otimes \epsilon)(\mathcal{R}) = 1$.

Applying S to the first tensor argument of (4.1.12), multiplying the first two arguments and using (4.1.15), one obtains

(4.1.17) $$(S \otimes \text{id})\mathcal{R} = \mathcal{R}^{-1} .$$

Similarly,

(4.1.18) $$(\text{id} \otimes S)\mathcal{R}^{-1} = \mathcal{R} .$$

Together, eqs. (4.1.17) and (4.1.18) imply

(4.1.19) $$(S \otimes S)\mathcal{R} = \mathcal{R} .$$

3. Some properties of the element u

An immediate consequence of (4.1.19) is

(4.1.20) $$v = S(u) .$$

Let $\mathcal{R}^{-1} = c \otimes d$. We have $1 = (\text{id} \otimes S)(\mathcal{R}\mathcal{R}^{-1}) = (\text{id} \otimes S)(ac \otimes bd) = ac \otimes S(d)S(b)$. Using (4.1.18), one can rewrite it in the form

(4.1.21) $$a'a \otimes bS(b') = 1 ,$$

where the prime means another copy, the full version of $a'a \otimes bS(b')$ is $\sum_{i,j} a_i a_j \otimes b_j S(b_i)$. Multiplying the tensor terms of (4.1.21) in the inverse order, we get $bua = 1$, or, by (4.1.7), $bS^2(a)u = 1$. On the other hand, $ubS^2(a) = S^2(b)uS^2(a)$, which, by (4.1.19), equals $bua = 1$. Thus, the element u is invertible,

(4.1.22) $$u^{-1} = bS^2(a) .$$

Exercise. Prove that the element u is invertible in the general almost cocommutative setting (i.e., without assuming the quasi-triangularity).

Using the invertibility of u, one can rewrite (4.1.7) in the form

(4.1.23) $$S^2(x) = uxu^{-1} .$$

In particular, the antipode S is invertible (since S^2 is invertible). Note that in the quasitriangular situation, eqs. (4.1.8), (4.1.9) and (4.1.10) follow from (4.1.7) because of (4.1.20) and the invertibility of S.

For $x = u$, eq. (4.1.23) gives

(4.1.24) $$S^2(u) = u .$$

For $x = S(u)$, eq. (4.1.7) gives $S^3(u)u = uS(u)$, which, in view of (4.1.24), implies

(4.1.25) $$uS(u) = S(u)u .$$

4. Coproduct of u

From quasi-triangularity properties (4.1.12) and (4.1.13) it follows that

(4.1.26) $$(\Delta \otimes \Delta)(\mathcal{R}) = \mathcal{R}_{14}\mathcal{R}_{24}\mathcal{R}_{13}\mathcal{R}_{23} ,$$

or

(4.1.27) $$a_{(1)} \otimes a_{(2)} \otimes b_{(1)} \otimes b_{(2)} = aa'' \otimes a'a''' \otimes b''b''' \otimes bb' ,$$

where, as usual, primes denote different copies.

Rewriting the Yang-Baxter equation (4.1.14) in the form
$$\mathcal{R}_{23}^{-1}\mathcal{R}_{12}\mathcal{R}_{13} = \mathcal{R}_{13}\mathcal{R}_{12}\mathcal{R}_{23}^{-1}$$
and using (4.1.17), we obtain
(4.1.28) $$a'a'' \otimes S(a)b' \otimes bb'' = a''a' \otimes b'S(a) \otimes b''b \ .$$
Now

(4.1.29)
$$\Delta(u) = S(b_{(2)})a_{(1)} \otimes S(b_{(1)})a_{(2)} \stackrel{(4.1.27)}{=} S(bb') \otimes aa'' \otimes S(b''b''')a'a'''$$
$$= S(b')ua'' \otimes S(b''b''')a'a''' \stackrel{(4.1.7)}{=} S(b')S^2(a'')u \otimes S(b''b''')a'a'''$$
$$= S(S(a'')b')u \otimes S(b''b''')a'a''' \stackrel{(4.1.28)}{=} S(b'S(a''))u \otimes S(b'''b'')a'''a'$$
$$= S^2(a'')S(b')u \otimes S(b'')ua' \stackrel{(4.1.19)}{=} S(a'')S(b')u \otimes b''ua'$$
$$\stackrel{(4.1.17)}{=} \mathcal{R}^{-1} \cdot S(b')u \otimes ua' \stackrel{(4.1.7)}{=} \mathcal{R}^{-1} \cdot S(b')u \otimes S^2(a')u$$
$$\stackrel{(4.1.19)}{=} \mathcal{R}^{-1} \cdot b'u \otimes S(a')u \stackrel{(4.1.17)}{=} \mathcal{R}^{-1}\mathcal{R}_{21}^{-1} \cdot u \otimes u \ .$$

A number over "=" refers to an equation which is used in the corresponding equality.

Denote the element $\mathcal{R}_{21}\mathcal{R} \in A \otimes A$ by ϕ, $\phi = \mathcal{R}_{21}\mathcal{R}$. We obtained
(4.1.30) $$\Delta(u) = \phi^{-1} \cdot u \otimes u \ .$$

Obviously, $\phi \Delta(x) = \Delta(x)\phi$ for any $x \in A$. The element ϕ plays in important role in the theory of quasi-triangular Hopf algebras; a map from A^* (a dual Hopf algebra) to A, $f \mapsto \langle \phi, f \rangle_2$ (the pairing with the second argument of ϕ) is called a factorization map. The algebra A is called factorizable if the factorization map is not degenerate (and A is called triangular if $\phi = 1$).

For $x = u$, eq. (4.1.4) gives (using (4.1.30))
(4.1.31) $$\mathcal{R} \cdot u \otimes u = u \otimes u \cdot \mathcal{R}$$
(note that this equality follows from eqs. (4.1.23) and (4.1.19) as well).

Using now that $\Delta(S(x)) = (S \otimes S)\Delta'(x)$ for any x, one obtains
(4.1.32) $$\Delta(S(u)) = \phi^{-1} \cdot S(u) \otimes S(u) \ .$$

Therefore, the element $g = uS(u)^{-1}$ is group-like, $\Delta(g) = g \otimes g$; the fourth power of the antipode is given by the conjugation by g, $S^4(x) = gxg^{-1}$.

For the central element $uS(u)$ we have $\Delta(uS(u)) = \phi^{-2} \cdot uS(u) \otimes uS(u)$. If there exists a central element $\rho \in A$ such that $\rho^2 = uS(u)$ and $\Delta(\rho) = \phi^{-1} \cdot \rho \otimes \rho$, one says that A is a ribbon Hopf algebra; the element ρ is then called the ribbon element.

Exercise. Show that $\epsilon(\rho) = 1$, $S(\rho) = \rho$ and $\mathcal{R} \cdot \rho \otimes \rho = \rho \otimes \rho \cdot \mathcal{R}$.

4.1.2. *Matrix picture.* Let t be a representation of A in a vector space V. The numerical R-matrix is
(4.1.33) $$R = (t \otimes t)(\mathcal{R})$$
or, in some basis of V, $R^{ij}_{kl} = t(a)^i_k t(b)^j_l$. As usual, P will denote the permutation matrix.

Eq. (4.1.21) produces the following matrix equation

(4.1.34) $$R^{ij}_{kl}\Psi^{sl}_{it} = \delta^s_k \delta^j_t ,$$

where $\Psi = (t \otimes t)(a \otimes S(b))$, $\Psi^{ij}_{kl} = t(a)^i_k t(S(b))^j_l$.

Thus for $\hat{R} = PR$, $\hat{R}^{de}_{af} = R^{ed}_{af}$, and $\hat{\Psi} = P\Psi$, $\hat{\Psi}^{ba}_{cd} = \Psi^{ab}_{cd}$, we have

(4.1.35) $$\hat{R}^{ba}_{cd}\hat{\Psi}^{de}_{af} = \delta^e_c \delta^b_f .$$

One can rewrite it without indices as

(4.1.36) $$\mathrm{tr}_2(\hat{R}_{12}\hat{\Psi}_{23}) = P_{13} .$$

We could have used instead of (4.1.21) an equivalent relation

$$1 = (\mathrm{id} \otimes S)(\mathcal{R}^{-1}\mathcal{R}) = (\mathrm{id} \otimes S)(ca \otimes db) = ca \otimes S(b)S(d) \stackrel{(4.1.18)}{=} a'a \otimes S(b)b'$$

to obtain in the matrix form

(4.1.37) $$\mathrm{tr}_2(\hat{\Psi}_{12}\hat{R}_{23}) = P_{13} .$$

Definition. Given an operator $\hat{R}$, a solution of eq. (4.1.36) (respectively, eq. (4.1.37)) is called a right (respectively, left) skew inverse of $\hat{R}$. The operator $\hat{R}$ is called skew-invertible if it has left and right skew inverses.

We are concerned only with a finite-dimensional case, in which the relations (4.1.36) and (4.1.37) are equivalent: $(A\hat{\odot}B)^{eb}_{cf} = A^{es}_{ct}B^{tb}_{sf}$ is an associative product on the space of tensors with two upper and two lower indices, the permutation $P^{ij}_{kl} = \delta^i_l \delta^j_k$ is a unit element for the operation $\hat{\odot}$ and eq. (4.1.36) (correspondingly, (4.1.37)) defines $\hat{\Psi}$ as the right (correspondingly, left) inverse of $\hat{R}$ with respect to $\hat{\odot}$. In a finite-dimensional algebra left and right inverses (when one of them exists) coincide.

This product reflects a product[3] $(\alpha \otimes \beta) \odot (\gamma \otimes \delta) = \gamma\alpha \otimes \beta\delta$ defined for elements of the tensor square of an arbitrary algebra: for $x, y \in A \otimes A$ and $z = x \odot y$ let $X = (t\otimes t)(x)$, $Y = (t\otimes t)(y)$ and $Z = (t\otimes t)(z)$ be their images for the representation t. Then $Z^{ij}_{kl} = X^{aj}_{kb}Y^{ib}_{al}$ or $\hat{Z} = \hat{X}\hat{\odot}\hat{Y}$.

Let $Q = t(u)$ be the image of the element u, $Q^i_j = t(S(b))^i_k t(a)^k_j = \Psi^{ki}_{jk}$, or

(4.1.38) $$Q_1 = \mathrm{tr}_2(\hat{\Psi}_{12}) .$$

Similarly, for $\tilde{Q} = t(S(u))$ we have

(4.1.39) $$\tilde{Q}_2 = \mathrm{tr}_1(\hat{\Psi}_{12}) .$$

Thus,

(4.1.40) $$\mathrm{tr}_2(\hat{R}_{12}Q_2) = I_1 \quad \text{and} \quad \mathrm{tr}_1(\tilde{Q}_1\hat{R}_{12}) = I_2 ,$$

where I stands for the identity operator in a corresponding space.

If the representation t is irreducible, the central element $uS(u)$ takes a constant value, the square of the value of the ribbon element. Thus, for an irreducible representation, the product $Q\tilde{Q}$ is proportional to unity.

---

[3] More generally, for an element $\mu = \mu_1 \otimes \ldots \mu_n \in A^{\otimes n}$ and an element $\xi = \alpha \otimes \beta \in A \otimes A$ one defines $\mu \odot \xi_{kl} := \mu_1 \otimes \ldots \otimes \alpha\mu_k \otimes \ldots \otimes \mu_l \beta \otimes \ldots \otimes \mu_n$ and $\xi_{kl} \odot \mu := \mu_1 \otimes \ldots \otimes \mu_k \alpha \otimes \ldots \otimes \beta\mu_l \otimes \ldots \otimes \mu_n$; then there are rules like $(x_{12}y_{13}) \odot z_{23} = (y_{13}z_{23}) \odot x_{12}$, $x_{12}x_{13}x_{23} = x_{13} \odot (x_{12}x_{23})$ etc.

Exercise. Show that the standard $\hat{R}$-matrix (4.0.1) is skew-invertible with

(4.1.41) $$\hat{\Psi}^{ab}_{cd} = q^{-\delta_{ab}} \delta^a_d \delta^b_c - (q - q^{-1})\Theta(d-c) q^{2c-2d} \delta^a_c \delta^b_d .$$

Show that

(4.1.42) $$Q^a_b = q^{-2N+2a-1} \delta^a_b \quad \text{and} \quad \tilde{Q}^a_b = q^{1-2a} \delta^a_b ,$$

so the value of the square of the ribbon element is q^{-2N}.

We now adopt another point of view and forget that there was a quasi-triangular Hopf algebra behind. We shall leave, as a trace of quasi-triangularity, only the assumption that the numerical matrix $\hat{R}$ is skew-invertible, and derive, purely in the matrix language, some consequences (for $\hat{\Psi}$) of the Yang-Baxter equation.

Below we constantly use the following simple fact:

(4.1.43) $$\mathrm{tr}_1(P_{12}) = I_2 .$$

Multiplying the Yang-Baxter equation $\hat{R}_{12}\hat{R}_{23}\hat{R}_{12} = \hat{R}_{23}\hat{R}_{12}\hat{R}_{23}$ from the left by $\hat{\Psi}_{a1}$, from the right by $\hat{\Psi}_{3b}$ (a and b should be understood as numbers of some copies of the space V), taking traces in the spaces 1 and 3 and using (4.1.36) and (4.1.37), we obtain (after relabeling spaces - we do it in order to avoid a redundancy of unnecessary symbols; the result is formulated for the spaces with numbers 1, 2 and 3)

(4.1.44) $$\mathrm{tr}_0(\hat{\Psi}_{10} \hat{R}_{02} \hat{R}_{03}) P_{23} = P_{12} \, \mathrm{tr}_0(\hat{R}_{10} \hat{R}_{20} \hat{\Psi}_{03}) .$$

Exercise. The Yang-Baxter equation implies that $\hat{R}_{12}\hat{R}_{23}\hat{R}^n_{12} = \hat{R}^n_{23}\hat{R}_{12}\hat{R}_{23}$ and $\hat{R}^n_{12}\hat{R}_{23}\hat{R}_{12} = \hat{R}_{23}\hat{R}_{12}\hat{R}^n_{23}$ for an arbitrary integer n. Show that

(4.1.45) $$\mathrm{tr}_0(\hat{\Psi}_{10} \hat{R}_{02} \hat{R}^n_{03}) P_{23} = P_{12} \, \mathrm{tr}_0(\hat{R}^n_{10} \hat{R}_{20} \hat{\Psi}_{03})$$

and

(4.1.46) $$\mathrm{tr}_0(\hat{\Psi}_{10} \hat{R}^n_{02} \hat{R}_{03}) P_{23} = P_{12} \, \mathrm{tr}_0(\hat{R}_{10} \hat{R}^n_{20} \hat{\Psi}_{03}) .$$

Deduce from (4.1.45) and (4.1.46) that

(4.1.47) $$\mathrm{tr}_0(\hat{\Psi}_{10} \hat{R}^{n+1}_{02}) = P_{12} \, \mathrm{tr}_0(\hat{R}^n_{10} \hat{R}_{20} Q_0) ,$$

(4.1.48) $$\mathrm{tr}_0(\hat{\Psi}_{10} \hat{R}^{n+1}_{02}) = P_{12} \, \mathrm{tr}_0(\hat{R}_{10} \hat{R}^n_{20} Q_0)$$

and then

(4.1.49) $$\mathrm{tr}_0(\tilde{Q}_0 \hat{R}^{n+1}_{01}) = \mathrm{tr}_0(\hat{R}^{n+1}_{10} Q_0) .$$

Since the permutation matrix P squares to the identity, we can rewrite (4.1.44) as

(4.1.50) $$P_{12} \, \mathrm{tr}_0(\hat{\Psi}_{10} \hat{R}_{02} \hat{R}_{03}) = \mathrm{tr}_0(\hat{R}_{10} \hat{R}_{20} \hat{\Psi}_{03}) P_{23} .$$

Multiplying (4.1.50) from the left by $\hat{\Psi}_{a1}$, from the right by $\hat{\Psi}_{3b}$, taking traces in the spaces 1 and 3 and using (4.1.36) and (4.1.37), we obtain

(4.1.51) $$\mathrm{tr}_{01}(\hat{\Psi}_{a1} P_{12} \hat{\Psi}_{10} \hat{R}_{02} P_{0b}) = \mathrm{tr}_{03}(P_{a0} \hat{R}_{20} \hat{\Psi}_{03} P_{23} \hat{\Psi}_{3b}) .$$

This equation can be simplified. The expression under the trace in the l.h.s. can be rewritten as $\hat{\Psi}_{a1} P_{12} \hat{\Psi}_{10} \hat{R}_{02} P_{0b} = \hat{\Psi}_{a1} P_{12} P_{0b} \hat{\Psi}_{1b} \hat{R}_{b2}$. Now the trace in the space

0 can be taken (by (4.1.43)), so the l.h.s. of (4.1.51) is $\text{tr}_1(\hat{\Psi}_{a1}P_{12}\hat{\Psi}_{1b}\hat{R}_{b2})$. We have

$$\text{tr}_1(\hat{\Psi}_{a1}P_{12}\hat{\Psi}_{1b}\hat{R}_{b2}) = \text{tr}_1(\hat{\Psi}_{a1}P_{12}\hat{\Psi}_{1b})\hat{R}_{b2}$$

(4.1.52)
$$= \text{tr}_1(P_{12}\hat{\Psi}_{a2}\hat{\Psi}_{1b})\hat{R}_{b2} = \text{tr}_1(P_{12}\hat{\Psi}_{1b}\hat{\Psi}_{a2})\hat{R}_{b2}$$

$$= \text{tr}_1(P_{12}\hat{\Psi}_{1b})\hat{\Psi}_{a2}\hat{R}_{b2} = \text{tr}_1(\hat{\Psi}_{2b}P_{12})\hat{\Psi}_{a2}\hat{R}_{b2}$$

$$= \hat{\Psi}_{2b}\hat{\Psi}_{a2}\hat{R}_{b2} .$$

In a similar way one simplifies the r.h.s. of (4.1.51) and obtains (after relabeling spaces)

(4.1.53) $$\hat{\Psi}_{23}\hat{\Psi}_{12}\hat{R}_{32} = \hat{R}_{21}\hat{\Psi}_{23}\hat{\Psi}_{12} .$$

Assume that an operator B has a left skew inverse A, $\text{tr}_2(A_{12}B_{23}) = P_{13}$ (or $A\hat{\odot}B = P$). Then for any operator X_1, which acts as the identity in the space 2, we have

$$\text{tr}_2(A_{12}X_1B_{21}) = \text{tr}_{02}(A_{12}X_1P_{10}B_{21})$$

(4.1.54)
$$= \text{tr}_{02}(A_{12}X_1B_{20}P_{10}) = \text{tr}_{02}(A_{12}B_{20}X_1P_{10})$$

$$= \text{tr}_0(P_{10}X_1P_{10}) = \text{tr}(X)I_1 ,$$

where I_1 is the identity operator in the second space.

Therefore, taking tr_3 of (4.1.53), one obtains

(4.1.55) $$\hat{R}_{21}Q_2\hat{\Psi}_{12} = Q_1I_2 ;$$

similarly, taking tr_1 of (4.1.53), one obtains

(4.1.56) $$\hat{\Psi}_{12}\tilde{Q}_1\hat{R}_{21} = I_1\tilde{Q}_2 .$$

Here Q and $\tilde{Q}$ are the operators defined in (4.1.38) and (4.1.39).

On the other hand, one can rewrite eq. (4.1.44) as $P_{23}\,\text{tr}_0(\hat{\Psi}_{10}\hat{R}_{03}\hat{R}_{02}) = \text{tr}_0(\hat{R}_{20}\hat{R}_{10}\hat{\Psi}_{03})P_{12}$ or

(4.1.57) $$\text{tr}_0(\hat{\Psi}_{10}\hat{R}_{03}\hat{R}_{02})P_{12} = P_{23}\,\text{tr}_0(\hat{R}_{20}\hat{R}_{10}\hat{\Psi}_{03}) .$$

Exercises.

1. Multiply (4.1.57) from the left by $\hat{\Psi}_{a1}$, from the right by $\hat{\Psi}_{3b}$, take traces in the spaces 1 and 3 and obtain

(4.1.58) $$\hat{\Psi}_{12}\hat{\Psi}_{23}\hat{R}_{21} = \hat{R}_{32}\hat{\Psi}_{12}\hat{\Psi}_{23} .$$

2. Assume that an operator B has a left skew inverse A, $\text{tr}_2(A_{12}B_{23}) = P_{13}$. Show, similarly to (4.1.54) that

(4.1.59) $$\text{tr}_2(A_{21}X_1B_{12}) = \text{tr}(X)I_1 .$$

3. Apply tr_1 or tr_3 to eq. (4.1.58) and deduce that

(4.1.60) $$\hat{\Psi}_{12}Q_2\hat{R}_{21} = Q_1I_2$$

and

(4.1.61) $$\hat{R}_{21}\tilde{Q}_1\hat{\Psi}_{12} = \tilde{Q}_2 I_1 \ .$$

4. Let $\psi = a \otimes S(b) \in A \otimes A$ ($\mathcal{R} = a \otimes b$ is the universal R-matrix). Show that

(4.1.62) $$\mathcal{R}_{23}\psi_{12}\psi_{13} = \psi_{13}\psi_{12}\mathcal{R}_{23} \ ,$$

(4.1.63) $$\mathcal{R}_{12}\psi_{23}\psi_{13} = \psi_{13}\psi_{23}\mathcal{R}_{12} \ .$$

Show that these equalities induce, on the level of a representation, equalities (4.1.53) and (4.1.58) respectively.

5. What are Hopf-algebraic counterparts of eqs. (4.1.55), (4.1.56), (4.1.60) and (4.1.61)?

Write equations (4.1.55), (4.1.56), (4.1.60) and (4.1.61) in the form

(4.1.64) $$Q_2\hat{\Psi}_{12} = \hat{R}_{21}^{-1}Q_1 \ ,$$

(4.1.65) $$\hat{\Psi}_{12}\tilde{Q}_1 = \tilde{Q}_2\hat{R}_{21}^{-1} \ ,$$

(4.1.66) $$\hat{\Psi}_{12}Q_2 = Q_1\hat{R}_{21}^{-1} \ ,$$

(4.1.67) $$\tilde{Q}_1\hat{\Psi}_{12} = \hat{R}_{21}^{-1}\tilde{Q}_2 \ .$$

A compatibility of these equations provides new relations[4].

Comparing tr_1 of eqs. (4.1.64) and (4.1.66):

(4.1.68) $$Q_2\tilde{Q}_2 = \mathrm{tr}_1(\hat{R}_{21}^{-1}Q_1) \quad \text{and} \quad \tilde{Q}_2 Q_2 = \mathrm{tr}_1(Q_1\hat{R}_{21}^{-1})$$

and using the cyclic property of trace to move Q_1, we conclude that

(4.1.69) $$Q\tilde{Q} = \tilde{Q}Q \ .$$

This is a matrix counterpart of eq. (4.1.25).

Using (4.1.64)-(4.1.67), we can express in two different ways combinations $Q_2\hat{\Psi}_{12}\tilde{Q}_1$, $Q_2\hat{\Psi}_{12}Q_2$, $Q_2\tilde{Q}_1\hat{\Psi}_{12}$, $\hat{\Psi}_{12}\tilde{Q}_1 Q_2$, $\tilde{Q}_1\hat{\Psi}_{12}\tilde{Q}_1$ and $\tilde{Q}_1\hat{\Psi}_{12}Q_2$. This results in

(4.1.70) $$\hat{R}_{21}^{-1}Q_1\tilde{Q}_1 = Q_2\tilde{Q}_2\hat{R}_{21}^{-1} \ ,$$

(4.1.71) $$\hat{R}_{21}^{-1}Q_1 Q_2 = Q_1 Q_2\hat{R}_{21}^{-1} \ ,$$

(4.1.72) $$\tilde{Q}_1\hat{R}_{21}^{-1}Q_1 = Q_2\hat{R}_{21}^{-1}\tilde{Q}_2 \ ,$$

(4.1.73) $$\tilde{Q}_2\hat{R}_{21}^{-1}Q_2 = Q_1\hat{R}_{21}^{-1}\tilde{Q}_1 \ ,$$

(4.1.74) $$\hat{R}_{21}^{-1}\tilde{Q}_1\tilde{Q}_2 = \tilde{Q}_1\tilde{Q}_2\hat{R}_{21}^{-1} \ ,$$

(4.1.75) $$\hat{R}_{21}^{-1}\tilde{Q}_2 Q_2 = \tilde{Q}_1 Q_1\hat{R}_{21}^{-1} \ .$$

---

[4] Eqs. (4.1.53) and (4.1.58) also have a nontrivial compatibility relation: $\hat{R}_{21}$ commutes with $\hat{\Psi}_{23}\hat{\Psi}_{12}^2\hat{\Psi}_{23}$.

Eqs. (4.1.71) and (4.1.74) reflect the fact that for a quasi-triangular Hopf algebra A, elements $u \otimes u$ and $S(u) \otimes S(u)$ commute with $\mathcal{R}$ (see eq. (4.1.31)).

It is interesting to compare eqs. (4.1.70) and (4.1.75) in the Hecke case, when the $\hat{R}$-matrix satisfies a quadratic equation $\hat{R}^2 = \lambda \hat{R} + 1$ with $\lambda \neq 0$. Rewriting eq. (4.1.75) as $\hat{R}_{21} Q_1 \tilde{Q}_1 = Q_2 \tilde{Q}_2 \hat{R}_{21}$ (we used that Q commutes with $\tilde{Q}$) and subtracting from (4.1.70) we obtain

(4.1.76) $$Q\tilde{Q} = const .$$

So, even if a representation t is not irreducible but the $\hat{R}$-matrix is of Hecke type, the value of the square of the ribbon element on all subrepresentations of t is the same.

If $\lambda = 0$ (i.e. $\hat{R}$ is triangular, $\hat{R}^2 = 1$), eq. (4.1.68) implies immediately that $Q\tilde{Q} = I$.

Exercise. Suppose that operators Q and $\tilde{Q}$ are invertible. Show, without taking skew inverses, that eqs. (4.1.72) and (4.1.73) follow from eqs. (4.1.70), (4.1.71), (4.1.74) and (4.1.75).

Use (4.1.48) (or multiply (4.1.37) from the left by Q_3 and use (4.1.64)) to obtain

(4.1.77) $$\mathrm{tr}_2(\hat{R}_{12} \hat{R}_{32}^{-1} Q_2) = Q_3 P_{13} .$$

Therefore, if Q is invertible then $\hat{R}^{-1}$ has a skew inverse $\hat{\Xi}$, $\hat{\Xi}_{12} = Q_1 \hat{R}_{21} Q_2^{-1}$.

On the other hand, assume that $\hat{R}^{-1}$ has a skew inverse $\hat{\Xi}$. Multiply (4.1.77) by $\hat{\Xi}_{03}$ and take tr_3 to obtain $\mathrm{tr}_0(\hat{\Xi}_{01})Q_1 = I_1$.

Therefore, Q is invertible iff $\hat{R}^{-1}$ has a skew inverse. Similarly, $\tilde{Q}$ is invertible iff $\hat{R}^{-1}$ has a skew inverse.

It follows then that Q is invertible iff $\tilde{Q}$ is invertible.

There is also an implication: Q is invertible $\Rightarrow$ $\hat{\Psi}$ is invertible (it follows immediately from, for example, (4.1.64).

Assuming that the operator $\hat{\Psi}$ is invertible, one can rewrite the Yang-Baxter equation entirely in terms of $\hat{\Psi}$. To this end, rewrite eq. (4.1.53) in the form

(4.1.78) $$\hat{\Psi}_{12} \hat{R}_{32} \hat{\Psi}_{12}^{-1} = \hat{\Psi}_{23}^{-1} \hat{R}_{21} \hat{\Psi}_{23} .$$

Multiplying (4.1.78) from the left by $\hat{\Psi}_{a3}$, from the right by $\hat{\Psi}_{1b}$, taking traces in the spaces 1 and 3 and using (4.1.36) and (4.1.37), we obtain

(4.1.79) $$P_{12} \, \mathrm{tr}_0(\hat{\Psi}_{01} \hat{\Psi}_{02}^{-1} \hat{\Psi}_{03}) = \mathrm{tr}_0(\hat{\Psi}_{10} \hat{\Psi}_{20}^{-1} \hat{\Psi}_{30}) P_{23} .$$

Note that on the way from the Yang-Baxter to eq. (4.1.79) we were making only reversible transformations, so eq. (4.1.79) is equivalent (assuming the skew-invertibility of $\hat{\Psi}$) to the original Yang-Baxter equation.

We conclude by a remark that from the Hopf-algebraic point of view the invertibility of $\hat{\Psi}$ is natural. The element $\psi = a \otimes S(b) \in A \otimes A$ has an inverse,

(4.1.80) $$\psi^{-1} = a \otimes S^2(b) .$$

Also, the element $\mathcal{R}^{-1}$ has a (left and right) skew-inverse ξ (that is, the inverse with respect to the multiplication $\odot$),

(4.1.81) $$\xi = S^2(a) \otimes b .$$

From the Hopf-algebraic perspective the matrix identities which we derived are quite transparent. However, for the construction of orthogonal and symplectic $\hat{R}$-matrices one needs the matrix form of the identities, so it is important to understand how much one can derive using only matrices.

Exercises.

1. Verify (4.1.80) and (4.1.81); show that

(4.1.82) $$\mathcal{R}_{12}\xi_{13}\xi_{23} = \xi_{23}\xi_{13}\mathcal{R}_{12},$$

(4.1.83) $$\mathcal{R}_{23}\xi_{13}\xi_{12} = \xi_{12}\xi_{13}\mathcal{R}_{23},$$

(4.1.84) $$\mathcal{R}_{13}\xi_{12}\psi_{23} = \psi_{23}\xi_{12}\mathcal{R}_{13},$$

(4.1.85) $$\mathcal{R}_{13}\xi_{23}\psi_{12} = \psi_{12}\xi_{23}\mathcal{R}_{13}.$$

2. What are Hopf-algebraic counterparts of eqs. (4.1.44) and (4.1.79)?

4.2. Ice $\hat{R}$-matrices. The standard $\hat{R}$-matrix (4.0.1) has two properties: it is of Hecke type (that is, it has two eigenvalues) and it satisfies the so-called "ice" condition which means that $\hat{R}^{ij}_{kl}$ can be different from zero only if the pair of the upper indices $\{i,j\}$ is a permutation of the pair of the lower ones, $\{i,j\} = \{k,l\}$ or $\{i,j\} = \{l,k\}$. Here we shall explain that these two properties (Hecke and ice) are not independent; we shall introduce the notion of indecomposable ice $\hat{R}$-matrix and demonstrate that such $\hat{R}$-matrices satisfy the Hecke condition[5]. Ideologically, this shows that the search of ice solutions of equations similar to the Yang-Baxter equation is justified only in the Hecke case (and then one imposes the Hecke condition first, as it is done in [31] for the dynamical Yang-Baxter equation).

Let $\hat{R}^{ij}_{kl} = a_{ij}\delta^i_l\delta^j_k + b_{ij}\delta^i_k\delta^j_l$ be an ice matrix. We fix $b_{ii} = 0$ for uniqueness. Let also $a_i = a_{ii}$.

We suppose that the matrix $\hat{R}$ is invertible and skew-invertible. It follows then (an easy exercise) that $a_i \neq 0$ and $a_{ij} \neq 0$ for all i and j.

Assume that $\hat{R}$ satisfies the Yang-Baxter equation, $Y^{ikj}_{abc} = 0$, where $Y^{ikj}_{abc} = (\hat{R}_{12}\hat{R}_{23}\hat{R}_{12} - \hat{R}_{23}\hat{R}_{12}\hat{R}_{23})^{ijk}_{abc}$.

When two indices among $\{i,j,k\}$ are different, the equation $Y^{ikj}_{abc} = 0$ gives (here $i \neq j$):

(4.2.1) $$a_{ij}b_{ij}b_{ji} = 0,$$

(4.2.2) $$b_{ij}(a_i^2 - a_ib_{ij} - a_{ij}a_{ji}) = 0,$$

(4.2.3) $$b_{ij}(a_j^2 - a_jb_{ij} - a_{ij}a_{ji}) = 0.$$

---

[5]The opposite is not true: there are many Hecke $\hat{R}$-matrices which cannot be brought to an ice form by a change of a basis.

For all three indices $\{i, j, k\}$ different, $i \neq j \neq k \neq i$, equations are

(4.2.4) $$(a_{ij}a_{ji} - a_{jk}a_{kj})b_{ik} + b_{ij}b_{jk}(b_{ij} - b_{jk}) = 0 ,$$

(4.2.5) $$a_{jk}(b_{ij}b_{ik} - b_{ij}b_{jk} - b_{ik}b_{kj}) = 0 ,$$

(4.2.6) $$a_{ij}(b_{jk}b_{ik} - b_{ij}b_{jk} - b_{ji}b_{ik}) = 0 .$$

Let Γ be a graph with vertices i. We draw an oriented edge $\vec{ij}$ from the vertex i to the vertex j if the number b_{ij} is not zero.

Since $a_{ij} \neq 0$, eq. (4.2.1) shows that two vertices can be joined by not more that one edge.

When the graph Γ is not connected, equations, corresponding to different connected components, do not notice each other. So, one has to study only the situation when the graph Γ is connected.

Definition. We say that the ice $\hat{R}$-matrix is indecomposable if its graph Γ is connected.

Proposition. Let $\hat{R}$ be an invertible and skew-invertible solution of the Yang-Baxter equation. Assume that $\hat{R}$ satisfies the ice condition and is indecomposable. Then $\hat{R}$ is of Hecke type (that is, it satisfies a quadratic equation).

Proof. Since $a_{ij} \neq 0$ for all i and j, eqs. (4.2.5) and (4.2.6) imply

(4.2.7) $$b_{ij}b_{ik} - b_{ij}b_{jk} - b_{ik}b_{kj} = 0 ,$$

(4.2.8) $$b_{jk}b_{ik} - b_{ij}b_{jk} - b_{ji}b_{ik} = 0 .$$

(i) Suppose that the graph Γ has edges $\vec{ij}$ and $\vec{jk}$. Then Γ has an edge $\vec{ik}$, as on the Figure:

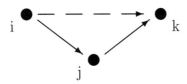

This is an immediate consequence of eq. (4.2.7): $b_{kj} = 0$ because, by assumption, $b_{jk} \neq 0$; therefore, $b_{ij}(b_{ik} - b_{jk}) = 0$ but, by assumption, $b_{ij} \neq 0$.

(ii) Suppose that the graph Γ has edges $\vec{ij}$ and $\vec{kj}$. Then Γ has either an edge $\vec{ik}$ or an edge $\vec{ki}$, as on the Figures:

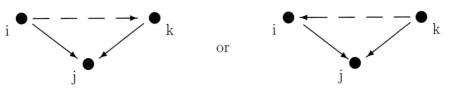

To prove this, interchange j and k, $j \leftrightarrow k$, in eq. (4.2.8):

(4.2.9) $$b_{kj}b_{ij} - b_{ik}b_{kj} - b_{ki}b_{ij} = 0$$

and note that either b_{ik} or b_{ki} is 0.

(*iii*) Situations when Γ has edges $\vec{ji}$ and $\vec{kj}$ or edges $\vec{ji}$ and $\vec{jk}$ are considered similarly.

We conclude:

(a) If Γ contains two sides of a triangle, it contains the third side. This immediately implies (since Γ is connected) that Γ is a full graph, that is, every two vertices are joined. In other words, for each pair (i, j) at least one number, b_{ij} or b_{ji}, is not zero, $\{b_{ij}, b_{ji}\} \neq \{0, 0\}$.

(b) An oriented triangle of Γ is never a cycle (see Figures above). By (a), Γ is the full graph; an easy exercise shows then that Γ has no cycles. Therefore, orientations of edges induce an order on the set of vertices and we can relabel vertices in such a way that Γ has an edge $\vec{ij} \in \Gamma$ if and only if $i < j$. In other words, $b_{ij} \neq 0$ if and only if $i < j$.

Consider a triangle with vertices i, j and k, $i < j < k$. Then the oriented edges are $\vec{ij}$, $\vec{ik}$ and $\vec{jk}$. We have $b_{ji} = b_{kj} = 0$; eq. (4.2.7) shows that $b_{ik} = b_{jk}$; eq. (4.2.8) shows that $b_{ik} = b_{ij}$. Therefore, for all i and j with $i < j$ the parameters b_{ij} take the same value, say b, $b_{ij} = b$.

At this stage, eqs. (4.2.1), (4.2.5) and (4.2.6) are solved and the $\hat{R}$-matrix has the form

(4.2.10) $$\hat{R}^{ij}_{kl} = a_{ij}\delta^i_l \delta^j_k + b\Theta(l-k)\delta^i_k \delta^j_l \ .$$

(*iv*) Eq. (4.2.4) simplifies now; it implies that all the products $a_{ij}a_{ji}$ take the same value. Denote this value by a, $a_{ij}a_{ji} = a$ for all i and j with $i \neq j$.

(*v*) The remaining two equations, (4.2.2) and (4.2.3) imply that for all i the parameters a_i satisfy a quadratic equation

(4.2.11) $$a_i^2 - ba_i - a = 0 \ .$$

Now it is immediate to verify that the matrix $\hat{R}$ satisfies the same quadratic equation

(4.2.12) $$\hat{R}^2 = a + b\hat{R} \ .$$

The proof of the Proposition is finished. $\square$

When $4a + b^2 = 0$, the matrix $\hat{R}$ has a nontrivial jordanian structure.

Assume that eq. $t^2 - bt - a = 0$ has two different roots μ_1 and μ_2. In this case the matrix $\hat{R}$ is diagonalizable and has two projectors. Let m (correspondingly n) be the number of those a_i which are equal to μ_1 (correspondingly μ_2). Then the ranks of the projectors are $\dfrac{m(m+1)}{2} + \dfrac{n(n-1)}{2} + mn$ and $\dfrac{m(m-1)}{2} + \dfrac{n(n+1)}{2} + mn$. These are exactly the ranks of the symmetrizer and the antisymmetrizer for the superspace of dimension $m|n$. The $\hat{R}$-matrices, constructed in the Proposition above, are called the multiparametric $\hat{R}$-matrices for the quantum supergroups $GL_q(m|n)$; we have shown that with the ice condition there are essentially no more solutions.

4.3. Construction of orthogonal and symplectic $\hat{R}$-matrices.

Let V be a vector space and V^* its dual. The natural pairing between V and V^* can be used to define either a symmetric or antisymmetric scalar product on the space $V \oplus V^*$. This scalar product is invariant under the natural action of the group $GL(V)$ of general linear transformations of the space V. Therefore, $GL(V)$ gets imbedded into a corresponding orthogonal or symplectic group.

Such logic goes very well for quantum spaces also. We shall model in this way quantum spaces for orthogonal and symplectic quantum groups.

1. Yang-Baxter equation and ordering

A quantum space, being defined by only a part of projectors of an $\hat{R}$-matrix, does not carry the whole information about the $\hat{R}$-matrix itself. It is not difficult to find a quantum space which can be defined by several different $\hat{R}$-matrices (*Exercise*: give an example).

However, there is a convenient way to encode an $\hat{R}$-matrix in a framework of quantum spaces. It requires several copies of a quantum space. Let x^i be coordinates of some quantum space. We shall not be interested in commutation relations between the elements x^i; rather, we introduce copies, say $x(1)^i$, $x(2)^i$ etc, and define commutation relations between different copies to be

$$(4.3.1) \qquad x(M)^i x(N)^j = \hat{R}^{ij}_{kl} x(N)^k x(M)^l$$

for $M < N$. The relations (4.3.1) allow to reorder any multilinear combination $x(M_1)^{i_1} x(M_2)^{i_2} \ldots x(M_p)^{i_p}$ with pairwise distinct labels $M_1, M_2, \ldots, M_p$ in the descending (with respect to the labels $M_1, M_2, \ldots, M_p$) order.

There are two ways to reorder a monomial $x(M_1)^{i_1} x(M_2)^{i_2} x(M_3)^{i_3}$, with $M_1 < M_2 < M_3$, in a descending way, starting from $x(M_1)^{i_1} x(M_2)^{i_2}$ or $x(M_2)^{i_2} x(M_3)^{i_3}$. The equality of two resulting ordered expressions is a compatibility condition. Assume that monomials $x(M_1)^{i_1} x(M_2)^{i_2} \ldots x(M_p)^{i_p}$, where $M_1 > M_2 > \cdots > M_p$, are linearly independent. Then the compatibility condition is precisely equivalent to the Yang-Baxter equation for the matrix $\hat{R}$. This interpretation of the Yang-Baxter equation is very useful especially in cases when the index i of coordinates x^i is composite (like, for instance, a pair of indices $\{\alpha, \beta\}$ if one wants to view the elements T^α_β of the quantum matrix as coordinates of a quantum space).

In the sequel, to avoid the cumbersome notation $x(M)^i$, we shall write $x^i y^j = \hat{R}^{ij}_{kl} y^k x^l$ instead of (4.3.1).

2. Assumptions

Our starting point is a solution $\hat{R}$ of the Yang-Baxter equation. We impose several conditions:

A1. $\hat{R}$ is invertible.
A2. $\hat{R}$ is skew-invertible with a skew inverse $\hat{\Psi}$.
A3. an operator Q defined by (4.1.38) is invertible (thus, $\tilde{Q}$ defined by (4.1.39) is invertible as well).

As any solution of the Yang-Baxter equation, $\hat{R}$ defines a quantum group; in this subsection it will be enough to understand it as an algebra generated by T^i_j

and $(T^{-1})^i_j$ with relations

(4.3.2) $$\hat{R}_{12}T_1T_2 = T_1T_2\hat{R}_{12}$$

and

(4.3.3) $$TT^{-1} = T^{-1}T = I,$$

or $T^i_j(T^{-1})^j_k = (T^{-1})^i_j T^j_k = \delta^i_k$.

We need one more assumption. The relations $\hat{R}_{12}T_1T_2 = T_1T_2\hat{R}_{12}$ imply that $W_{12}T_1T_2 = T_1T_2W_{12}$ for any polynomial W in $\hat{R}$, $W = \sum_\alpha c_\alpha \hat{R}^\alpha$. We shall say that $\hat{R}$ is rigid if every W for which $W_{12}T_1T_2 = T_1T_2W_{12}$ is a polynomial in $\hat{R}$. And this is our assumption:

A4. $\hat{R}$ is rigid.

3. Auxiliary formulas

Here are some immediate consequences from (4.3.2) and (4.3.3). First,

(4.3.4) $$T_1^{-1}\hat{R}_{12}T_1 = T_2\hat{R}_{12}T_2^{-1},$$

(4.3.5) $$\hat{R}_{12}T_2^{-1}T_1^{-1} = T_2^{-1}T_1^{-1}\hat{R}_{12}.$$

Multiplying (4.3.4) by $\hat{\Psi}_{a1}$ from the left, by $\hat{\Psi}_{2b}$ from the right and taking traces in the spaces 1 and 2, we obtain (as usual, after relabeling spaces)

(4.3.6) $$\text{tr}_0(\hat{\Psi}_{10}T_0^{-1}P_{02}T_0) = \text{tr}_0(T_0P_{10}T_0^{-1}\hat{\Psi}_{02}).$$

Attention: one cannot move T_0 cyclically under the tr_0 in (4.3.6) because the matrix elements of T_0 do not commute with matrix elements of other operators in the expression.

Tracing (4.3.6) in the spaces 1 or 2 gives

(4.3.7) $$\text{tr}_0(T_0P_{10}T_0^{-1}Q_0) = Q_1,$$

(4.3.8) $$\text{tr}_0(\tilde{Q}_0T_0^{-1}P_{01}T_0) = \tilde{Q}_1.$$

Sometimes it is more transparent to write eq. (4.3.6), as well as eqs. (4.3.7) and (4.3.8), in indices:

(4.3.9) $$\hat{\Psi}^{ua}_{vb}(T^{-1})^b_i T^j_a = T^s_v(T^{-1})^u_t \hat{\Psi}^{tj}_{si}$$

and

(4.3.10) $$T^a_i(T^{-1}Q)^b_a = Q^j_i \quad \text{and} \quad (\tilde{Q}T^{-1})^a_i T^j_a = \tilde{Q}^j_i.$$

Operators Q and $\tilde{Q}$ are invertible, so we can rewrite (4.3.10) in terms of an associative operation $(X \circ Y)^i_j := X^a_j Y^i_a$ (or $X \circ Y = (X^t Y^t)^t$, where t means the transposition) as

(4.3.11) $$T \circ (Q^{-1}T^{-1}Q) = I \quad \text{and} \quad (\tilde{Q}T^{-1}\tilde{Q}^{-1}) \circ T = I,$$

where I is the identity (with respect to the usual multiplication as well as to the multiplication $\circ$). Left and right inverses coincide, so

(4.3.12) $$TQ\tilde{Q} = Q\tilde{Q}T.$$

It follows from (4.3.11) that T^{-1} also has an inverse with respect to $\circ$:

(4.3.13) $$(QTQ^{-1}) \circ T^{-1} = I .$$

Note that (4.3.6) can be rewritten as $(\hat{\Psi}_{12} T_2^{-1}) \circ T_2 = T_1 \circ (T_1^{-1} \hat{\Psi}_{12})$ or

(4.3.14) $$T_2^{-1} \circ \hat{\Psi}_{12} \circ T_2 = T_1 \circ \hat{\Psi}_{12} \circ T_1^{-1} ,$$

since, if the matrix elements of an operator X commute with the matrix elements of an operator Y then $X \circ Y = YX$.

4. Covariance

As explained in the part 1 of this subsection, the operator $\hat{R}$ provides a consistent set of relations

(4.3.15) $$x^i y^j = \hat{R}^{ij}_{kl} y^k x^l .$$

The relations (4.3.15) are covariant under the following (co)action of the quantum group (generators T^i_j commute with x^i):

(4.3.16) $$x^i \to T^i_j x^j ,$$

and the same for y^i.

We are going to build a quantum analog of the direct sum $V \oplus V^*$, so we need, in addition to x^i, another set of generators, x_i. To mimic that the generators x_i describe a dual space, we require their transformation law to be

(4.3.17) $$x_i \to x_j (T^{-1})^j_i$$

(the same for y_i).

A little later we will restrict ourselves to the case when $\hat{R}$ has only two eigenvalues. But already now we can partly analyze possible ordering relations. We have two "multiplets", $x^\bullet = \{x^i\}$ and $x_\bullet = \{x_i\}$. For the moment, let $S \in End(V \otimes V)$ be an arbitrary operator. Let us say that a matrix element $S^{\alpha\beta}_{\gamma\delta}$ is "ice" if either $\alpha = \gamma$ and $\beta = \delta$ or $\alpha = \delta$ and $\beta = \gamma$. If all non-vanishing matrix elements of S are ice then we have an ice matrix in the sense of the subsection 4.2. We shall apply the same terminology to the whole multiplets $x^\bullet$ and $x_\bullet$: if x belongs to a multiplet A (A can be $\bullet$ or $_\bullet$) and y belongs to a multiplet B then the parts in the ordered expression for xy, which contain the same multiplets will be called "ice". We shall see that the ice part of ordering relations is strongly governed by the covariance.

We fix the ordering relations for $x^i y^j$ to be as in (4.3.15).

In the ordered expression for $x_i y_j$ the ice terms are $y_k x_l$,

(4.3.18) $$x_i y_j = E^{kl}_{ij} y_k x_l + \ldots ,$$

where dots stand for terms with other structures of indices. Then the covariance under the transformations (4.3.17) requires $T_1^{-1} T_2^{-1} E_{21} = E_{21} T_1^{-1} T_2^{-1}$ so, by rigidity, E_{12} is a polynomial in $\hat{R}_{21}$, $E_{12} = e(\hat{R}_{21})$.

In the ordered expression for $x^i y_j$ we may have terms like $y_k x^l$ and $y^k x_l$,

(4.3.19) $$x^i y_j = A^{ik}_{jl} y_k x^l + B^{il}_{jk} y^k x_l + \ldots ,$$

dots stand for terms with other structures of indices. Then the covariance under the transformations (4.3.16) and (4.3.17) requires

(4.3.20) $$T_1 A_{12} T_1^{-1} = T_2^{-1} A_{12} T_2 ,$$

(4.3.21) $$T_1 B_{12} T_1^{-1} = \text{tr}_0(B_{10} T_0 P_{02} T_0^{-1}) .$$

Similarly, if, in the ordered expression for $x_i y^j$, we have terms like $y^k x_l$ and $y_k x^l$,

(4.3.22) $$x_i y^j = C_{ik}^{jl} y^k x_l + D_{il}^{jk} y_k x^l + \cdots ,$$

where dots stand for terms with other structures of indices, then the covariance under the transformations (4.3.16) and (4.3.17) requires

(4.3.23) $$\text{tr}_0(T_0^{-1} P_{01} T_0 C_{02}) = \text{tr}_0(C_{10} T_0 P_{02} T_0^{-1}) ,$$

(4.3.24) $$\text{tr}_0(T_0^{-1} P_{01} T_0 D_{02}) = T_2^{-1} D_{12} T_2 .$$

Due to rigidity of $\hat{R}$, it follows from eq. (4.3.20) that A_{12} is a polynomial in $\hat{R}_{21}$, $A_{12} = a(\hat{R}_{21})$.

Multiply (4.3.21) by $\hat{R}_{a2}$ from the right and take tr_2. The l.h.s. becomes $T_1 \tilde{B}_{1a} T_1^{-1}$, where $\tilde{B}_{1a} = \text{tr}_2(B_{12} \hat{R}_{a2})$. The r.h.s. becomes

$$\text{tr}_{02}(B_{10} T_0 P_{02} T_0^{-1} \hat{R}_{a2}) = \text{tr}_{02}(B_{10} T_0 P_{02} \hat{R}_{a2} T_0^{-1})$$

(4.3.25)
$$= \text{tr}_{02}(P_{02} B_{12} T_2 \hat{R}_{a2} T_0^{-1}) \stackrel{(c)}{=} \text{tr}_{02}(B_{12} T_2 \hat{R}_{a2} T_0^{-1} P_{02})$$

$$= \text{tr}_{02}(B_{12} T_2 \hat{R}_{a2} P_{02} T_2^{-1}) \stackrel{\text{tr}_0}{=} \text{tr}_2(B_{12} T_2 \hat{R}_{a2} T_2^{-1})$$

$$\stackrel{(4.3.4)}{=} \text{tr}_2(B_{12} T_a^{-1} \hat{R}_{a2} T_a) = T_a^{-1} \tilde{B}_{1a} T_a .$$

We used: the cyclic property of the trace to move P_{02}, it is indicated by (c) over "="; we took tr_0 (it is indicated over "="); and we used eq. (4.3.4).

Therefore, $\tilde{B}_{1a}$ is, by rigidity of $\hat{R}$, a polynomial in $\hat{R}_{a1}$, $\text{tr}_2(B_{12} \hat{R}_{a2}) = b(\hat{R}_{a1})$ with some polynomial b. Multiplying by $\hat{\Psi}_{ba}$ from the left and taking tr_a, we find

(4.3.26) $$B_{12} = \text{tr}_0(\hat{\Psi}_{20} b(\hat{R}_{01})) .$$

Similarly, multiplying (4.3.24) by $\hat{R}_{1a}$ from the left and taking tr_1, we find

(4.3.27) $$T_2^{-1} \tilde{D}_{a2} T_2 = T_a \tilde{D}_{a2} T_a^{-1} ,$$

where $\tilde{D}_{a2} = \text{tr}_1(\hat{R}_{1a} D_{12})$. Therefore, $\tilde{D}_{a2}$ is a polynomial in $\hat{R}_{2a}$. Thus,

(4.3.28) $$D_{12} = \text{tr}_0(d(\hat{R}_{20}) \hat{\Psi}_{01})$$

for some polynomial d.

Finally, multiply eq. (4.3.23) from the left by $\hat{R}_{1a}$, from the right by $\hat{R}_{b2}$ and take tr_{12} to obtain

(4.3.29) $$T_a \tilde{C}_{ab} T_a^{-1} = T_b^{-1} \tilde{C}_{ab} T_b ,$$

where $\tilde{C}_{ab} = \text{tr}_{12}(\hat{R}_{1a} C_{12} \hat{R}_{b2})$. Therefore, $\tilde{C}_{ab}$ is a polynomial in $\hat{R}_{ba}$. Thus,

(4.3.30) $$C_{12} = \text{tr}_{03}(\hat{\Psi}_{23} c(\hat{R}_{30}) \hat{\Psi}_{01})$$

for some polynomial c.

5. Ansatz

We keep in mind that the multiplets $x^\bullet$ and $x_\bullet$ are associated to the group GL_N. For general N, the only invariant tensors with four indices are the permutation and the identity in $V \otimes V$. This motivates the following Ansatz:

$$(4.3.31) \qquad x^i y^j = \hat{R}^{ij}_{kl} y^k x^l ,$$

$$(4.3.32) \qquad x^i y_j = A^{ik}_{jl} y_k x^l + B^{il}_{jk} y^k x_l ,$$

$$(4.3.33) \qquad x_i y^j = C^{jl}_{ik} y^k x_l + D^{jk}_{il} y_k x^l ,$$

$$(4.3.34) \qquad x_i y_j = E^{lk}_{ij} y_l x_k .$$

Here

$$A_{12} = a(\hat{R}_{21}) , \quad C_{12} = \mathrm{tr}_{03}(\hat{\Psi}_{23} c(\hat{R}_{30}) \hat{\Psi}_{01}) , \quad B_{12} = \mathrm{tr}_0(\hat{\Psi}_{20} b(\hat{R}_{01})) ,$$

$$(4.3.35) \quad D_{12} = \mathrm{tr}_0(d(\hat{R}_{20}) \hat{\Psi}_{01}) , \quad E_{12} = e(\hat{R}_{21})$$

with some polynomials a, b, c, d and e.

The original matrix $\hat{R}$ is a matrix of the size $N^2 \times N^2$, where N is the dimension of the space V, the range of indices of multiplets x^i and x_i. A solution $\hat{\mathbf{R}}^{IJ}_{KL}$ of the consistency conditions for the ordering relations (4.3.31)-(4.3.34) is a matrix of a bigger size $(2N^2) \times (2N)^2$, each of four indices of $\hat{\mathbf{R}}$ runs from 1 to $2N$. The new index is the union of upper and lower indices of the original multiplets. To remember it, we shall write, for the new index I, $I = (\underline{k})$ for a value of the original index from the multiplet x^k or $I = (\overline{k})$ for a value of the original index from the multiplet x_k. In this notation, the nonzero matrix elements of $\hat{\mathbf{R}}$ are

$$(4.3.36) \quad \begin{array}{l} \hat{\mathbf{R}}^{(\underline{i})(\underline{j})}_{(\underline{k})(\underline{l})} = \hat{R}^{ij}_{kl} , \quad \hat{\mathbf{R}}^{(\underline{i})(\overline{j})}_{(\overline{k})(\underline{l})} = A^{ik}_{jl} , \quad \hat{\mathbf{R}}^{(\underline{i})(\overline{j})}_{(\underline{k})(\overline{l})} = B^{il}_{jk} , \\[6pt] \hat{\mathbf{R}}^{(\overline{i})(\underline{j})}_{(\underline{k})(\overline{l})} = C^{jl}_{ik} , \quad \hat{\mathbf{R}}^{(\overline{i})(\underline{j})}_{(\overline{k})(\underline{l})} = D^{jk}_{il} , \quad \hat{\mathbf{R}}^{(\overline{i})(\overline{j})}_{(\overline{l})(\overline{k})} = E^{lk}_{ij} . \end{array}$$

We are looking for a skew-invertible $\hat{\mathbf{R}}$. In the notation, as in (4.3.36), it is easy to see that if A is zero then the matrix $\hat{\mathbf{R}}^{IJ}_{KL}$ has a zero eigenvector with respect to the skew multiplication, that is, a quantity v^I_K which satisfies $v^I_K \hat{\mathbf{R}}^{IJ}_{KL} = 0$: one may take any v whose non-zero elements are only $v^{(\overline{k})}_{(\underline{i})}$; so, such $\hat{\mathbf{R}}$ cannot be skew invertible. In fact, this argument shows that the skew invertibility of $\hat{\mathbf{R}}$ requires that the operator A is invertible (with respect to the usual multiplication). Similarly, C must be invertible. The conditions

$$(4.3.37) \qquad A \text{ and } C \text{ are invertible}$$

we will use in the process of solving the Yang-Baxter equation for $\hat{\mathbf{R}}$.

6. Yang-Baxter equation for $\hat{\mathbf{R}}$

224   O. OGIEVETSKY

As explained in the beginning of this subsection, the Yang-Baxter equation for $\hat{\mathbf{R}}$ we obtain by ordering in two different ways expressions $x^A y^B z^C$, where the indices A, B and C can belong now to any of multiplets, $^\bullet$ or $_\bullet$.

Ordering $x^\bullet y^\bullet z_\bullet$:

(4.3.38) $\hat{R}_{23} A_{21} A_{32} = A_{21} A_{32} \hat{R}_{12}$,

(4.3.39) $\hat{R}_{21} B_{23} A_{12} = P_{23}\, \text{tr}_0(A_{02} B_{10} \hat{R}_{30}) + P_{23}\, \text{tr}_0(A_{10} D_{02} P_{03} B_{03})$,

(4.3.40) $\hat{R}_{12} B_{23} \hat{R}_{13} = \text{tr}_0(B_{20} \hat{R}_{10} B_{03}) + \text{tr}_0(P_{01} B_{01} A_{20} C_{03})$.

Ordering $x^\bullet y_\bullet z^\bullet$:

$$P_{12}\, \text{tr}_0(A_{10} \hat{R}_{02} C_{03}) + P_{12}\, \text{tr}_0(B_{10} B_{03} P_{02} D_{02})$$

(4.3.41) $= \text{tr}_0(C_{10} \hat{R}_{20} A_{03}) P_{23} + \text{tr}_0(D_{10} D_{03} P_{02} B_{02}) P_{23}$,

(4.3.42) $A_{23} D_{12} \hat{R}_{32} = P_{12}\, \text{tr}_0(\hat{R}_{02} D_{03} A_{10}) + P_{12}\, \text{tr}_0(B_{10} A_{03} P_{02} D_{02})$,

(4.3.43) $C_{23} B_{12} \hat{R}_{23} = P_{12}\, \text{tr}_0(\hat{R}_{20} B_{03} C_{10}) + P_{12}\, \text{tr}_0(D_{10} C_{03} P_{02} B_{02})$.

Ordering $x_\bullet y^\bullet z^\bullet$:

(4.3.44) $C_{23} C_{12} \hat{R}_{23} = \hat{R}_{12} C_{23} C_{12}$,

(4.3.45) $\hat{R}_{12} D_{23} C_{12} = P_{23}\, \text{tr}_0(C_{02} D_{10} \hat{R}_{03}) + P_{23}\, \text{tr}_0(C_{10} B_{02} P_{03} D_{03})$,

(4.3.46) $\hat{R}_{13} D_{12} \hat{R}_{23} = \text{tr}_0(D_{10} \hat{R}_{03} D_{02}) + \text{tr}_0(C_{10} A_{02} P_{03} D_{03})$.

Exercise. Verify eqs. (4.3.38)-(4.3.46).

Equations, arising from ordering $x^\bullet y_\bullet z_\bullet$, $x_\bullet y^\bullet z_\bullet$ and $x_\bullet y_\bullet z^\bullet$, can be quickly obtained by noticing that the system (4.3.31)-(4.3.34) is invariant under a substitution $x^\bullet \leftrightarrow x_\bullet$, $y^\bullet \leftrightarrow y_\bullet$, $\hat{R} \leftrightarrow E^t$, $A \leftrightarrow C^t$ and $B \leftrightarrow D^t$, where t stands for the transposition. We have:

for $x^\bullet y_\bullet z_\bullet$:

(4.3.47) $A_{12} A_{23} E_{12} = E_{23} A_{12} A_{23}$,

(4.3.48) $A_{12} B_{23} E_{12} = \text{tr}_0(E_{03} B_{10} A_{02}) P_{23} + \text{tr}_0(P_{03} B_{30} D_{02} A_{10}) P_{23}$,

(4.3.49) $E_{23} B_{12} E_{13} = \text{tr}_0(B_{02} E_{03} B_{10}) + \text{tr}_0(P_{03} B_{30} C_{02} A_{10})$;

for $x_\bullet y^\bullet z_\bullet$:

$$\text{tr}_0(A_{03} E_{02} C_{10}) P_{12} + \text{tr}_0(P_{02} B_{20} D_{03} D_{10}) P_{12}$$

(4.3.50) $= P_{23}\, \text{tr}_0(C_{03} E_{20} A_{10}) + P_{23}\, \text{tr}_0(P_{02} D_{20} B_{03} B_{10})$,

(4.3.51) $E_{23} D_{12} A_{23} = \text{tr}_0(A_{10} D_{03} E_{20}) P_{12} + \text{tr}_0(P_{02} D_{20} A_{03} B_{10}) P_{12}$,

(4.3.52) $E_{32} B_{12} C_{23} = \text{tr}_0(C_{10} B_{03} E_{02}) P_{12} + \text{tr}_0(P_{02} B_{20} C_{03} D_{10}) P_{23}$;

for $x_\bullet y_\bullet z^\bullet$:

(4.3.53) $\quad E_{12}C_{32}C_{21} = C_{32}C_{21}E_{23}$,

(4.3.54) $\quad C_{12}D_{23}E_{21} = \mathrm{tr}_0(E_{30}D_{10}C_{02})P_{23} + \mathrm{tr}_0(P_{03}D_{30}B_{02}C_{10})P_{23}$,

(4.3.55) $\quad E_{12}D_{32}E_{13} = \mathrm{tr}_0(D_{02}E_{10}D_{30}) + \mathrm{tr}_0(P_{10}D_{10}A_{02}C_{30})$.

Finally, ordering $x_\bullet y_\bullet z_\bullet$ implies the Yang-Baxter equation for E:

(4.3.56) $\quad\quad\quad\quad\quad\quad E_{12}E_{23}E_{12} = E_{23}E_{12}E_{23}$.

7. Specifying to the Hecke case

We shall solve the system (4.3.38)-4.3.56) in the Hecke case - when the matrix $\hat{R}$ satisfies a quadratic equation $\hat{R}^2 = \lambda\hat{R} + 1$. Note that $\hat{R}$ cannot be proportional to a constant, it would contradict the skew invertibility.

As we have seen in the subsection 4.1.2 (see eq. (4.1.76)), in the Hecke case the product $Q\tilde{Q}$ is proportional to a unity, $Q\tilde{Q} = r^2 I$ (r corresponds to the ribbon element in the quasi-triangular case). Due to the assumption **A3**, $r \neq 0$. Therefore, by (4.1.68) (and (4.1.49))

(4.3.57) $\quad\quad\quad\quad 1 - \lambda\,\mathrm{tr}(Q) = 1 - \lambda\,\mathrm{tr}(\tilde{Q}) \neq 0$.

Because of Hecke condition, the polynomials in (4.3.35) contain only constant and linear terms.

8. Block triangularity

The standard $\hat{R}$-matrix (4.0.1) has a following property:

(4.3.58) $\quad\quad\quad\quad \hat{R}^{ij}_{kl} = 0 \quad \text{if} \quad ji < kl$,

where $<$ is the lexicographic ordering (*i.e.*, $ji < kl$ when $j < k$ or $j = k$ and $i < l$). This means that the matrix $P\hat{R}$ is lower triangular.

The standard $\hat{R}$-matrix (4.0.1) has also another triangularity property: $\hat{R}^{ij}_{kl} = 0$ if $ij > lk$; this means that the matrix $\hat{R}P$ is upper triangular.

For the ordering relations $x^i y^j = \hat{R}^{ij}_{kl}$, the property (4.3.58) says that the ordered expression for $x^i y^j$ can contain only monomials which are lexicographically not bigger than $y^j x^i$.

As a first step towards a solution of the system (4.3.38)-4.3.56) in the Hecke case, we shall prove that the relations (4.3.31)-(4.3.34) are "block triangular": say, block upper triangularity means that we define an order on the set $\mathcal{S} = \{^\bullet, _\bullet\}$ of multiplets $x^\bullet$ and $x_\bullet$, $x^\bullet > x_\bullet$ and then the ordered expression for $x^\mathcal{I} y^\mathcal{J}$, $\mathcal{I}, \mathcal{J} \in \mathcal{S}$, contains only monomials which are not bigger than $y^\mathcal{J} x^\mathcal{I}$.

In the simple situation of eqs. (4.3.31)-(4.3.34), the block triangularity means that either $B = 0$ or $D = 0$.

To prove the block triangularity, it is enough to consider two equations, (4.3.38) and (4.3.39). Eqn. (4.3.38) implies that A_{12} is proportional to either $\hat{R}_{21}$ or $\hat{R}^{-1}_{21}$. We shall write it as $A_{12} \propto \hat{R}_{12} - \epsilon\lambda I_1 I_2$, where $\epsilon = 0$ or 1. The coefficient of proportionality is different from 0 due to (4.3.37).

The expressions (4.3.35) for B and D reduce in the Hecke case to

(4.3.59) $$B_{12} = \mu I_1 Q_2 + \nu P_{12} ,$$

(4.3.60) $$D_{12} = \alpha \tilde{Q}_1 I_2 + \beta P_{12}$$

with some constants μ, ν, α and β.

Substituting the expressions for A, B and D into (4.3.39), we obtain, after using identities from the subsection 4.1.2, an equality

(4.3.61)
$$(\mu\lambda(1-\epsilon) + \alpha\nu(1 - \epsilon\lambda \operatorname{tr}(\tilde{Q})) - \epsilon\lambda\beta\nu - \epsilon\lambda\alpha\mu)P_{23}I_1$$
$$+(\beta\mu - \mu(1-\epsilon)\lambda)\hat{R}_{21}Q_3 + \alpha\mu r^2 P_{23}\hat{R}_{31}$$
$$+\beta\nu P_{23}\hat{R}_{21} - \epsilon\lambda\beta\mu I_1 I_2 Q_3 = 0 .$$

The tensors $P_{23}I_1$, $\hat{R}_{21}Q_3$, $P_{23}\hat{R}_{31}$, $P_{23}\hat{R}_{21}$ and $I_1 I_2 Q_3$, entering eqn. (4.3.61) are linearly independent: to see it, multiply them from the right by $\hat{\Psi}_{14}$ and take tr_1; the tensors become $P_{23}\tilde{Q}_4$, $P_{24}Q_3$, $P_{23}P_{34}$, $P_{23}P_{24}$ and $I_2 Q_3 \tilde{Q}_4$, which are obviously independent. Thus, the coefficients must vanish,

(4.3.62)
$$\mu\lambda(1-\epsilon) + \alpha\nu(1 - \epsilon\lambda \operatorname{tr}(\tilde{Q})) - \epsilon\lambda\beta\nu - \epsilon\lambda\alpha\mu = 0 ,$$
$$\beta\mu - \mu(1-\epsilon)\lambda = 0 , \ \alpha\mu = 0 , \ \beta\nu = 0 , \ \epsilon\lambda\beta\mu = 0 .$$

For $\epsilon = 0$, it follows from eqs. (4.3.62) that $\alpha\mu = 0$, $\beta\nu = 0$ and $\beta\mu + \alpha\nu = 0$, which implies that either B or D is zero.

For $\epsilon = 1$, it follows from eqs. (4.3.62) that $\beta\mu = 0$, $\alpha\nu(1 - \lambda \operatorname{tr}(\tilde{Q})) = 0$, $\alpha\mu = 0$ and $\beta\nu = 0$; in view of (4.3.57) we conclude again that either B or D is zero.

It is enough to consider the case $B = 0$; another case can be reduced to it by considering the opposite ordering (if we read (4.3.31)-(4.3.34) from the right to the left, as instructions to order yx to the form xy).

9. Solution

With $B = 0$ the system (4.3.38)-4.3.56) simplifies drastically, can be fully analyzed and one can write down all solutions. It is lengthy and we shall not do it here.

It turns out that solutions which give rise to the orthogonal and symplectic quantum groups are those for which the coefficient α in (4.3.60) is different from 0.

Proposition. Let $\hat{R}$ be a solution of the Yang-Baxter equation with $\hat{R}^2 = \lambda\hat{R} + 1$. If $\hat{R}$ satisfies assumptions **A1-A4** then the ordering relations

(4.3.63) $$x^i y^j = \hat{R}^{ij}_{kl} y^k x^l ,$$

(4.3.64) $$x_i y_j = \hat{R}^{kl}_{ji} y_l x_k ,$$

(4.3.65) $$x^i y_j = \kappa(\hat{R}^{-1})^{ki}_{lj} y_k x^l ,$$

(4.3.66) $$x_i y^j = \kappa^{-1}\hat{\Psi}^{uj}_{vi} y^v x_u + \lambda y_i x^j + \nu\lambda\tilde{Q}^j_i y_k x^k ,$$

where κ is an arbitrary non-zero number, provide an invertible and skew invertible solution $\hat{\mathbf{R}}$ of the Yang-Baxter equation when $\nu^2 + \lambda\nu - 1 = 0$.

If $\hat{R}$ is of GL_N-type then $\hat{\mathbf{R}}$ is of SO_{2N} type for $\nu = -q$ and of Sp_{2N} type for $\nu = q^{-1}$.

10. $SO(2N+1)$

Without going into details we shall describe the situation with the odd-dimensional orthogonal groups.

One has to add a new generator x^0 to the multiplets $x^\bullet$ and $x_\bullet$. The matrix $\hat{\mathbf{R}}$ again turns out to be block-triangular; we will write the answer for the order $x^\bullet > x^0 > x_\bullet$. Relations (4.3.63), (4.3.64) and (4.3.65) are the same. Relation (4.3.66) has to be replaced by

(4.3.67) $\qquad x_i y^j = \hat{\Psi}^{uj}_{vi} y^v x_u + \lambda y_i x^j - \lambda \tilde{Q}^j_i y_k x^k - q^{-1/2} \lambda \tilde{Q}^j_i y^0 x^0 \ .$

Finally, when one of generators has an index 0, the ordering relations are

$$\begin{aligned} x^i y^0 &= y^0 x^i \ , \\ x^0 y^0 &= y^0 x^0 - q^{1/2} \lambda y_l x^l \ , \\ x_i y^0 &= y^0 x_i + \lambda y_i x^0 \ , \\ x^0 y^i &= y^i x^0 + \lambda y^0 x^i \ , \\ x^0 y_i &= y_i x^0 \ . \end{aligned}$$

(4.3.68)

Proposition. Under the same conditions as in the Proposition above, the ordering relations (4.3.63)-(4.3.65), (4.3.67) and (4.3.68) provide an invertible and skew invertible solution $\hat{\mathbf{R}}$ of the Yang-Baxter equation.

If $\hat{R}$ is of GL_N-type then $\hat{\mathbf{R}}$ is of SO_{2N+1} type.

Remarks. 1. For a standard $\hat{R}$ for GL, it was noted in [**32**], that the commutation relations between coordinates and derivatives (even or odd) can be given by projectors of $\hat{R}$ for Sp and SO. Our propositions in this subsection generalize it to the construction of the whole $\hat{R}$-matrix for SO and Sp from the $\hat{R}$-matrix for GL, which works in all cases, not only for the standard deformation.

2. If one starts with $\hat{R}$ corresponding to a supergroup $GL(M|N)$, the constructions of the propositions from this subsection produce Yang-Baxter matrices for the quantum supergroups of OSp type.

5. Real forms

In this subsection we explain how to classify real forms of RTT-algebras using quantum spaces [**33, 34**][6].

---

[6]The description of real forms of the dual algebra for a generic q is given in [**35**]. Our description is more precise, it requires only that $q^4 \neq 1$ in SL case and $q^8 \neq 1$ in the SO and Sp cases

5.1. General linear quantum groups.

We shall start with a standard Drinfeld-Jimbo $\hat{R}$ matrix (4.0.1) for the quantum group $GL_q(N)$. We shall assume that $q^4 \neq 1$.

Exercise. Show that the $\hat{R}$-matrix (4.0.1) satisfies the Yang-Baxter equation. Show that the spectral decomposition of $\hat{R}$ is $\hat{R} = qS - q^{-1}A$ (S and A are projectors, $S^2 = S$, $A^2 = A$) with $\operatorname{rk} S = \frac{N(N+1)}{2}$ and $\operatorname{rk} A = \frac{N(N-1)}{2}$.

Let $*$ be an involution on the RTT-algebra, that is, an antilinear operation, satisfying $*(ab) = *(b) * (a)$ and $(* \otimes *) \circ \Delta = \Delta \circ *$. Then $*x^i$ form a comodule for the $SL_q(N)$. There are two comodules of dimension N: one is generated by x^i, another one is generated by x_i. So we may have two different types of conjugations, $*$ can map A_+^l to itself or to A_+^r.

We shall consider in some details the first possibility. So, we assume

$$(5.1.1) \qquad *x^i = J_j^i x^j \ .$$

Since the matrix T coacts on the vector x, we have $*(Tx) = JTx$; on the other hand, $*(Tx)^i = *(T_j^i x^j) = *x^j * T_j^i = *T_j^i * x^j = *T_j^i J_k^j x^k$ (we used that T_j^i commutes with x^k). It follows then that

$$(5.1.2) \qquad *T = JTJ^{-1} \ .$$

Conjugate now the relation $\hat{R}T_1 T_2 = T_1 T_2 \hat{R}$:

$$\overline{\hat{R}} * T_2 * T_1 = *T_2 * T_1 \overline{\hat{R}}$$

(here ̄ is the complex conjugate), or

$$\overline{\hat{R}}_{21} * T_1 * T_2 = *T_1 * T_2 \overline{\hat{R}}_{21} \ .$$

Substituting $*T$ from (5.1.2) we find

$$(5.1.3) \qquad \Phi T_1 T_2 = T_1 T_2 \Phi \ ,$$

where $\Phi = J_1^{-1} J_2^{-1} \overline{\hat{R}}_{21} J_1 J_2$.

Proposition. Let $\hat{R}$ be the standard $SL_q(N)$ $\hat{R}$-matrix (4.0.1). If an operator $\Phi = \Phi_{12}$ satisfies an equality

$$(5.1.4) \qquad [\Phi, T_1 T_2] = 0 \ ,$$

then Φ is a polynomial in $\hat{R}$.

Sketch of the proof. Take a 1-dimensional representation for T, $T_j^i \mapsto \mu_j \delta_j^i$ with some commuting variables μ_i. Then it follows from (5.1.4) that Φ is of "ice" type, that is, Φ_{kl}^{ij} can be different from zero only if $i = k$, $j = l$ or $i = l$, $j = k$.

Take now another representation, $(T_j^i)_b^a = \hat{R}_{jb}^{ai}$. Writing (5.1.4) in this representation with an ice Φ, one arrives at the statement of the proposition. $\square$

$\hat{R}$ satisfies the Yang-Baxter equation (YBe) $\Rightarrow \hat{R}_{21}$ satisfies YBe $\Rightarrow \overline{\hat{R}}_{21}$ satisfies YBe $\Rightarrow \Phi = J_1^{-1} J_2^{-1} \overline{\hat{R}}_{21} J_1 J_2$ satisfies YBe.

The following proposition is easy:

Proposition. A non-constant polynomial in $\hat{R}$ which satisfies YBe is either $\alpha\hat{R}$ or $\alpha\hat{R}^{-1}$ for some constant α.

The operator $\hat{R}_{21} = P\hat{R}P$ (P is the permutation) has the same spectrum as $\hat{R}$. Therefore, the spectrum of $\Phi = J_1^{-1}J_2^{-1}\hat{R}_{21}J_1J_2$ contains an eigenvalue $\bar{q}$ with the multiplicity $\frac{N(N+1)}{2}$ and the eigenvalue $(-\bar{q}^{-1})$ with the multiplicity $\frac{N(N-1)}{2}$.

According to the Proposition, we have to consider two possibilities, $\Phi = \alpha\hat{R}$ or $\Phi = \alpha\hat{R}^{-1}$.

Comparing spectra, we find that if $\Phi = \alpha\hat{R}$ then $\bar{q} = \alpha q$ and $-\bar{q}^{-1} = -\alpha q^{-1}$. Therefore, $\alpha^2 = 1$ and $\bar{q} = \pm q$.

Similarly, if $\Phi = \alpha\hat{R}^{-1}$ then $\bar{q} = \alpha q^{-1}$ and $-\bar{q}^{-1} = -\alpha q$. Therefore, $\alpha^2 = 1$ and $\bar{q} = \pm q^{-1}$.

We have four cases. Let us see which equations we have to solve. For example, for $\bar{q} = q$ we have $\Phi = \hat{R}$; for q real, $\overline{\hat{R}} = \hat{R}$ and we have therefore equations $\hat{R}_{21}J_1J_2 = J_1J_2\hat{R}$. This is a system of quadratic equations and it turns out that for the $\hat{R}$-matrix (4.0.1) one can completely solve the system. One can solve the corresponding system in the other three cases as well.

For the other type of conjugation (when $*$ of a quantum vector is a quantum covector), the operator J has two lower indices, $*x_i = J_{ij}x^j$. Again at the end one arrives at a system of quadratic equations for j which admits a complete solution.

The last step is to impose the condition that the square of the conjugation is the identity, $** = \text{Id}$; this produces a further restriction on the operator J.

The final result is presented below. We use a notation $\Omega(c_1,\ldots,c_N)$ for an antidiagonal matrix $\begin{pmatrix} & & & & c_1 \\ & & & c_2 & \\ & & \ldots & & \\ & c_{N-1} & & & \\ c_N & & & & \end{pmatrix}$.

In the formulation of the theorem below, a letter "a" appears sometimes in the name of a real form. The letter "a" stands for "alternative"; it signifies that there are several real forms having the same classical limit.

Theorem. (i) There are no real forms in the nonquasiclassical cases $\bar{q} = -q^{\pm 1}$; all real forms admit the classical limit.

(ii) For $\bar{q} = q^{-1}$ the real forms are:

$SL_q(N,\mathbb{R})$; here $J = 1$.

$SU_q^a(N - [N/2], [N/2])$; $J_{ij} = \delta_{i+j}^{N+1}$.

(iii) For $\bar{q} = q$ the real forms are:

$SL_q^a(N,\mathbb{R})$; $J_j^i = \delta_{i+j}^{N+1}$.

$SU_q^*(2n)$, $N = 2n$; $J = \text{antidiag}(\underbrace{1,\ldots,1}_{n \text{ times}},\underbrace{-1,\ldots,-1}_{n \text{ times}})$.

$SU_q(\mu_1,\ldots,\mu_N)$; $J_{ij} = \mu_i\delta_i^j$, $\mu_i = \pm 1$.

In the last case the sequences $\{\mu_i\}$ and $\{-\mu_i\}$ produce equivalent real forms. What is more interesting is that the sequences $\{\mu_i\}$ and $\{\mu_i'\}$ where $\mu_i' = \mu_{i'}$, where $i' = N + 1 - i$ produce equivalent real forms as well. An explanation: classically, there is an outer automorphism $T \mapsto (T^{-1})^t$ of the algebra, corresponding to the symmetry $T_\leftrightarrow$ of the Dynkin diagram A_l. For the quantum T, we have $(T^{-1})^i_k T^k_j = \delta^i_j = T^i_k(T^{-1})^k_j$ but $(T^{-1})^k_i T^j_k \neq \delta^j_i$. The correct version is[7] $(T^{-1})^k_i (QTQ^{-1})^j_k = \delta^j_i$ where the numerical matrix Q is defined by (4.1.38) (we remind that the standard $\hat{R}$-matrix (4.0.1) is skew-invertible, see (4.1.41)). It is, up to a factor, the same Q which cyclically rotated the E-tensor.

Set $\tau(T^i_j) = (T^{-1})^{j'}_{i'}$.

Proposition. The map τ preserves the RTT-relations.

The proof follows from the fact that $\hat{R}Q_1Q_2 = Q_1Q_2\hat{R}$ and $\hat{R}^{i'j'}_{k'l'} = \hat{R}^{ji}_{lk}$ for the standard $\hat{R}$. Moreover, $\tau((T^{-1})^j_k) = (Q^{-1})^{k'}_v T^v_u Q^u_{j'}$. The effect of τ on the sequence $\{\mu_i\}$ is exactly $\{\mu_i\} \mapsto \{\mu'_i\}$.

5.2. Orthogonal and symplectic quantum groups. I shall very shortly list the real forms for orthogonal and symplectic quantum groups.

The answer below is written in the basis, in which the ordering relations for the quantum planes have the form as in (4.3.63)-(4.3.66), with $\kappa = 1$, for $SO_q(2N)$ and $Sp_q(2N)$, or (4.3.63)-(4.3.65) and (4.3.67)-(4.3.68) for $SO_q(2N+1)$.

Let $B = \begin{pmatrix} 1 & & & & & & \\ & \ddots & & & & & \\ & & 1 & & & & \\ & & & & 1 & & \\ & & & 1 & \cdot & & \\ & & & & & 1 & \\ & & & & & \ddots & \\ & & & & & & 1 \end{pmatrix}$ (the nondiagonal 2 by 2 block is in

the middle); in the formulation of the theorem below, a letter "b" in the name of a real form signifies that the matrix J involves the matrix B.

Theorem. (i) Again, all real forms admit the classical limit.

(ii) For $\bar{q} = q^{-1}$ the real forms are:
$SO_q([N/2], N - [N/2])$; $J = 1$;
$SO^b_q(n+1, n-1)$, $N = 2n$; $J = B$.
$Sp_q(N, \mathbb{R})$; $J = 1$.

(iii) For $\bar{q} = q$ the real forms are ($\mu_i = \pm 1$):
$SO_q(\mu_1, \ldots, \mu_N)$; $J = \Omega(\mu_1, \ldots, \mu_N)$ with $J^t = J$.
$SO^b_q(\mu_1, \ldots, \mu_N)$; $J = B\Omega(\mu_1, \ldots, \mu_N)$ with $\mu_{i'} = \mu_i$.
$SO^*_q(\mu_1, \ldots, \mu_N)$; $J = \Omega(\mu_1, \ldots, \mu_N)$ with $J^t = -J$.
$USp_q(\mu_1, \ldots, \mu_N)$; $J = \Omega(\mu_1, \ldots, \mu_N)$ with $J^t = J$.
$Sp_q(\mu_1, \ldots, \mu_N; \mathbb{R})$; $J = \Omega(\mu_1, \ldots, \mu_N)$ with $J^t = -J$.

I shall end the lectures by a comparison with the classical (Cartan) way of classifying the real forms (see, eg. [**36**]).

---

[7] see eqs. (4.3.7), (4.3.8) and (4.3.12).

1. One proves that there exists a unique compact real form u; denote the corresponding $*$ by τ.

2. For an arbitrary real form σ one proves that there exists an equivalent to it real form $\tilde{\sigma}$ such that the automorphism $\theta = \tilde{\sigma}\tau$ is involutive, $\theta^2 = 1$. For a description of involutive automorphisms one should analyze each Cartan data concretely.

3. The automorphism θ acts on u; under this action, u decomposes according to the eigenvalues of θ, $u = u_1 \oplus u_{-1}$. The real form corresponding to $\tilde{\sigma}$ is $u_1 \oplus \sqrt{-1}u_{-1}$.

In the classification of real forms of quantum groups given above, these steps become hidden because quantum spaces are more "rigid" (they admit less automorphisms).

Acknowledgements. I am deeply grateful to my friends and colleagues R. Coquereaux and A. Isaev for inspiration and numerous valuable comments.

References

[1] V. Chari and A. Pressley, *A guide to quantum groups*, Cambridge University Press (1994).
[2] J. Dixmier, *Algèbres enveloppantes*, Gauthier-Villars (1974); English translation: J. Dixmier, *Enveloping algebras*, North-Holland (1977).
[3] S. Montgomery, *Hopf algebras and their actions on rings*, CBMS Regional Conference Series in Mathematics, 82; published by the American Mathematical Society, Providence, RI, 1993
[4] V. Drinfeld, *On some unsolved problems in quantum group theory*, in: Lect. Notes in Math. **1510**, Springer 1992.
[5] A. Medina and Ph. Revoy, *Algèbres de Lie et produit scalaire invariant*, Ann. Scient. Éc. Norm. Sup. **18** (1985).
[6] N. Jacobson, *Lie Algebras*, New York – London, Interscience, 1962.
[7] P. Etingof, T. Schedler and O. Schiffman, *Explicit quantization for dynamical r-matrices for finite dimensional semisimple Lie algebras*, J. Amer. Math. Soc. **13** (2000).
[8] A. Isaev and O. Ogievetsky, *On quantization of r-matrices for Belavin-Drinfeld triples*, Yadernaya Fizika **64** (2001).
[9] A. Belavin and V. Drinfeld, *Triangle equations and simple Lie algebras*, Sov. Sci. Rev. **C 4** (1984).
[10] L. Faddeev, N. Reshetikhin and L. Takhtajan, *Quantization of Lie groups and Lie algebras*, Algebra i Analiz **1** (1989); English translation: Leningrad Math. J. **1** (1990).
[11] V. Drinfeld, *Quasi-Hopf algebras*, Algebra i Analiz **1** (1989); English translation: Leningrad Math. J. **1** (1990).
[12] Yu. Manin, *Quantum groups and noncommutative geometry*, Centre de Recherches Mathématiques, Univ. de Montréal, 1988.
[13] L. Bokut', *Imbeddings into simple associative algebras* (Russian), Algebra i Logika **15** (1976).
[14] G. Bergman, *The diamond lemma for ring theory*, Adv. Math. **29** (1978).
[15] J. Wess and B. Zumino, *Covariant differential calculus on the quantum hyperplane*, Nucl. Phys. B (Proc. Suppl.) **18 B** (1990).
[16] D. Gurevich, *Algebraic aspects of the quantum Yang-Baxter equation*, Algebra i Analiz **2** (1990); English translation: Leningrad Math. J. **2** (1991).
[17] H. Ewen, O. Ogievetsky and J. Wess, *Quantum matrices in 2 dimensions*, Lett. Math. Phys. **22** (1991).
[18] M. Artin and W. Schelter, *Graded algebras of global dimension 3*, Adv. in Math. **66** (1987).
[19] H. Ewen and O. Ogievetsky, *Classification of quantum matrix groups in 3 dimensions*, Preprint MPI-Ph/94-93.
[20] G. Lusztig, *Introduction to quantum groups*, Birkhäuser (1993).

[21] V. Ufnarovsky, *Combinatorial and asymptotic methods in algebra*, Current problems in mathematics: Fundamental directions, Vol. **57** (Russian), 5–177, Itogi Nauki i Tekhniki, Akad. Nauk SSSR, Vsesoyuz. Inst. Nauchn. i Tekhn. Inform., Moscow, 1990.

[22] O. Ogievetsky, *Matrix Structure of $SL_q(2)$ for q a root of unity*, Preprint CPT 96/P3390.

[23] N. Reshetikhin and V. Turaev, *Invariants of 3-manifolds via link polynomials and quantum groups*, Invent. Math. **103** (1991).

[24] A. Alekseev, D. Glushchenkov and A. Lyakhovskaya, *Regular representation of the quantum group $sl_q(2)$ (q is a root of unity)*, Algebra i Analiz **6** (1994); English translation: St. Petersburg Math. J. **6** (1995).

[25] H.-J. Schneider, *Lectures on Hopf algebras*; notes by Sonia Natale. Trabajos de Matemática [Mathematical Works], 31/95. Universidad Nacional de Córdoba, Facultad de Matemática, Astronomia y Fisica, Córdoba, 1995.

[26] C. De Concini and V. Kac, *Representations of quantum groups at roots of 1*, in: *Operator algebras, unitary representations, enveloping algebras, and invariant theory* (Paris, 1989); Progr. Math., **92**, Birkhäuser Boston, Boston, MA, 1990.

[27] V. Drinfeld, *Quantum groups*, Proc. Int. Congr. Math. **1** (1986).

[28] M. Jimbo, *A q-difference analogue of $U_q(g)$ and the Yang-Baxter equation*, Lett. Math. Phys. **1** (1985).

[29] V. Drinfeld, *Almost cocommutative Hopf algebras*, Algebra i Analiz **1** (1989); English translation: Leningrad Math. J. **1** (1990).

[30] S. L. Woronowicz, *Solutions of the braid equation related to a Hopf algebra*, Lett. Math. Phys. **23** (1991).

[31] A. P. Isaev, *Twisted Yang-Baxter equations for linear quantum (super)groups*, J. Phys. A: Math. Gen. **29** (1996).

[32] B. Zumino, *Deformation of the quantum mechanical phase space with bosonic or fermionic coordinates*, Modern Phys. Lett. A**6** (1991).

[33] O. Ogievetsky, *Real forms of special linear quantum groups*, Preprint MPI-Ph/92-49.

[34] O. Ogievetsky, *Real forms of symplectic and orthogonal quantum groups*, Preprint MPI-Ph/92-50.

[35] E. Twietmeyer, *Real forms of $U_q(g)$*, Lett. Math. Phys. **24** (1992).

[36] V. Gorbatsevich, A. Onishchik and E. Vinberg, *Foundations of Lie theory and Lie transformation groups*, Encyclopaedia Math. Sci., 20, Springer, Berlin, 1993.

CENTRE DE PHYSIQUE THÉORIQUE, LUMINY, 13288 MARSEILLE, FRANCE. ON LEAVE OF ABSENCE FROM P. N. LEBEDEV PHYSICAL INSTITUTE, THEORETICAL DEPARTMENT, LENINSKY PR. 53, 117924 MOSCOW, RUSSIA.

E-mail address: oleg@cpt.univ-mrs.fr

CFT, BCFT, ADE and all that

J.-B. Zuber

ABSTRACT. These pedagogical lectures present some material, classical or more recent, on (Rational) Conformal Field Theories and their general setting "in the bulk" or in the presence of a boundary. Two well posed problems are the classification of modular invariant partition functions and the determination of boundary conditions consistent with conformal invariance. It is shown why the two problems are intimately connected and how graphs –ADE Dynkin diagrams and their generalizations– appear in a natural way.

Crimen Ade Quantum
N. Porpora (1686-1768)

0. Introduction

These lectures aim at presenting some curious features encountered in the study of 2 D conformal field theories. The key words are graphs, or Dynkin diagrams, and indeed we shall encounter new avatars of the ADE Dynkin diagrams, and some generalizations thereof. The first lecture is devoted to a lightning review of Conformal Field Theory (CFT), essentially to recall essential notions and to establish basic notations. The study of modular invariant partition functions for theories related to the simplest Lie algebra $sl(2)$ leads to an ADE classification, as has been known for more than ten years. A certain frustration comes from the fact that we have no good reason to explain why this ADE classification appears, or no definite way to connect it to another existing classification (Lecture 2). Or, which amounts to the same, we have too many: depending on the way we look at these $sl(2)$ theories –their topological counterparts, their lattice realization– the reason looks different. Moreover, when we turn to higher rank algebras, the situation is even more elusive: it had been guessed long ago that classification should involve again graphs, though for reasons not very well understood, and a list of graphs had been proposed in the case of $sl(3)$. Recent progress has confirmed these expectations: as will be discussed in the final lecture 3, through the study of boundary conditions we now understand why graphs are naturally associated with

---

1991 *Mathematics Subject Classification.* 47Lxx, 81Rxx, 47N50, 47N55, 16W30, 20G42, 81R50, 81R60 .

© 2002 American Mathematical Society

CFTs and which properties they must satisfy. In the case of $sl(2)$, this leads in a straightforward way (but up to a little subtlety) to the ADE diagrams. Moreover, the classification of the graphs relevant in the case of $sl(3)$ has just been completed, see A. Ocneanu's lectures at this school.

1. A crash course on CFT

This section is devoted to a fast summary of concepts and notations in conformal field theories (CFTs).

1.1. On CFTs

A conformal field theory is a quantum field theory endowed with covariance properties under conformal transformations. We shall restrict our attention to two dimensions, and in a first step to the Euclidean plane, where the conformal transformations are realized by any analytic change of coordinates $z \mapsto \zeta(z)$, $\bar{z} \mapsto \bar{\zeta}(\bar{z})$. In fact, it appears that the contributions of the variables z and $\bar{z}$ decouple [1] and that z and $\bar{z}$ may be regarded as independent variables. The zz component of the energy-momentum tensor $T_{\mu\nu}$ is analytic as a consequence of its tracelessness and conservation, and denoted $T(z)$, $T(z) := T_{zz}(z)$, and likewise $\bar{T}(\bar{z}) := T_{\bar{z}\bar{z}}(\bar{z})$ is antianalytic. $T(z)$ is the generator of the infinitesimal change $z \mapsto \zeta = z + \epsilon(z)$ and likewise $\bar{T}(\bar{z})$ for $\bar{z} \mapsto \bar{\zeta}(\bar{z})$, in the sense that under such a change a correlation function of fields undergoes the change

$$\delta \langle \phi_{i_1}(z_1, \bar{z}_1) \cdots \phi_{i_n}(z_n, \bar{z}_n) \rangle = \frac{1}{2\pi i} \oint_C dw\, \epsilon(w) \langle T(w) \phi_{i_1}(z_1, \bar{z}_1) \cdots \phi_{i_n}(z_n, \bar{z}_n) \rangle + \text{c.c.} \tag{1.1}$$

with an integration contour encircling all points $z_1, \cdots, z_n$.

Fields of a CFT are assigned an explicit transformation under these changes of variables. In particular, *primary fields* transform as $(h, \bar{h})$ forms, i.e. according to

$$\tilde{\phi}(\zeta, \bar{\zeta}) = \left(\frac{\partial z}{\partial \zeta}\right)^h \left(\frac{\partial \bar{z}}{\partial \bar{\zeta}}\right)^{\bar{h}} \phi(z, \bar{z}) \tag{1.2}$$

where the real numbers $(h, \bar{h})$ are the *conformal weights* of the field ϕ (see below for a representation-theoretic interpretation). For an infinitesimal change, $\zeta = z + \epsilon(z)$, $\bar{\zeta} = \bar{z} + \bar{\epsilon}(\bar{z})$, if we set $\tilde{\phi}(z, \bar{z}) = \phi(z, \bar{z}) - \delta\phi(z, \bar{z})$

$$\delta\phi = (h\epsilon' + \epsilon\partial_z)\phi + \text{c.c.} \tag{1.3}$$

Here and in the following, "c.c." denotes the formal complex conjugate, $(z, h) \to (\bar{z}, \bar{h})$, "formal" because $\bar{z}$ is at this stage independent of z and h and $\bar{h}$ are also a priori independent. When the condition that $\bar{z} = z^*$ is imposed, $h + \bar{h}$ turns out to be the scaling dimension of the field ϕ and $h - \bar{h}$ its spin: locality (singlevaluedness of the correlators) imposes only on $h - \bar{h}$ to be an integer or half-integer.

In contrast to primary fields, the change of $T(z)$ itself has the form

$$\tilde{T}(\zeta) = \left(\frac{\partial z}{\partial \zeta}\right)^2 T(z) + \frac{c}{12}\{z, \zeta\} \tag{1.4}$$

where $\{z, \zeta\}$ denotes the schwarzian derivative

$$\{z, \zeta\} = \frac{\frac{\partial^3 z}{\partial \zeta^3}}{\frac{\partial z}{\partial \zeta}} - \frac{3}{2} \left(\frac{\frac{\partial^2 z}{\partial \zeta^2}}{\frac{\partial z}{\partial \zeta}} \right)^2 \tag{1.5}$$

and where the parameter c in front of the anomalous schwarzian term is the *central charge* (of the Virasoro algebra, to come soon!). In other words, T transforms almost as a primary field of conformal weights $(h, \bar{h}) = (2, 0)$ –a conserved current of scaling dimension 2– up to the schwarzian anomaly.

Exercise : derive the form of the variation of T under an infinitesimal transformation, i.e. the analog of (1.3).

In the spirit of local field theory, we assume that correlation functions as above are well defined in the complex plane with possible singularities only at coinciding points $z_i = z_j$ or $w = z_i$. Close to these points, there is a short distance expansion of products of fields. As an exercise, using Cauchy theorem, show that equation (1.3) (and its analog for T) may be rephrased as a statement on the expansions

$$T(w)\phi(z, \bar{z}) = \frac{h\phi(z, \bar{z})}{(w-z)^2} + \frac{\partial \phi(z, \bar{z})}{(w-z)} + \text{regular} \tag{1.6a}$$

$$T(w)T(z) = \frac{\frac{c}{2}}{(w-z)^4} + \frac{2T(z)}{(w-z)^2} + \frac{\partial T(z)}{(w-z)} + \text{regular} \tag{1.6b}$$

with similar expressions for the products with $\bar{T}(\bar{z})$. Eqs (1.6) are meant to describe the singular behaviour of correlation functions $\langle T(w)\phi(z, \bar{z}) \cdots \rangle$, $\langle T(w)T(z) \cdots \rangle$, in the presence of spectator fields, as $z \to w$.

As usual in Quantum Field Theory, it is good to have two dual pictures at hand: the one dealing with correlation (Green) functions, as we have done so far, and in the spirit of quantum mechanics, the operator formalism, which describes the system by "states", i.e. vectors in a Hilbert space. In CFT, it is appropriate to think of a radial quantization in the plane: surfaces of equal "time" are circles centered at the origin, and the Hamiltonian is the dilatation operator. The origin in the plane plays the rôle of remote past, the remote future lies on the circle at infinity. On any circle, there is a description of the system in terms of a Hilbert space $\mathcal{H}$ of states, on which field *operators* act. If we expand the energy momentum tensor on its Laurent modes

$$T(z) = \sum_{n=-\infty}^{\infty} z^{-n-2} L_n \tag{1.7}$$

it is a good exercise (making use again of Cauchy theorem) to check that the expansion (1.6b), now regarded as an "operator product expansion" (OPE), may be rephrased as a commutation relation between the L's

$$[L_n, L_m] = (n - m)L_{n+m} + \frac{c}{12} n(n^2 - 1) \delta_{n+m, 0} . \tag{1.8}$$

This is the celebrated **Virasoro algebra**, in which the *central charge c* appears indeed as the coefficient of the central term. Note that L_0 is the generator of dilatations, L_{-1} the one of translations. Together with L_1, they form a subalgebra.

What is the interpretation of L_1 and of that subalgebra?

Exercise : derive the commutation relation of L_n with the field operator ϕ, as a consequence of the equation (1.6a).

1.2. Extension to other chiral algebras.

A frequently encountered situation in CFT is that there is a larger ("extended") *chiral* algebra $\mathcal{A}$ encompassing the Virasoro algebra, and acting on fields of the theory. The latter thus fall again in representations of $\mathcal{A}$. The most common cases are those involving a current algebra, i.e. the affine extension $\hat{\mathfrak{g}}$ of a Lie algebra $\mathfrak{g}$, or the so-called W algebras, or the various superconformal algebras, etc.

In the affine algebra $\hat{\mathfrak{g}}$, the important objects from our standpoint are the current $J(z)$ or its moments J_n, with values in the adjoint representation of $\mathfrak{g}$. In some basis, they satisfy the commutation relation

$$[J_n^a, J_m^b] = if^{ab}{}_c J_{n+m}^c + \hat{k}\, n\, \delta_{n+m,\,0}\, \delta_{ab}\,, \qquad (1.9)$$

with $\hat{k}$ a central element.

The Virasoro algebra is either a proper subalgebra (for ex, in the superconformal cases), or contained in the enveloping algebra of this chiral algebra. For example, in affine algebras, the energy momentum tensor and Virasoro generators are obtained through the Sugawara construction as quadratic forms in the currents: $T(z) = \text{const.} : (\mathbf{J}(z))^2 :$. (The colons refer to a specific regularization of this ill-defined product of two currents at coinciding points, and the constant is fixed in any irreducible representation ...). For many cases of such an extended chiral algebra, the representation theory has been developed (maybe in less detail for W algebras, even less for other, more exotic, extended algebras ...).

1.3. Elements of Representation theory of the Virasoro algebra

The representations that we shall consider are the *highest weight representations (or Verma modules)*. They are parametrized by a real number h: the Verma module $\mathfrak{V}_h$ is generated from a highest weight (h.w.) vector denoted $|h\rangle$ satisfying

$$L_0|h\rangle = h|h\rangle \qquad L_n|h\rangle = 0, \quad \forall n \in \mathbb{N} \qquad (1.10)$$

by the action of the L_{-n}, $n > 0$

$$\mathfrak{V}_h = \text{Span}\{L_{-p_1} L_{-p_2} \cdots L_{-p_r}|h\rangle\}, \quad 1 \leq p_1 \leq p_2 \leq \cdots \leq p_r\,. \qquad (1.11)$$

Such a representation has the important property of being graded for the action of the Virasoro generator L_0. This means that the spectrum of L_0 in $\mathfrak{V}_h$ is of the form $\{h, h+1, h+2, \cdots\}$. The subspace of eigenvalue $h+N$ is called the eigenspace of level N.

Whether this module is or is not irreducible is the object of a theorem (Kac, Feigin-Fuchs)[2]

Theorem Let $c = 1 - 6/x(x+1)$, where $x \in \mathbb{C}$. Then $\mathfrak{V}_h$ is reducible iff there exist two positive integers r and s such that

$$h = h_{rs} := \frac{((r(x+1) - sx)^2 - 1}{4x(x+1)} \,. \qquad (1.12)$$

If h takes one of these values, then $\mathfrak{V}_h$ is reducible in the sense that it contains a "singular" vector, i.e. a vector satisfying the axioms (1.10) of a h.w. vector. This vector thus supports itself a h.w. module, which is a submodule of $\mathfrak{V}_h$. Moreover the theorem asserts that this "degeneracy" occurs at level $r.s$. In fact $\mathfrak{V}_h$ may contain several such submodules, with a non trivial intersection: this is what happens if the parameter x is rational. One constructs the irreducible representation $\mathcal{V}_h$ by quotienting out this (or these) submodule(s) of $\mathfrak{V}_h$.

Assume that the parameter x in (1.12) is of the form $p'/(p-p')$, with p, p' two coprime integers. Show that $h = h_{r\,s} = h_{p'-r\,p-s}$ so that there are degeneracies at the two levels $r.s$ and $(p'-r).(p-s)$ and thus two distinct submodules in $\mathfrak{V}_h$.

Highest weight representations of other chiral algebras may also be constructed. For example, let us sketch the results for an affine algebra $\hat{\mathfrak{g}}$ associated with a simple algebra $\mathfrak{g}$ [3]. Let $\alpha_1, \cdots, \alpha_r$ and θ be the ordinary simple roots and the highest root of the finite algebra $\mathfrak{g}$, $\alpha_i^\vee = 2\alpha_i/(\alpha_i, \alpha_i)$ the corresponding coroots; θ is normalised by $(\theta, \theta) = 2$. Then the so-called *integrable* representations of the algebra $\hat{\mathfrak{g}}$ are labelled by a pair $(\bar{\lambda}, k)$ where k is a non negative integer (the "level" of the representation) and where $\bar{\lambda} \in P$, (P the weight lattice of $\mathfrak{g}$), is subject to the inequalities

$$(\bar{\lambda}, \alpha_i^\vee) \in \mathbb{N}, \quad i = 1, \cdots, r, \qquad (\bar{\lambda}, \theta) \leq k \,. \qquad (1.13)$$

Many formulae simplify if expressed in terms of the shifted weight $\lambda = \bar{\lambda} + \rho$, with ρ the Weyl vector $\rho = \frac{1}{2} \sum_{\alpha > 0} \alpha$. In the case of $\mathfrak{g} = sl(N)$, $r = N-1$, these formulae reduce to $\theta = \sum_1^{N-1} \alpha_i$, the shifted weight $\lambda = \sum_{i=1}^{N-1} \lambda_i \Lambda_i$, with Λ_i the dominant fundamental weights of $sl(N)$, satisfies the inequalities $\lambda_i \geq 1, \sum_{i=1}^{N-1} \lambda_i \leq k+N-1$. The simplest example is of course that of $\widehat{sl}(2)$ for which the representations are labelled by the pair of integers (λ, k), with $1 \leq \lambda \leq k+1$, (λ may be thought of as $2j+1$, i.e. the dimension of the corresponding finite-dimensional spin j representation of $sl(2)$).

If $\mathcal{V}_i$ is some representation of a chiral algebra, we shall use the label i^* to denote the complex conjugate representation. It may happen that $\mathcal{V}_{i^*}$ is identical or equivalent to $\mathcal{V}_i$, like for Vir, or $\widehat{sl}(2)$. We shall in general label by $i = 1$ the identity representation; its conformal weight (eigenvalue of L_0 on the highest weight vector) vanishes.

As in the case of Vir discussed above, these representations are graded for the action of the Virasoro generator L_0. The spectrum of L_0 in $\mathcal{V}_i$ is of the form $\{h_i, h_i + 1, h_i + 2, \cdots\}$, with non-trivial multiplicities $\#_n = \dim$ (subspace of eigenvalue $h + n$). It is thus natural to introduce a generating function of these multiplicities, i.e. a function of a dummy variable q, the *character* of the representation $\mathcal{V}_i$

$$\chi_i(q) = \mathrm{tr}_{\mathcal{V}_i} q^{L_0 - \frac{c}{24}} = q^{h_i - \frac{c}{24}} \sum_{n=0}^{\infty} \#_n q^n \,. \qquad (1.14)$$

Show that in the original h.w. Verma module $\mathfrak{V}_h$ of Vir, the character is simply $\chi_h(q) = \frac{q^{h-\frac{c}{24}}}{\prod_1^\infty (1-q^n)}$. Assuming that for $c = 1$, the representations with a singular vector have a conformal weight given by the limit $x \to \infty$ of (1.12), namely $h = \ell^2/4$, $\ell \in \mathbb{N}$, $r = \ell+1$, $s = 1$, show that $\chi_h(q) = \frac{q^{\frac{\ell^2}{4}}(1-q^{\ell+1})}{\eta(q)}$ with Dedekind's eta function: $\eta(q) = q^{\frac{1}{24}} \prod_1^\infty (1-q^n)$. Another interesting case is for $c < 1$, when the parameter x is rational: one writes

$$c = 1 - \frac{6(p-p')^2}{pp'}, \quad p, p' \in \mathbb{N} \tag{1.15}$$

and one concentrates on the irreducible representations ("irrep") with h of the form

$$h_{rs} = h_{p'-r,p-s} = \frac{(rp-sp')^2 - (p-p')^2}{4pp'}, \quad 1 \le r \le p'-1, \ 1 \le s \le p-1. \tag{1.16}$$

The character of this irrep reads, with $\lambda := (rp - sp')$, $\lambda' := (rp + sp')$

$$\chi_{rs} = \frac{1}{\eta(q)} \sum_{n \in \mathbb{Z}} \left(q^{\frac{(2npp'+\lambda)^2}{4pp'}} - q^{\frac{(2npp'+\lambda')^2}{4pp'}} \right). \tag{1.17}$$

In representations of the affine algebra $\hat{\mathfrak{g}}$, we may also consider the characters $\mathrm{tr}\, q^{L_0-c/24}$. They are called "specialized characters" since they count states according to their L_0 grading only. Non-specialized characters can be introduced, which are sensitive to generators of the Cartan subalgebra $\mathbf{J}_0$

$$\chi(q, \mathbf{u}) = \mathrm{tr}\, q^{L_0 - \frac{c}{24}} e^{2\pi i (\mathbf{u}, \mathbf{J}_0)}. \tag{1.18}$$

For the case of $\widehat{sl}(2)$, and for the representations (λ, k) discussed above

$$\chi_\lambda(q) = \frac{1}{\eta^3(q)} \sum_{p=-\infty}^{\infty} (2(k+2)p + \lambda) q^{\frac{(2(k+2)p+\lambda)^2}{4(k+2)}}. \tag{1.19}$$

The expressions of non-specialized characters and for general algebras may be found in [3].

Similar considerations apply to other chiral algebras ...

1.4. Modular properties of characters

If the multiplicity $\#_n$ doesn't grow too fast in (1.14), this sum converges for $|q| < 1$: it is thus natural to write $q = \exp 2i\pi\tau$, with τ a complex number in the upper half-plane.

It is a remarkable fact that such functions χ enjoy beautiful transformation properties under the action of the modular group on the variable τ

$$\begin{pmatrix} a & b \\ c & d \end{pmatrix} \in PSL(2, \mathbb{Z}) \quad : \tau \mapsto \frac{a\tau + b}{c\tau + d} \tag{1.20}$$

(i.e. a, b, c, d integers defined up to a global sign, with $ad - bc = 1$).

By definition, Rational Conformal Field Theories (RCFT) are CFTs that are consistently described by a finite set $\mathcal{I}$ of representations $\mathcal{V}_i$, $i \in \mathcal{I}$, of a certain chiral algebra $\mathcal{A}$. Moreover the corresponding characters $\chi_i(q)$ form a finite dimensional unitary representation of the modular group (in fact of its double covering, see below): they transform among themselves linearly (and unitarily) under the action of (1.20). It is well known that the modular group is generated by the two transformations

$$T : \tau \mapsto \tau + 1 \qquad S : \tau \mapsto -\frac{1}{\tau} . \tag{1.21}$$

It is clear from the definition (1.14) that under T

$$\chi_i(q) \to \chi_i(q\, e^{2i\pi}) = e^{2i\pi(h_i - \frac{c}{24})} \chi_i(q) \tag{1.22}$$

and the non-trivial part of the above statement is that, if $\tilde{q} := \exp -\frac{2i\pi}{\tau}$ there exists a unitary $|\mathcal{I}| \times |\mathcal{I}|$ matrix S such that

$$\chi_i(q) = \sum_{j \in \mathcal{I}} S_{ij} \chi_j(\tilde{q}) . \tag{1.23}$$

Moreover the matrix S satisfies $S^T = S$, $S^\dagger = S^{-1}$, $(S_{ij})^* = S_{i^*j} = S_{ij^*}$, $S^2 = C =$ the conjugation matrix defined by $C_{ij} = \delta_{ij^*}$, $S^4 = I$.[1]

1st Example: for the $c < 1$ minimal representations,

$$\mathcal{I} = \{(r,s) \equiv (p'-r, p-s) ; \quad 1 \leq r \leq p'-1,\ 1 \leq s \leq p-1\}$$

with h_{rs} given in (1.16), the S matrix reads

$$S_{rs,r's'} = \sqrt{\frac{8}{pp'}} (-1)^{(r+s)(r'+s')} \sin \pi r r' \frac{p-p'}{p'} \sin \pi s s' \frac{p-p'}{p} \tag{1.24}$$

2nd Example: the $\widehat{sl}(2)$ affine algebra

At level k, we have seen that the integrable representations are labelled by the set $\mathcal{I} = \{1, 2, \cdots, k+1\}$. The conformal weight of representation (λ) reads $h_\lambda = (\lambda^2 - 1)/4(k+2)$, $\lambda \in \mathcal{I}$. Then one finds

$$S_{\lambda\mu} = \sqrt{\frac{2}{k+2}} \sin \frac{\pi \lambda \mu}{k+2} , \quad \lambda, \mu \in \mathcal{I} . \tag{1.25}$$

For non-specialised characters, the transformation reads:

$$\chi_\lambda(q, \mathbf{u}) = e^{-ik\pi(\mathbf{u},\mathbf{u})/\tau} \sum_{\mu \in \mathcal{I}} S_{\lambda\mu} \chi_\mu(\tilde{q}, -\mathbf{u}/\tau) . \tag{1.26}$$

The expression for more general affine algebras may be found in [3].

One notes that the S matrix of the minimal case (1.24) is "almost" the tensor product of two matrices of the form (1.25), at two different levels $k = p - 2$ and $k' = p' - 2$. This would be true for $|p - p'| = 1$, and if one could omit the identification $(r, s) \equiv (p' - r, p - s)$. This is of course not a coincidence but reflects the "coset construction" of $c < 1$ representations of Vir out of the affine algebra $\widehat{sl}(2)$ [4].

---

[1] The fact that $S^2 = C$ rather than $S^2 = I$ as expected from the transformation (1.21) signals that we are dealing with a representation of a double covering of the modular group.

1.5. Notion of Fusion Algebra

The last concept of crucial importance for our discussion is that of fusion algebra. Fusion is an operation among representations of chiral algebras of RCFTs, inherited from the operator product algebra of Quantum Field Theory. It looks similar to the usual tensor product of representations, but contrary to the latter, it is consistent with the finiteness of the set $\mathcal{I}$ and it preserves the central elements (instead of adding them). I shall refer to the literature [3,5] for a systematic discussion of this concept, and just introduce a notation $\star$ to denote it and distinguish it from the tensor product. It is natural to decompose the fusion of two representations of a chiral algebra on the irreps, thus defining multiplicities, or "fusion coefficients"

$$\mathcal{V}_i \star \mathcal{V}_j = \oplus_k \mathcal{N}_{ij}{}^k \mathcal{V}_k, \qquad \mathcal{N}_{ij}{}^k \in \mathbb{N} \ . \tag{1.27}$$

There is a remarkable formula, due to Verlinde [6], expressing these multiplicities in terms of the unitary matrix S:

$$\mathcal{N}_{ij}{}^k = \sum_{\ell \in \mathcal{I}} \frac{S_{i\ell} S_{j\ell} (S_{k\ell})^*}{S_{1\ell}} \tag{1.28}$$

Note that the fact that the r.h.s. of (1.28), computed with the matrices (1.25) or (1.24) yields non negative integers is a priori non-trivial!

This completes our review of the basics of RCFTs. The data c (or k, etc), $\mathcal{V}_i, h_i, i \in \mathcal{I}$, $S_i{}^j$, $\mathcal{N}_{ij}{}^k$, form what I call the "chiral data": they are relative to a "chiral half" of the conformal theory (in the plane), which means they refer only to the holomorphic variables z or to their antiholomorphic counterparts $\bar{z}$, rather than to the pair $(z, \bar{z})$.

Our task is now to use these ingredients to construct physically sensible theories.

2. Modular Invariant Partition Functions

In the plane punctured at the origin, equipped with the coordinate z, or equivalently on the cylinder of perimeter L with the coordinate w, with the conformal mapping from the latter to the former $z = \exp -2\pi i w/L$, a given RCFT is described by a Hilbert space $\mathcal{H}_P$. This Hilbert space is decomposable into a *finite* sum of irreps of **two** copies of the chiral algebra (Vir or else), associated with the holomorphic and anti-holomorphic "sectors" of the theory:

$$\mathcal{H}_P = \oplus N_{j\bar{j}} \mathcal{V}_j \otimes \mathcal{V}_{\bar{j}} \ , \tag{2.1}$$

with (non negative integer) multiplicities $N_{j\bar{j}}$. On the cylinder, it is natural to think of the Hamiltonian as the operator of translation along its axis (the imaginary axis in w), or along any helix, defined by its period τL in the w plane, with $\Im m\, \tau > 0$. If L^{cyl}_{-1} and $\bar{L}^{\text{cyl}}_{-1}$ are the two generators of translation in w and $\bar{w}$, $H^{\text{cyl},\tau} = (\tau L^{\text{cyl}}_{-1} + \bar{\tau} \bar{L}^{\text{cyl}}_{-1})$. Mapped back in the plane using the transformation law of the energy-momentum tensor (1.4), L^{cyl}_{-1} reads

$$L^{\text{cyl}}_{-1} = -\frac{2\pi i}{L}(L_0 - \frac{c}{24}) \tag{2.2}$$

where the term $c/24$ comes from the schwarzian derivative of the exponential mapping. The evolution operator of the system, i.e. the exponential of L times the Hamiltonian is thus

$$e^{-H^{\text{cyl},\tau}L} = e^{2\pi i(\tau(L_0-\frac{c}{24})-\bar{\tau}(\bar{L}_0-\frac{c}{24}))} \ . \tag{2.3}$$

A convenient way to encode the information (2.1) is to look at the partition function of the theory on a torus $\mathcal{T}$. Up to a global dilatation, irrelevant here, a torus may be defined by its modular parameter τ, $\Im m\, \tau > 0$, such that its two periods are 1 and τ. Equivalently, it may be regarded as the quotient of the complex plane by the lattice generated by the two numbers 1 and τ:

$$\mathcal{T} = \mathbb{C}/(\mathbb{Z} \oplus \tau\mathbb{Z}) \ , \tag{2.4}$$

in the sense that points in the complex plane are identified according to $w \sim w' = w + n + m\tau$, $n, m \in \mathbb{Z}$. There is, however, a redundancy in this description of the torus: the modular parameters τ and $M\tau$ describe the same torus, for any modular transformation $M \in PSL(2, \mathbb{Z})$. The partition function of the theory on this torus is just the trace of the evolution operator (2.3), with the trace taking care of the identification of the two ends of the cylinder into a torus

$$Z = \text{tr}_{\mathcal{H}_P} e^{2\pi i[\tau(L_0-\frac{c}{24})-\bar{\tau}(\bar{L}_0-\frac{c}{24})]} \ . \tag{2.5}$$

Using (2.1) and the definition (1.14) of characters, this trace may be written as

$$Z = \sum N_{j\bar{j}} \chi_j(q)\chi_{\bar{j}}(\bar{q}) \qquad q = e^{2\pi i\tau} \quad \bar{q} = e^{-2\pi i\bar{\tau}}. \tag{2.6}$$

Let's stress that in these expressions, $\bar{\tau}$ is the complex conjugate of τ, and $\bar{q}$ that of q, and therefore, $Z = \sum N_{j\bar{j}}\chi_j(q)(\chi_{\bar{j}}(q))^*$ is a *sesquilinear* form in the characters. Finally this partition function must be intrinsically attached to the torus, and thus be invariant under modular transformations. This key observation, together with the expression of Z as a sesquilinear form in the characters, is due to Cardy [7]. As we shall now show, it opens a route to the classification of RCFTs. We have been led indeed to the ...

Classification Problem : *find all possible sesquilinear forms (2.6) with non negative integer coefficients that are modular invariant, and such that $N_{11} = 1$.*

The extra condition $N_{11} = 1$ expresses the unicity of the identity representation (i.e. of the "vacuum").[2] As explained in the previous section, the finite set of characters of any RCFT, labelled by $\mathcal{I}$, supports a unitary representation of the

---

[2] Notice that the property of modular invariance is just a necessary condition of consistency of the theory. It may be –and it seems to happen– that some modular invariants do not correspond to any consistent CFT. The general conditions to be fulfilled by a CFT to be fully consistent have been spelled out by Sonoda [8] and by Moore and Seiberg [9]. They amount essentially to the consistency (duality equations) of the 4-point function in the plane, and modular invariance (or covariance) of the 0- and 1-point functions on the torus. Nobody has been able, however, to analyse systematically these conditions besides the simplest cases of $sl(2)$-related theories.

modular group. This implies that any diagonal combination of characters $Z = \sum_{i \in \mathcal{I}} \chi_i(q)\chi_i(\bar{q})$ is modular invariant. Are there other solutions to the problem?

The problem has been completely solved only in a few cases: for the RCFTs with an affine algebra, the $\widehat{sl}(2)$ [10] and $\widehat{sl}(3)$ [11] theories at arbitrary level, plus a host of cases with constraints on the level, e.g. the general $\widehat{sl}(N)$ for $k = 1$ [12], etc; associated coset theories have also been fully classified, including all the minimal $c < 1$ theories, $N = 2$ "minimal" superconformal theories, etc. A good review on the current state of the art is provided by T. Gannon [13].

In the case of CFTs with a current algebra, it is in fact better to look at the same problem of modular invariants after replacing in (2.6) all specialized characters by non-specialized ones, v.i.z. $\sum N_{j\bar{j}} \chi_j(q, \mathbf{u}) \left(\chi_{\bar{j}}(q, \mathbf{u}) \right)^*$. Because these non-specialized characters are linearly independent, there is no ambiguity in the determination of the multiplicities $N_{j\bar{j}}$ from Z. This alternative form of the partition function may be seen as resulting from a modification of the energy-momentum tensor $T(z) \to T(z) - \frac{2\pi i}{L}(\mathbf{u}, \mathbf{J}(z)) - \frac{k}{2}\left(\frac{2\pi}{L}\right)^2(\mathbf{u}, \mathbf{u})$, see [14].

Table 1
List of modular invariant partition functions in terms of $SU(2)$ Kac–Moody characters χ_λ.

$k \geq 0$	$\sum_{\lambda=1}^{k+1} \|\chi_\lambda\|^2$	(A_{k+1})
$k = 4\rho \geq 4$	$\sum_{\lambda \text{ odd} = 1}^{2\rho-1} \|\chi_\lambda + \chi_{4\rho+2-\lambda}\|^2 + 2\|\chi_{2\rho+1}\|^2$	$(D_{2\rho+2})$
$k = 4\rho - 2 \geq 6$	$\sum_{\lambda \text{ odd} = 1}^{4\rho-1} \|\chi_\lambda\|^2 + \|\chi_{2\rho}\|^2 + \sum_{\lambda \text{ even} = 2}^{2\rho-2} (\chi_\lambda \chi^*_{4\rho-\lambda} + \text{c.c.})$	$(D_{2\rho+1})$
$k = 10$	$\|\chi_1 + \chi_7\|^2 + \|\chi_4 + \chi_8\|^2 + \|\chi_5 + \chi_{11}\|^2$	(E_6)
$k = 16$	$\|\chi_1 + \chi_{17}\|^2 + \|\chi_5 + \chi_{13}\|^2 + \|\chi_7 + \chi_{11}\|^2 + \|\chi_9\|^2$ $+ [(\chi_3 + \chi_{15})\chi_9^* + \text{c.c.}]$	(E_7)
$k = 28$	$\|\chi_1 + \chi_{11} + \chi_{19} + \chi_{29}\|^2 + \|\chi_7 + \chi_{13} + \chi_{17} + \chi_{23}\|^2$	(E_8)

2.1. The $\widehat{sl}(2)$ cases

This was the first case fully solved. With the notations introduced in the previous section for the representations of $\widehat{sl}(2)$, the complete list of modular invariant partition functions is as listed in Table 1. A remarkable feature appears, namely an unexpected connection with *ADE* Dynkin diagrams. By this I mean that if we concentrate on the *diagonal* terms of these expressions, their labels λ turn out to be the Coxeter *exponents* of these Dynkin diagrams. I recall that for such a diagram, the eigenvalues of its adjacency matrix G are of the form $2\cos\frac{\pi\ell}{h}$, h the Coxeter number, and ℓ is an exponent taking rank(G) (= number of vertices of G) values between 1 and $h - 1$, with possible multiplicities. Alternatively, the Cartan matrix

$C = 2\mathbb{I} - G$ has eigenvalues $4\sin^2\frac{\pi\ell}{2h}$. These Coxeter number and exponents are listed in Table 2 (do not pay attention for the time being to the last entry denoted T_n). Anticipating a little on the following, let's notice that the off-diagonal terms in the partition functions of Table 1 may also be determined in terms of the data of the ADE diagrams.

Table 2: The graphs A-D-E-T, their Coxeter number and their Coxeter exponents

	h	exponents
A_n	$n+1$	$1, 2, \cdots, n$
$D_{\ell+2}$	$2(\ell+1)$	$1, 3, \cdots, 2\ell+1, \ell+1$
E_6	12	$1, 4, 5, 7, 8, 11$
E_7	18	$1, 5, 7, 9, 11, 13, 17$
E_8	30	$1, 7, 11, 13, 17, 19, 23, 29$
T_n	$2n+1$	$1, 3, 5, \cdots, 2n-1$

2.2. On ADE classifications

The content of this section is not essential for what follows (except point (vii) below). The subject is so intriguing and so fascinating, however, that I cannot resist presenting it.

It is well known that there are many mathematical objects that fall in an ADE classification [15]. The list includes
  (i) simple simply-laced Lie algebras, i.e. with roots of equal length [16];
  (ii) finite reflection groups of cristallographic and of simply-laced type [17];
  (iii) finite subgroups of $SO(3)$ or of $SU(2)$, (or the associated platonic solids);
  (iv) Kleinian singularities [18];
  (v) "simple" singularities, i.e. with no modulus [19];
  (vi) finite type quivers [20];
  (vii) symmetric matrices with eigenvalues between -2 and $+2$;
  (viii) algebraic solutions to the hypergeometric equation ([21], p 385);
  (ix) subfactors of finite index [22], see also D. Evans' lectures at this school;
  and presumably others ...

To elaborate a little on these various items:
(i) Simple Lie algebras have been classified by Killing and Cartan; restricting ourselves to the simply laced ones, i.e. with roots of equal length, leaves us with the ADE cases.

(ii) Reflection groups are groups generated by reflections in hyperplanes orthogonal to vectors $\{\alpha_a\}$ in the Euclidean space $\mathbb{R}^n$ called roots: $S_a : x \mapsto x - 2\alpha_a(\alpha_a, x)/(\alpha_a, \alpha_a)$. The group is of finite order iff the bilinear form (α_a, α_b) is positive definite. This leads to a list A_n, $B_n \equiv C_n$, D_n, E_6, E_7, E_8, F_4, G_2, H_3, H_4, $I_2(k)$, where the subscript gives the space dimension n [17]. If moreover, the condition that the root system is crystallographic is imposed, (i.e. that for all pairs of roots α, β, $2(\alpha, \beta)/(\beta, \beta) \in \mathbb{Z}$), only the cases A to G_2 are left: they are of course the Weyl groups of the Lie algebras of (i). Imposing also the condition of simple lacedness leaves only ADE.

(iii) Finite subgroups of $SU(2)$ form two infinite series and three exceptional cases: the cyclic groups $\mathcal{C}_n$, the binary dihedral groups $\mathcal{D}_n$, the binary tetrahedral group $\mathcal{T}$, the binary octahedral group $\mathcal{O}$ and the binary icosahedral group $\mathcal{I}$. It is natural to label them by ADE as we shall see (Table 3). A related classification is that of the 5 regular solids in three-dimensional Euclidean space: this may be the oldest ADE classification, since it goes back to the school of Plato; strictly speaking, only the exceptional cases E_6, E_7, E_8 appear there, since E_6 is associated with the group of the tetrahedron, E_7 with the group of the octahedron or of the cube, and E_8 with the group of the dodecahedron or of the icosahedron. The cyclic and dihedral groups may be thought of respectively as the rotation invariance group of a pyramid and of a prismus of base a regular n-gon, but those are not regular platonic solids.

(iv) Kleinian singularities: Let Γ be a finite subgroup of $SU(2)$. It acts on $(u, v) \in \mathbb{C}^2$. The algebra of Γ-invariant polynomials in u, v is generated by three polynomials X, Y, Z subject to one relation $W(X, Y, Z) = 0$. The quotient variety $\mathbb{C}^2/\Gamma$ is parametrized by these polynomials X, Y, Z and is thus embedded into the hypersurface $W(x, y, z) = 0$, $x, y, z \in \mathbb{C}^3$. This variety $\mathcal{S}$ is singular at the origin [18], see Table 3.

(v) Simple singularities, i.e. polynomials $W(x_1, x_2, \cdots, x_p)$ with $\partial W/\partial x_j|_0 = 0$ for all $i = 1, \cdots, p$, and with no modulus (up to regular changes of the x), are also given by the same Kleinian polynomials W (up to the addition or subtraction of quadratic terms) [19].

(vi) Quivers are oriented graphs, with vertices a and edges e. A representation of a quiver is the assignment to each vertex a of a non-negative integer d_a and of a vector space $V_a \equiv \mathbb{C}^{d_a}$ and to each edge $e = (a \to b)$ of a linear map $f_e : V_a \mapsto V_b$. Two such representations (V_a, f_e) and (W_a, g_e) are equivalent if there are linear maps $\varphi_a : V_a \mapsto W_a$ such that $\varphi_b \circ f_e = g_e \circ \varphi_a$ for all edges $e = (a, b)$. One proves that quivers of finite type, i.e. such that they admit only a finite number of inequivalent indecomposable representations, are ADE diagrams ! [20],[18].

(vii) Symmetric matrices with non negative integer entries and eigenvalues between -2 and 2 are the adjacency matrices of the $ADET$ graphs of Table 2. The "tadpoles" $T_n = A_{2n}/\mathbb{Z}_2$ may be ruled out if the condition of 2-colourability (or "bipartiteness") is imposed [23].

Γ	$\mathcal{C}_n$	$\mathcal{D}_n$	$\mathcal{T}$	$\mathcal{O}$	$\mathcal{I}$
$\|\Gamma\|$	n	$4n$	24	48	120
W	$X^n - YZ$	$X^{n+1} + XY^2 + Z^2$	$X^4 + Y^3 + Z^2$	$X^3 + XY^3 + Z^2$	$X^5 + Y^3 + Z^2$
G	A_{n-1}	D_{n+2}	E_6	E_7	E_8

Table 3: Finite subgroups of $SU(2)$, their orders, the Kleinian singularity and the associated Dynkin diagram.

The most obvious manifestation of the ADE classification of these objects is provided by the Dynkin diagram and its exponents. In cases (i) and (ii) the Dynkin diagram encodes the geometry of the root system $\{\alpha_a\}$: $(\alpha_a, \alpha_b) = C_{ab}$ the Cartan matrix $= 2\mathbb{I} - G_{ab}$, G the adjacency matrix, while the exponents shifted by 1 give the degrees of the invariant polynomials. Also in case (ii), the product over all the simple roots of the reflections S_a defines the Coxeter element, unique up to conjugation, whose eigenvalues are $\exp 2i\pi m_i/h$, m_i running over the exponents. In case (iii), as found by McKay [24], things are subtle and beautiful: the corresponding *affine* Dynkin diagrams of $\hat{A}$-$\hat{D}$-$\hat{E}$ type describe the decomposition into irreducible representations of the tensor products of the representations of Γ by a two dimensional representation. By removing the vertex corresponding to the identity representation, one recovers the standard ADE Dynkin diagrams in accordance with Table 3. See Appendix A for more elements on the McKay correspondence. In the case (iv) of Kleinian singularities, one considers the *resolution* $\tilde{S}$ of the singular surface $\mathcal{S}$: this is a smooth variety with a projection $\pi : \tilde{\mathcal{S}} \to \mathcal{S}$ which is one-to-one everywhere except above the singularity at 0: one proves that the *exceptional divisor* $\pi^{-1}(0)$ is a connected union of spheres, $\pi^{-1}(0) = C_1 \cup \cdots \cup C_r$, $C_i \cong \mathbb{P}^1\mathbb{C}$. The Dynkin diagram of ADE type listed on the last line of Table 3 (or more precisely the negative of its Cartan matrix) describes the intersection form of these components C_i. In the case (v) of a simple singularity, one may consider its deformation $W = \epsilon$ and look at the intersection of the homology cycles of its level set $\{x \in \mathbb{C}_p, |x| < \delta \,|\, W(x) = \epsilon\}$, or at their monodromy as ϵ circles around the origin: the intersection is again encoded in the Dynkin diagram, while the monodromy is given by the Coxeter element of the associated Coxeter group. Shifted by -1 the exponents give the degrees of the homogeneous polynomials of the local ring $\mathbb{C}[x_1, \cdots, x_p]/(\partial_{x_i} W)$ of the simple singularity [19] etc, etc.

This is just a sample of all the fascinating properties and crossrelations between these problems.

In many cases, the classification follows from the spectral condition (vii). In some others, however, the key point is the determination of triplets of integers (p, q, r) such that $\frac{1}{p} + \frac{1}{q} + \frac{1}{r} > 1$. (Prove that the ADE list below includes all the solutions except $(p = 1, q \neq r)$.) Note also that for the D and E cases, these integers give the length of the three branches of the Dynkin diagram counted from the vertex of valency 3. (The A entry may look slightly artificial at this stage).

G	A_{2n-1}	D_{n+2}	E_6	E_7	E_8
(p,q,r)	$(1,n,n)$	$(2,2,n)$	$(2,3,3)$	$(2,3,4)$	$(2,3,5)$

The integers (p, q, r) also appear in the definition of the binary polyhedral groups by generators and relations : $R^p = S^q = T^q = RST$, $(RST)^2 = 1$. The relationship between these two occurrences of (p, q, r) is a nice manifestation of the "dual McKay correspondence" of [25].

To the above list, we have now added one more item: the $\widehat{sl}(2)$ modular invariant partition functions. A natural question is: does this new item connect to any previously known case? If it is so, this may give us a hint about what may be expected in other cases. For example, is there some spectral property on matrices that would be related to the modular invariants of $\widehat{sl}(3)$ type? Or would the subgroups of $SU(3)$ be of relevance? Or a certain class of singularities beyond the "simple" ones?

I think it is fair to say that the question has not yet received a clear answer. We do see connections between existing ADE classifications and some consequences of the ADE classification of $\widehat{sl}(2)$ modular invariants, but we do not see how they extend to higher rank cases. Or at least, in no direct and systematic way ... This is what we shall show by discussing next the case of $\widehat{sl}(3)$ theories (sect. 2.3). Then in section 3, we shall see that the study of CFT in the presence of boundaries ("BCFT") gives us some new insight in these questions.

In the $\widehat{sl}(2)$ case, there are two related classes of CFTs for which another ADE classification appears from another standpoint: the $c < 1$ (unitary) minimal models, and the "simple" $\mathcal{N} = 2$ superconformal field theories or their topological cousins [26]. Both may be obtained by the coset construction from the $\widehat{sl}(2)$ models, and inherit from them a variant of the ADE classification. In the former case, it is known that $c < 1$ minimal models admit a lattice integrable realisation [27]: the configuration space of these lattice models is the space of paths on a graph, and demanding that this space supports a representation of the Temperley-Lieb algebra (a quantum deformation of the symmetric group algebra and a known way to achieve integrability) and that the model is critical (in the sense of statistical mechanics) forces us to restrict to graphs with eigenvalues of their adjacency matrix between -2 and 2, hence of ADE type. On the other hand, the $\mathcal{N} = 2$ superCFTs have been argued to admit a description of their "chiral sector" by a Landau-Ginsburg superpotential which must be one of the simple singularities described above, whence again of type ADE [28]. Thus in these two cases, we have an alternative way to see why and how ADE appears. Unfortunately, these alternative standpoints are of little help in the higher rank cases: the class of graphs supporting a representation of the Hecke algebra associated with $sl(N)$, $N > 3$, is not yet known, (for $N = 3$, see [29,30,31,32]) and the $\mathcal{N} = 2$ theories related to higher $sl(N)$ are not all described by a Landau-Ginsburg potential.

There is a finer subdivision of items classified by ADE into two classes: those classified by $A, D_{2\ell}, E_6$, or E_8 and those classified by $D_{2\ell+1}$ or E_7. The distinction appears in many cases when one looks at positivity properties of some numbers, coefficients, etc. For example, in the list of Table 1, the modular invariants of the first class may be written as sums of squares of linear combinations of χ with non negative coefficients. I am not sure that the relevant positivity property has been identified in all cases listed above (see [33] for a discussion of some aspects of this issue). Ocneanu has shown that Dynkin diagrams of the first subclass admit a "flat connection" [34]. Another manifestation of the distinction is the existence or non-existence of a fusion-like algebra, called the Ocneanu-Pasquier algebra, attached to the Dynkin diagram [35,36,33]. From the point of view of CFT, this apparently innocent looking distinction reveals a different structure of the theory. The block-diagonal modular invariant partition functions classified by $A, D_{2\ell}, E_6, E_8$ may be regarded as diagonal in terms of characters of some extended chiral algebra; the others are obtained from the latter by some twisting procedure [9,37]. All these considerations extend beyond the case of $\widehat{sl}(2)$ theories.

2.3. The sl(3) case, the associated graphs

Let us turn to the case of $\widehat{sl}(3)$ for which complete results are now available. According to what was said above on the affine algebras $\widehat{sl}(N)$ at a given level k, each integrable representation of $\widehat{sl}(3)_k$ is labelled by a weight $\lambda = \lambda_1\Lambda_1 + \lambda_2\Lambda_2$ subject to inequalities $\lambda_i \geq 1, \lambda_1 + \lambda_2 \leq k + 2$. With these notations, the complete list of modular invariants is given in Table 4 [11]. It includes four infinite series and six exceptional cases. In the same way as the modular invariants of $\widehat{sl}(2)$ were associated with Dynkin diagrams, with the exponents of the latter giving the diagonal terms of the former, we find here the

Fact With each modular invariant Z of $\widehat{sl}(3)$ one may associate (at least) one graph, whose spectrum is encoded in the diagonal terms of Z. More precisely, the adjacency matrix of this graph has eigenvalues of the form: $S_{(2,1),\lambda}/S_{(1,1),\lambda}$, with a multiplicity equal to the diagonal element $N_{\lambda,\lambda}$ of Z.

This fact was originally proposed as an Ansatz and graphs were found by empirical methods [29]. Later, more systematic techniques to determine the graph were developed in a variety of cases [38], but the physical interpretation of the graph itself remained unclear until recently, when it took a new perspective in the light of boundary conformal field theory (Lecture 3). On a more abstract level, this also inspired developments by Ocneanu, by Xu and by Böckenhauer, Evans and Kawahigashi [39,31,40]. The list of $\widehat{sl}(3)$ graphs displayed in Fig. 1 and 2 were recapitulated in a recent work with Behrend, Pearce and Petkova [14].

Let us discuss briefly the properties of these graphs (Fig. 1 and 2).
⋆ Most of them turn out to be 3-colourable: a colour, black (b), gray (g) or white (w), may be assigned to each vertex, and it is understood that the oriented edges go as: $b \to g \to w \to b$. Some of the graphs, however, are not 3-colourable, and then the orientation of the edges has been explicitly displayed whenever necessary, while the absence of arrow in these cases means that the edge carries both orientations.
⋆ All the graphs listed in Fig. 1 and 2 satisfy the spectral property stated above, but many others also do, which are not associated with any modular invariant [29].
⋆ There are a few cases for which two graphs are associated with the same modular invariant, for example $\mathcal{D}^{(6)}$ and $\mathcal{D}^{(6)*}$, or $\mathcal{E}_1^{(12)}$ and $\mathcal{E}_2^{(12)}$. The graph $\mathcal{E}_3^{(12)}$ which is isospectral with the two latter and was believed so far to be also associated with the same modular invariant, is discarded by A. Ocneanu on the basis that it does not support a system of "triangular cells" [32], i.e., presumably, that the corresponding CFT has no consistent Operator Algebra. It is thus marked with a question mark on Fig. 2.
⋆ Many of these graphs may be obtained by the following procedure, inspired by the McKay correspondence for $SU(2)$. Given a finite subgroup Γ of $SU(3)$, the decomposition into irreducibles (b) of the tensor product of each irreducible representation (a) of Γ by the restriction to Γ of the defining three-dimensional representation of $SU(3)$ yields a matrix $\hat{G}$: $(a) \otimes (3) = \oplus_b \hat{G}_{ab}(b)$. Some appropriate truncation of the graph of $\hat{G}$ may then yield a graph adequate for our problem. Contrary to the case of $SU(2)$, however, things are neither systematic –which vertices/edges have to be deleted is not clear a priori– nor exhaustive: some graphs like $\mathcal{E}^{(24)}$ in Fig. 2 are not reached by this procedure.
⋆ It may be interesting, in view of point (v) of the ADE list above, to note that some of these graphs enable one to construct a reflection group associated with a singularity [41],[42].
⋆ For a proof that the list of graphs is complete from the subfactor perspective see [32]. See also the discussion in [31],[40], where several of these graphs have been reproduced.

Of course, there is nothing special with $\widehat{sl}(3)$ at this stage, and everything could be repeated for higher rank, except that no complete list of modular invariants nor of graphs is known in these cases. Also, for $\widehat{sl}(N)$, it is in fact a collection of $N-1$ graphs, labelled by the fundamental representations of $SU(N)$, which must

Table 4. List of $\widehat{sl(3)}_k$ modular invariants and associated graphs.
The superscript n on the graph must equal $k + 3$.

$(\mathcal{A}^{(n)}) \quad Z = \sum_{\lambda \in P^{(n)}_{++}} |\chi_\lambda|^2$

$(\mathcal{A}^{(n)*}) \quad Z = \sum_{\lambda \in P^{(n)}_{++}} \chi_\lambda \chi_{\lambda^*}$

$(\mathcal{D}^{(n)}) \quad Z = \begin{cases} \frac{1}{3} \sum_{\lambda \in \mathcal{Q} \cap P^{(n)}_{++}} |\chi_\lambda + \chi_{\sigma\lambda} + \chi_{\sigma^2\lambda}|^2 & \text{if 3 divides } n \\ \sum_{\lambda \in \mathcal{Q} \cap P^{(n)}_{++}} |\chi_\lambda|^2 + \sum_{\lambda \in P^{(n)}_{++} \backslash \mathcal{Q}} \chi_\lambda \chi_{\sigma^{n+r}\lambda} & \text{if 3 does not divide } n \end{cases}$

$(\mathcal{D}^{(n)*}) \quad Z = \begin{cases} \frac{1}{3} \sum_{\lambda \in \mathcal{Q} \cap P^{(n)}_{++}} \chi_\lambda \chi_{\lambda^*} + \sum_{\lambda \in \mathcal{Q} \cap P^{(n)}_{++}} \left(\sum_{\ell=0}^{2} \chi_{\sigma^\ell \lambda} \right) \left(\sum_{\ell=0}^{2} \chi_{\sigma^\ell \lambda^*} \right)^* & \text{if 3 divides } n \\ \sum_{\lambda \in P^{(n)}_{++} \backslash \mathcal{Q}} \chi_\lambda \chi_{\sigma^{-n\tau}\lambda^*} \end{cases}^*$ if 3 does not divide n

$(\mathcal{E}^{(8)}) \quad Z = |\chi_{(1,1)} + \chi_{(3,3)}|^2 + |\chi_{(3,2)} + \chi_{(1,6)}|^2 + |\chi_{(2,3)} + \chi_{(6,1)}|^2 + |\chi_{(4,1)} + \chi_{(1,4)}|^2 + |\chi_{(1,3)} + \chi_{(4,3)}|^2 + |\chi_{(3,1)} + \chi_{(3,4)}|^2$

$(\mathcal{E}^{(8)*}) \quad Z = |\chi_{(1,1)} + \chi_{(3,3)}|^2 + (\chi_{(3,2)} + \chi_{(1,6)})(\chi_{(2,3)} + \chi_{(6,1)})^* + \text{c.c.} + |\chi_{(4,1)} + \chi_{(1,4)}|^2 + (\chi_{(1,3)} + \chi_{(4,3)})(\chi_{(3,1)} + \chi_{(3,4)})^* + \text{c.c.}$

$(\mathcal{E}_i^{(12)}, i = 1, 2 \text{ (and 3?)}) \quad Z = |\chi_{(1,1)} + \chi_{(10,1)} + \chi_{(1,10)}|^2 + |\chi_{(3,3)} + \chi_{(3,6)} + \chi_{(6,3)}|^2 + |\chi_{(5,5)} + \chi_{(5,2)} + \chi_{(2,5)}|^2 + |\chi_{(4,7)} + \chi_{(7,1)} + \chi_{(1,4)}|^2 + |\chi_{(7,4)} + \chi_{(1,7)} + \chi_{(4,1)}|^2 + 2|\chi_{(2,2)} + \chi_{(8,2)} + \chi_{(2,8)}|^2$

$(\mathcal{E}_5^{(12)}) \quad Z = |\chi_{(1,1)} + \chi_{(10,1)} + \chi_{(1,10)}|^2 + 2|\chi_{(4,4)}|^2 + (\chi_{(2,2)} + \chi_{(8,2)} + \chi_{(2,8)}) \chi_{(4,4)}^* + \text{c.c.}$

$(\mathcal{E}_4^{(12)*}) \quad Z = |\chi_{(1,1)} + \chi_{(10,1)}|^2 + |\chi_{(3,3)} + \chi_{(6,3)} + \chi_{(5,5)}|^2 + |\chi_{(2,5)} + \chi_{(2,8)}|^2 + 2|\chi_{(4,4)}|^2 + (\chi_{(4,1)} + \chi_{(1,4)})(\chi_{(7,4)} + \chi_{(1,7)} + \chi_{(4,1)})^* + \text{c.c.}$

$(\mathcal{E}^{(24)}) \quad Z = |\chi_{(1,1)} + \chi_{(22,1)} + \chi_{(1,22)} + \chi_{(5,5)} + \chi_{(14,5)} + \chi_{(5,14)} + \chi_{(11,11)} + \chi_{(11,2)} + \chi_{(2,11)} + \chi_{(7,7)} + \chi_{(7,10)} + \chi_{(10,7)}|^2 + |\chi_{(7,1)} + \chi_{(16,7)} + \chi_{(1,16)} + \chi_{(1,7)} + \chi_{(7,16)} + \chi_{(16,1)} + \chi_{(5,8)} + \chi_{(11,5)} + \chi_{(8,11)} + \chi_{(8,5)} + \chi_{(5,11)} + \chi_{(11,8)}|^2$

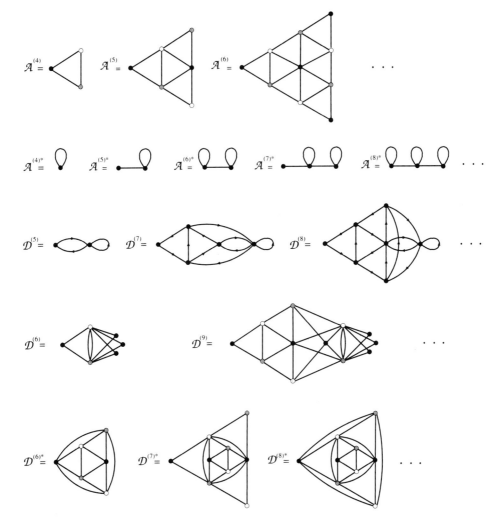

Fig. 1: The graphs of $\widehat{sl}(3)$ (continuation on Fig. 2)

be provided. Complex conjugate representations give graphs with all orientations reversed, and the first $[N/2]$ are thus sufficient.

2.4. General case

In general, we expect that a graph (or a collection of graphs) will be associated with any "rational" conformal field theory, (see section 1.4), with the spectrum of its adjacency matrix determined by the diagonal terms of the partition function. Among the pairs $(j, \bar{j})$ appearing in Z, a special role will therefore be played by the diagonal subset

$$\mathcal{E} = \{j | j = \bar{j}, N_{jj} \neq 0\} \,, \tag{2.7}$$

the elements of which, the "exponents" of the theory, will be counted with the multiplicity N_{jj}. Note that $\mathcal{E}$ is stable under conjugation: j and j^* occur with the same multiplicity. The justification of this association of one (or several) graph(s) with a CFT will appear in the next section.

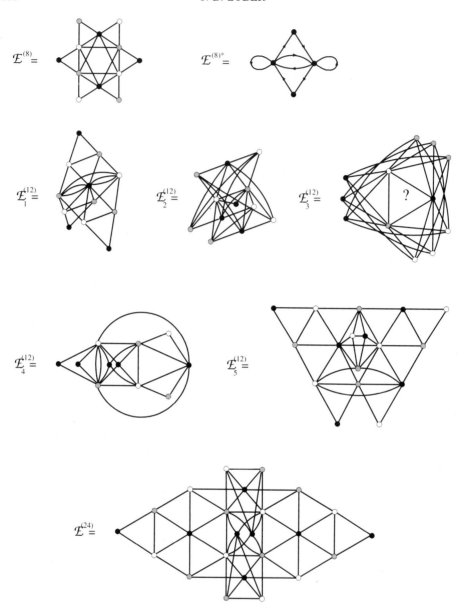

Fig. 2: The graphs of $\widehat{sl}(3)$ (continued)

3. Boundary Conformal Field Theory

3.1. RCFT in the half-plane, boundary conditions and operator content

We now turn to the study of RCFTs in a half-plane. There are several physical reasons to look at this problem, –critical systems in the presence of a boundary, open strings and generalized D-branes, one-dimensional electronic systems with "quantum impurities" etc. Here we shall only look at the new information and perspective that this situation gives us in the classification problem of RCFT.

In a half-plane, the admissible diffeomorphisms must respect the boundary, taken as the real axis: thus only real analytic changes of coordinates, satisfying $\epsilon(z) = \bar{\epsilon}(\bar{z})$ for $z = \bar{z}$ real, are allowed. The energy momentum itself has this property:

$$T(z) = \bar{T}(\bar{z})|_{\text{real axis}}, \qquad (3.1)$$

which expresses simply the absence of momentum flow across the boundary and which enables one to extend the definition of T to the lower half-plane by $T(z) := \bar{T}(z)$ for $\Im m\, z < 0$. There is thus only one copy of the Virasoro algebra $L_n = \bar{L}_n$. This continuity equation (3.1) on T extends to more general chiral algebras and their generators, at the price however of some complication. In general, the continuity equation on generators of the chiral algebra involves some automorphism of that algebra:

$$W(z) = \Omega \bar{W}(\bar{z})|_{\text{real axis}} \qquad (3.2)$$

(see [14] and further references therein).

The half-plane, punctured at the origin, (which introduces a distinction between the two halves of the real axis), may also be conformally mapped on an infinite horizontal strip of width L by $w = \frac{L}{\pi} \log z$. Boundary conditions, loosely specified at this stage by labels a and b, are assigned to fields on the two boundaries z real > 0, < 0 or $\Im m\, w = 0, L$. For given boundary conditions on the generators of the algebra and on the other fields of the theory, i.e. for given automorphisms Ω and given a, b, we may again use a description of the system by a Hilbert space of states $\mathcal{H}_{ba}$ (we drop the dependence on Ω for simplicity). On the half-plane or on the finite-width strip, only **one copy** of the Virasoro algebra, or of the chiral algebra $\mathcal{A}$ under consideration, acts on $\mathcal{H}_{ba}$, and this space decomposes on representations of Vir or $\mathcal{A}$ according to

$$\mathcal{H}_{ba} = \oplus n_{ib}{}^a \mathcal{V}_i, \qquad (3.3)$$

with a new set of multiplicities $n_{ib}{}^a \in \mathbb{N}$. The natural Hamiltonian on the strip is the translation operator in $\Re e\, w$, hence, mapped back in the half-plane

$$H_{b|a} = \frac{\pi}{L}\left(L_0 - \frac{c}{24}\right). \qquad (3.4)$$

To summarize, in order to fully specify the operator content of the theory in various configurations, we need not only determine the multiplicities "in the bulk" $N_{j\bar{j}}$ of (2.1), but also the possible boundary conditions a, b on a half-plane and the associated multiplicities $n_{ib}{}^a$. This will be our task in the following, and as we shall see, a surprising fact is that the latter have some bearing on the former.

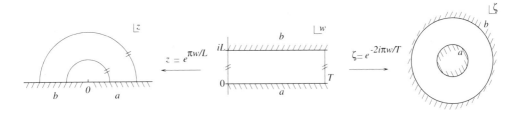

Fig. 3: The same domain seen in different coordinates: a semi-circular annulus, with the two half-circles identified, a rectangular domain with two opposite sides identified, and a circular annulus.

3.2. Boundary states

In the same way that we found useful to chop a finite segment of the infinite cylinder and identify its ends to make a torus, it is suggested to consider a finite segment of the strip – or a semi-annular domain in the half-plane– and identify its edges, thus making a cylinder. This cylinder can be mapped back into an annular domain in the plane, with open boundaries. More explicitly, consider the segment $0 \leq \Re w \leq T$ of the strip –i.e. the semi-annular domain in the upper half-plane comprised between the semi-circles of radii 1 and $e^{\pi T/L}$, the latter being identified. It may be conformally mapped into an annulus in the complex plane by $\zeta = \exp(-2i\pi w/T)$, of radii 1 and $e^{2\pi L/T}$, see Fig. 3. By working out the effect of this change of coordinates on the energy-momentum T, using (1.4), one finds that (3.1) implies

$$\zeta^2 T(\zeta) = \bar{\zeta}^2 \bar{T}(\bar{\zeta}) \quad \text{for } |\zeta| = 1, \; e^{2\pi \frac{L}{T}} . \tag{3.5}$$

Exercise : assuming that W transforms as a primary field of conformal weight $(h_W, 0)$, find the corresponding condition on $W(\zeta)$.

After radial quantization, this translates into a condition on *boundary states* $|a\rangle, |b\rangle \in \mathcal{H}_P$ which describe the system on these two boundaries.

$$(L_n - \bar{L}_{-n})|a\rangle = 0 \tag{3.6}$$

(and likewise for $|b\rangle$). Analogously $(W_n - (-1)^{h_W} \Omega(\bar{W}_{-n}))|a\rangle = 0$, with h_W = spin of W.

We shall now look for a basis of states, solutions of this linear system of boundary conditions. One may look for solutions of these equations in each $\mathcal{V}_j \otimes \mathcal{V}_{\bar{j}} \subset \mathcal{H}_P$, since these spaces are invariant under the action of the two copies of Vir or of the chiral algebra $\mathcal{A}$. Consider only for simplicity the case of the Virasoro generators.

Lemma There is an independent "Ishibashi state" $|j\rangle\rangle$, solution of (3.6), for each $j = \bar{j}$, i.e. $j \in \mathcal{E}$, the set of exponents.

Proof (G. Watts): Use the identification between states $|a\rangle \in \mathcal{V}_j \otimes \mathcal{V}_{\bar{j}}$ and operators $X_a \in \text{Hom}(\mathcal{V}_{\bar{j}}, \mathcal{V}_j)$, namely

$$|a\rangle = \sum_{n,\bar{n}} a_{n,\bar{n}} |j,n\rangle \otimes |\bar{j},\bar{n}\rangle \leftrightarrow X_a = \sum_{n,\bar{n}} a_{n,\bar{n}} |j,n\rangle \langle \bar{j},\bar{n}|$$

Here we make use of the scalar product in $\mathcal{V}_{\bar{j}}$ for which $\bar{L}_{-n} = \bar{L}_n^\dagger$, hence (3.6) means that $L_n X_a = X_a L_n$, i.e. X_a intertwines the action of Vir in the two irreps $\mathcal{V}_j$ and $\mathcal{V}_{\bar{j}}$. By Schur's lemma, this implies that they are equivalent, $\mathcal{V}_j \sim \mathcal{V}_{\bar{j}}$, i.e. that their labels coincide $j = \bar{j}$ and that X_a is proportional to P_j, the projector in $\mathcal{V}_j$. We shall denote $|j\rangle\rangle$ the corresponding state, solution to (3.6).

Since "exponents" $j \in \text{Exp}$ may have some multiplicity, an extra label should be appended to our notation $|j\rangle\rangle$. We omit it for the sake of simplicity. The previous considerations extend with only notational complications to more general chiral algebras and their possible gluing automorphisms Ω. See [14] for more details on these points. Also, in this discussion, I have been a bit cavalier on some points: the fact that these Ishibashi states have no finite norm and thus do not really belong to $\mathcal{H}$, and the use of Schur's lemma in this context would require some justification: See [43] for an alternative and more precise discussion.

The normalization of this Ishibashi state requires some care. One first notices that, for $\tilde{q}$ a real number between 0 and 1,

$$\langle\!\langle j|\tilde{q}^{\frac{1}{2}(L_0+\bar{L}_0-\frac{c}{12})}|j'\rangle\!\rangle = \delta_{jj'}\chi_j(\tilde{q}) \tag{3.7}$$

up to a constant that we choose equal to 1. It would seem natural to then define the norm of these states by the limit $\tilde{q} \to 1$ of (3.7). This limit diverges, however, and the adequate definition is rather

$$\langle\!\langle j\|j'\rangle\!\rangle = \delta_{jj'}S_{1j} \tag{3.8}$$

This comes about in the following way: a natural regularization of the above limit is:

$$\langle\!\langle j\|j'\rangle\!\rangle = \lim_{\tilde{q}\to 1} q^{c/24}\langle\!\langle j'|\tilde{q}^{\frac{1}{2}(L_0+\bar{L}_0-\frac{c}{12})}|j\rangle\!\rangle \tag{3.9}$$

where q is the modular transform of $\tilde{q} = e^{-2\pi i/\tau}$, $q = e^{2\pi i\tau}$. In a ("unitary") theory in which the identity representation (denoted 1) is the one with the smallest conformal weight, show that in the limit $q \to 0$, the r.h.s. of (3.9) reduces to (3.8). In non unitary theories, this limiting procedure fails, but we keep (3.8) as a definition of the new norm.

At the term of this study, we have found a basis of solutions to the constraint (3.6) on boundary states, and it is thus legitimate to expand the two states attached to the two boundaries of our domain as

$$|a\rangle = \sum_{j\in\mathcal{E}} \frac{\psi_a^j}{\sqrt{S_{1j}}} |j\rangle\!\rangle \tag{3.10}$$

with coefficients denoted ψ_a^j, and likewise for $|b\rangle$. We define an involution $a \to a^*$ on the boundary states by $\psi_{a^*}^j = \psi_a^{j^*} = (\psi_a^j)^*$, (recall that $j \to j^*$ is an involution in $\mathcal{E}$). One may show [44] that it is natural to write for the conjugate state

$$\langle b| = \sum_{j\in\mathcal{E}} \langle\!\langle j| \frac{\psi_{b^*}^j}{\sqrt{S_{1j}}}. \tag{3.11}$$

As a consequence

$$\langle b\|a\rangle = \sum_{j\in\mathcal{E}} \frac{\psi_a^j \left(\psi_b^j\right)^*}{S_{1j}} \langle\!\langle j\|j\rangle\!\rangle = \sum_{j\in\mathcal{E}} \psi_a^j \left(\psi_b^j\right)^* \tag{3.12}$$

so that the orthonormality of the boundary states is equivalent to that of the ψ's.

3.3. Cardy equation

Let us return to the annulus $1 \leq |\zeta| \leq e^{2\pi L/T}$ considered in last subsection, or equivalently to the cylinder of length L and perimeter T, with boundary conditions (b.c.) a and b on its two ends. Following Cardy [45], we shall compute its partition function $Z_{b|a}$ in two different ways. If we regard it as resulting from the evolution between the boundary states $|a\rangle$ and $\langle b|$, with $\tilde{q}^{\frac{1}{2}} = e^{-2\pi L/T}$, we find

$$Z_{b|a} = \langle b|(\tilde{q}^{\frac{1}{2}(L_0+\bar{L}_0-\frac{c}{12})}|a\rangle = \sum_{j,j'\in\mathcal{E}} \frac{\left(\psi_b^j\right)^* \psi_a^{j'}}{S_{1j}} \langle\!\langle j|\tilde{q}^{\frac{1}{2}(L_0+\bar{L}_0-\frac{c}{12})}|j'\rangle\!\rangle$$

$$= \sum_{j\in\mathcal{E}} \psi_a^j \left(\psi_b^j\right)^* \frac{\chi_j(\tilde{q})}{S_{1j}}. \tag{3.13}$$

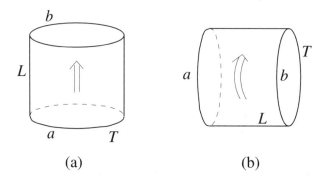

Fig. 4: Two alternative computations of the partition function $Z_{b|a}$: (a) on the cylinder, between the boundary states $|a\rangle$ and $\langle b|$, (b) as a periodic time evolution on the strip, with boundary conditions a and b.

On the other hand, if we regard it as resulting from the periodic "time" evolution on the strip with b.c. a and b, using the decomposition (3.3) of the Hilbert space $\mathcal{H}_{ba}$, and with $q = e^{-\pi T/L}$

$$Z_{b|a}(q) = \sum_{i \in \mathcal{I}} n_{ib}{}^a \chi_i(q) \,. \tag{3.14}$$

(Note that string theorists would refer to these two situations as (a): the tree approximation of the propagation of a closed string; (b) the one-loop evolution of an open string). Performing a modular transformation on the characters $\chi_j(\tilde{q}) = \sum_i S_{ji^*}\chi_i(q)$ in (3.13), and identifying the coefficients of χ_i yields

$$n_{ia}{}^b = \sum_{j \in \mathcal{E}} \frac{S_{ij}}{S_{1j}} \psi_a^j \left(\psi_b^j\right)^* , \tag{3.15}$$

a fundamental equation for our discussion that we refer to as Cardy equation [45]. In deriving this form of Cardy equation, we have made use of the first of the following symmetry properties of $n_{ia}{}^b$

$$n_{ia}{}^b = n_{i^*b}{}^a \tag{3.16}$$

which follow from the previous relations on ψ and of the symmetries of S.

(Comment: this identification of coefficients of specialized characters is in general not justified, as the $\chi_i(q)$ are not linearly independent. As in sect. 2, it is better to generalize the previous discussion, in a way which introduces non-specialized –and linearly independent– characters. This has been done in [14] for the case of CFTs with a current algebra. Unfortunately, little is known about other chiral algebras and their non-specialized characters.)

Let us stress that in (3.15), the summation runs over $j \in \mathcal{E}$, i.e. this equation incorporates some information on the spectrum of the theory "in the bulk", i.e. on the modular invariant partition function (2.6).

Cardy equation (3.15) is a non linear constraint relating a priori unknown complex coefficients ψ^j to integer multiplicities $n_{ia}{}^b$. We need additional assumptions to exploit it.

We shall thus assume that

- we have found an orthonormal set of boundary states $|a\rangle$, i.e. satisfying

$$(n_1)_a{}^b = \sum_{j\in\mathcal{E}} \psi_a{}^j (\psi_b{}^j)^* = \delta_{ab} ; \qquad (3.17)$$

- we have been able to construct a *complete* set of such boundary states $|a\rangle$

$$\sum_a \psi_a{}^j (\psi_a{}^{j'})^* = \delta_{jj'} . \qquad (3.18)$$

Note that the second assumption –which looks far from obvious to me– implies that

$$\text{\# boundary states} = \text{\# independent Ishibashi states} = |\mathcal{E}| .$$

3.4. Representations of the fusion algebra and graphs

Return to Cardy equation (3.15) and observe that it gives a decomposition of the matrices n_i, defined by $(n_i)_a{}^b = n_{ia}{}^b$, into their orthonormal eigenvectors ψ and their eigenvalues S_{ij}/S_{1j}. Observe also that as a consequence of Verlinde formula (1.28), these eigenvalues form a one-dimensional representation of the fusion algebra

$$\frac{S_{i\ell}}{S_{1\ell}} \frac{S_{j\ell}}{S_{1\ell}} = \sum_{k\in\mathcal{I}} \mathcal{N}_{ij}{}^k \frac{S_{k\ell}}{S_{1\ell}}, \qquad \forall i,j,\ell \in \mathcal{I} . \qquad (3.19)$$

Hence the matrices n_i also form a representation of the fusion algebra

$$n_i \, n_j = \sum_{k\in\mathcal{I}} \mathcal{N}_{ij}{}^k \, n_k \qquad (3.20)$$

and they thus commute. Moreover, as we have seen above, they satisfy $n_1 = I$, $n_i^T = n_{i^*}$.

Conversely, consider any $\mathbb{N}$-valued matrix representation of the Verlinde fusion algebra n_i, such that $n_i^T = n_{i^*}$. Since the algebra is commutative, $[n_i, n_j^T] = [n_i, n_{i^*}] = 0$. The $\{n_i\}$ form a set of *normal* matrices, hence are diagonalizable in a common orthonormal basis. Their eigenvalues are known to be of the form S_{ij}/S_{1j}. They may thus be written as in (3.15). Thus any such $\mathbb{N}$-valued matrix representation of the Verlinde fusion algebra gives a (complete orthonormal) solution to Cardy's equation.

Conclusion:

> $\mathbb{N}$-valued matrix representation of the fusion algebra
> $\Longleftrightarrow$ complete, orthonormal solution of Cardy equation

Moreover, since $\mathbb{N}$-valued matrices are naturally interpreted as graph adjacency matrices, graphs appear naturally!

The relevance of the fusion algebra in the solution of Cardy equation had been pointed out by Cardy himself for diagonal theories [45] and foreseen in general in [29] with no good justification; the importance of the assumption of completeness of boundary conditions was first stressed by Pradisi et al [46].

3.5. The case of $\widehat{sl}(2)$ WZW theories

Problem: Classify all $\mathbb{N}$-valued matrix reps of $\widehat{s\ell}(2)_k$ fusion algebra with k fixed. The algebra is generated recursively by n_2

$$n_1 = I, \qquad n_2 \, n_i = n_{i+1} + n_{i-1}, \quad i = 2, \ldots, k \qquad (3.21)$$

$$S \text{ real} \Rightarrow n_i = n_i^T \, .$$

Even though ψ^j and $\mathcal{E}$ are yet unknown we know from (1.25) that n_2 has eigenvalues of the form

$$\gamma_j = \frac{S_{2j}}{S_{1j}} = 2 \cos \frac{\pi j}{k+2}, \quad j \in \mathcal{E} \, . \qquad (3.22)$$

But as discussed already in sect. 2.2, $\mathbb{N}$-valued matrices G with spectrum $|\gamma| < 2$ have been classified. They are the adjacency matrices either of the A-D-E Dynkin diagrams or of the "tadpoles" $T_n = A_{2n}/\mathbb{Z}_2$. Thus as a consequence of equation (3.15) alone, for a $\widehat{sl}(2)$ theory at level k, the possible boundary conditions are in one-to-one correspondence with the vertices of one of these diagrams G, with Coxeter number $h = k + 2$. If we remember, however, that the set $\mathcal{E}$ must appear in one of the modular invariant torus partition functions, the case $G = T_n$ has to discarded, and we are left with ADE. (Up to this last step, this looks like the simplest route leading to the ADE classification of $\widehat{sl}(2)$ theories.) We thus conclude that for each $\widehat{sl}(2)$ theory classified by a Dynkin diagram G of ADE type

$$\mathcal{E} = \text{Exp}(G), \qquad \dim(n_i) = |\mathcal{E}| = |G|$$

complete orthonormal b. c. $= a, b, \cdots$: vertices of G

$n_2 =$ adjacency matrix of G

$n_i =$ "i-th fused adjacency matrix" of G

$\psi^j =$ eigenvector of n_2 with eigenvalue γ_j

One checks indeed that the matrices n_i, given by equation (3.15), together with (1.25), have only non negative integer elements. Let us pause a little to examine the remarkable properties of these matrices which seem to play an ubiquitous role ...

3.6. Side remark: the $\widehat{sl}(2)$ "intertwiners"

Let G be a given Dynkin diagram of ADE type, with Coxeter number h. We want to look more closely at properties of the matrices n_i just defined. Explicitly, using (1.25),

$$n_{ia}{}^b = \sum_{\ell \in \text{Exp}(G)} \frac{\sin \frac{\pi \ell i}{h}}{\sin \frac{\pi \ell}{h}} \psi_a^\ell \psi_b^{\ell *} \, . \qquad (3.23)$$

As stated above, all their matrix elements are non negative integers, and they form a representation of the fusion algebra. But also, regarded as rectangular matrices for fixed a, they satisfy an intertwining property $An = nG$ with the adjacency matrix A of the Dynkin diagram A_{h-1}. More explicitly

$$\sum_{j \in A_{h-1}} A_{ij} n_{ja}{}^b = \sum_{c \in G} n_{ia}{}^c G_{cb} \qquad (3.24)$$

These matrices n have made repeated appearances in various contexts.
(1) In CFT: For an appropriate choice of a subset T of the vertices of $G^{(3)}$ and in particular of a special vertex denoted 1, $1 \in T$, one may write the torus partition functions of type $A, D_{\text{even}}, E_6, E_8$ in the form

$$Z_{\text{torus}} = \sum_{a \in T} |\hat{\chi}_a|^2 \qquad (3.25)$$

where $\hat{\chi}_a := \sum_i n_{i1}{}^a \chi_i$ are combinations of characters of the original algebra, interpreted as characters of a larger "extended" algebra (see sect. 2.2 in fine). This formula, originally found empirically in [29] and justified later in a variety of cases in [38], has been extended to the missing cases $D_{2\ell+1}, E_7$ by use of a relative twist between the right and left characters $\hat{\chi}$ pertaining to the $A_{4\ell-1}$ and D_{10} cases, respectively [39], in agreement with the general result of [37,9] recalled in sect 2.2 in fine. The general formula is thus

$$Z_{\text{torus}} = \sum_{a \in T} \hat{\chi}_a(q) \left(\hat{\chi}_{\zeta(a)}(q)\right)^* . \qquad (3.26)$$

Here ζ is an automorphism of the fusion algebra of the representations of the extended algebra labelled by $a \in T$. This formula has also received a new interpretation in the light of the work of [31][40]: see D. Evans' lectures at this school, and compare his expression $\langle \alpha_i^+, \alpha_j^- \rangle$ for the matrix N_{ij}, with that coming from (3.26) $N_{ij} = \sum_{a \in T} n_{a1}{}^i n_{\zeta(a)1}{}^j$.

(2) Lattice models, graphs, operator algebras. These same matrices n also give the decomposition of representations of the Temperley-Lieb algebra on the space of paths from a to b on the graph G onto the irreducible ones given by the paths from 1 to i on graph A_{h-1} [47]

$$R_a^{(G)b} = \oplus_i n_{ia}{}^b R_1^{(A)i}$$

Another manifestation of this is that they give the counting of "essential" paths [39]. See Appendix B for details.

(3) Kostant polynomials in McKay's correspondence: More surprisingly, maybe, these matrices also appear in the explicit expressions of the so-called Kostant polynomials, in the context of McKay correspondence, [48]: see Appendix A. In that context too, the matrices n have received a very neat group theoretical interpretation by Dorey, in terms of the action of the Coxeter element on simple roots [49].

(4) Finally, they have lately made repeated appearances in the context of integrable theories, e.g. in the S-matrices of affine Toda theories [50], or in the excitation spectrum of integrable lattice models [51].

3.7. Other cases

It should be clear that the situation that we have described in detail for $sl(2)$ extends to all RCFTs. The matrices n_i solutions to Cardy equation are the adjacency matrices of graphs. In the case of $\widehat{sl}(N)$, it is sufficient to supply the $(N-1)$

---

[3] the set of ambichiral vertices in the language of A. Ocneanu

fundamental matrices $n_{\square \atop \vdots}\bigg\}_p$, $p = 1, \cdots, N-1$, to determine all of them. The fact that all n_i then have non negative integer elements is non trivial. By Cardy equation again, they satisfy a very restrictive spectral property: their eigenvalues must be of the form S_{ij}/S_{1j}, when j runs over the set $\mathcal{E}$, i.e. the diagonal part of the modular invariant. We have thus found a justification of the empirical association between RCFTs and graphs, (see sect 2.4), and we have found a physical interpretation of the matrix element $n_{ia}{}^b$ as the multiplicity of representation i in the presence of the boundary conditions a and b.

The program of classifying these graphs/boundary conditions has been completed only in a few cases: $\widehat{sl}(2)$ as discussed above, $\widehat{sl}(3)$ through a combination of Gannon's work and the recent work of Ocneanu [32], see sect. 2.3; $\widehat{sl}(N)_1$ [14], where the results match those obtained in the study of modular invariants [12]: the graphs turn out to be star polygons. The case of minimal ($c < 1$) models has also been fully analysed [14].

3.8. Other algebraic features

The identification of the allowed boundary conditions with the determination of the multiplicities and of the associated graphs is just the beginning of the story. A more elaborate discussion of BCFT should include a study of the operator algebra in the presence of a boundary. This is an important, interesting and lively subject, a general understanding of which is still missing. It is also beyond the scope of these introductory lectures. Let me only mention that in that study, it appears that a triplet of algebras (n_i, M_i, N_a) plays a key role. Let

$$n_{ia}{}^b = \sum_{\ell \in \mathcal{E}} \frac{S_{i\ell}}{S_{1\ell}} \psi_a^\ell (\psi_b^\ell)^*$$

$$M_{ij}{}^k = \sum_{a \in G} \frac{\psi_a^i \psi_a^j (\psi_a^k)^*}{\psi_a^1} \qquad (3.27)$$

$$\hat{N}_{ab}{}^c = \sum_{\ell \in \mathcal{E}} \frac{\psi_a^\ell \psi_b^\ell (\psi_c^\ell)^*}{\psi_1^\ell} \ .$$

The first has already been encountered. In the definition of the second ("Pasquier algebra"), the positivity of the components of the Perron-Frobenius eigenvector ψ^1 is crucial. For the third, one assumes as above the existence of a special vertex denoted 1, such that all $\psi_1^\ell \neq 0$. Note that as matrices, the n, M and $\hat{N}$ satisfy

$$n_i n_j = \sum_k \mathcal{N}_{ij}{}^k n_k$$

$$M_i M_j = \sum_k M_{ij}{}^k M_k \qquad (3.28)$$

$$\hat{N}_a \hat{N}_b = \sum_c \hat{N}_{ab}{}^c \hat{N}_c \ .$$

The role of the first has just been discussed. In most known cases the $\hat{N}_{ab}{}^c$ turn out to be integers. In the type I cases, where they are non negative, they seem to describe a certain fusion algebra, generalised to a class of a yet ill-defined class of "twisted" representations of the extended algebra. Restricted to the subset $a \in T$ (see sect. 3.6 (1)), it reduces to the ordinary fusion of ordinary representations of the extended algebra. One may show that the graph adjacency matrices n_i are linear combinations of the $\hat{N}_a$, and the algebra of the latter may be called a graph fusion algebra. This graph fusion algebra is the dual of the Pasquier algebra in the sense of the theory of C-algebras [52],[36],[38]. In the context of BCFT, the Pasquier algebra is deeply connected with the properties of the so-called bulk-boundary coefficients, which describe the coupling of bulk and boundary operators. For more details, I refer the interested reader to [14] and to the many references quoted there.

I have been particularly sloppy on references in this last section. I should mention that over the last ten years, this subject of boundary CFT has received important contributions by many, in particular Saleur and Bauer, Cardy and Lewellen, Affleck and Ludwig, Affleck and Oshikawa, and Saleur, Pradisi, Recknagel and Schomerus, Sagnotti and Stanev, Fuchs and Schweigert, and Runkel: these references may be found in [14]. Additional recent references are by Gannon, by Huiszoon, Schellekens and Sousa, by Felder, Fröhlich, Fuchs and Schweigert, and many others.

4. Lattice integrable realizations

It should also be mentionned that parallel to the conformal field theoretic discussion sketched in these notes, there exists a discussion of lattice integrable models, the so-called face, or height, or RSOS, models. There the Yang-Baxter is realised through a representation of the Temperley-Lieb algebra, or of some other quotient of the Hecke algebra, on the space of paths on a graph: for example the Pasquier models [27] in the simplest case of $sl(2)$, or their higher rank generalizations. Finally, boundaries may be introduced without spoiling integrability, through a careful determination of the boundary Boltzmann weights, satisfying the Boundary Yang-Baxter Equation [53].

Through these different approaches we can see the various facets of a beautiful common algebraic structure ...

Acknowledgements

It is a pleasure to thank the two Roberts, Robert Coquereaux and Roberto Trinchero, for their invitation and hospitality in beautiful Patagonia and for offering me the opportunity of extremely stimulating talks with many participants, in particular Adrian Ocneanu. I also want to thank Robert Coquereaux and David Evans for comments on the first draft, Ariel Garcia for his assistance in the preparation of my typuscript and Valya Petkova for a careful and critical reading of these lecture notes and for thoughtful suggestions.

Appendix A. The McKay correspondence

The following presents details on the McKay correspondence, the Kostant polynomials and their connection with "intertwiners" of ADE type. The proof of the latter is a slightly expanded version of what appeared in a review article by Di Francesco [36].

Let Γ be a subgroup of $SU(2)$. We denote by (a) its irreducible representations, among which (0) is the identity representation; (f), the "fundamental", is the two-dimensional representation of Γ inherited from that of $SU(2)$. (It may be irreducible or reducible, depending on Γ). The fundamental observation of McKay [24] is that if we tensor product (a) by (f) and decompose it on irreps

$$(f) \otimes (a) = \oplus_b \widehat{G}_{ab} (b) \tag{A.1}$$

we find that $\widehat{G}_{ab}$ is the adjacency matrix of an affine Dynkin diagram $\widehat{G}$, thus canonically associated with Γ, according to Table 3 of sect 2.2. In the following, we shall restrict to the subgroups Γ such that $\hat{G}$ is bi-colourable, namely $\mathcal{C}_{2n}, \mathcal{D}_n, \mathcal{T}, \mathcal{O}, \mathcal{I}$.

We want to see how the irreps of $SU(2)$ decompose onto the irreps of Γ: let $[n]$ be the $(n+1)$-dimensional representation of $SU(2)$ restricted to Γ (in particular $[0]$ is the identity representation, and $[1] = (f)$, see above. Let

$$[n] = \oplus_a N_{na} (a) \tag{A.2}$$

with a generating function of the multiplicities N written as

$$\begin{aligned} F(t) &= \sum_{n=0}^{\infty} t^n [n] = \sum_{a,n} t^n N_{na} (a) \\ &= \sum_{\substack{a \\ \text{irreps of } \Gamma}} F_a(t) (a) . \end{aligned} \tag{A.3}$$

One writes easily recursion formulae

$$[1] \otimes F(t) = \sum_a F_a(t) (a) \otimes (f)$$

$$= \sum_{n=0}^{\infty} t^n ([n+1] + [n-1]) = \left(t + \frac{1}{t}\right) F(t) - \frac{(0)}{t} \tag{A.4}$$

We evaluate this by taking its character on conjugation classes C_i of Γ:

$$\sum_a F_a(t)\chi_a(C_i)\chi_f(C_i) = \left(t + \frac{1}{t}\right)\sum_a F_a(t)\chi_a(C_i) - \frac{1}{t} \qquad (A.5)$$

hence

$$\sum_a F_a(t)\chi_a(C_i) = \frac{1}{1 + t^2 - t\chi_f(C_i)}, \qquad (A.6)$$

or, using the orthogonality of characters $\sum_i |C_i|\chi_a(C_i)\chi_b^*(C_i) = |\Gamma|\delta_{ab}$:

$$F_a(t) = \sum_i \frac{|C_i|}{|\Gamma|} \frac{\chi_a^*(C_i)}{1 - t\chi_f(C_i) + t^2} \cdot \qquad (A.7)$$

The explicit result has been worked out by Kostant [48]. He found

$$F_a(t) = \frac{p_a(t)}{(1 - t^{\mathbf{a}})(1 - t^{\mathbf{b}})} \qquad (A.8)$$

where $p_a(t)$ is a polynomial in t of degree less or equal to h (h is the Coxeter number of the *finite* Dynkin diagram G associated with $\widehat{G}$, **a** and **b** are two integers satisfying $\mathbf{ab} = 2|\Gamma|$, $\mathbf{a} + \mathbf{b} = h + 2$, for example 6 and 8 for E_6).
Exercise : prove that in (A.7) only $|\Gamma|$-th roots of unity appear as poles in t, hence that **a** and **b** must divide $|\Gamma|$.

Γ	$\mathcal{C}_{2n}$	$\mathcal{D}_n$	$\mathcal{T}$	$\mathcal{O}$	$\mathcal{I}$
$|\Gamma|$	$2n$	$4n$	24	48	120
$(\mathbf{a}, \mathbf{b})$	$(2, 2n)$	$(4, 2n)$	$(6, 8)$	$(8, 12)$	$(12, 20)$
h	$2n$	$2n + 2$	12	18	30
G	A_{2n-1}	D_{n+2}	E_6	E_7	E_8

Let us plug the form above in the recursion formula, getting rid of the denominator $(1 - t^{\mathbf{a}})(1 - t^{\mathbf{b}})$

$$\left(t + \frac{1}{t}\right)\sum_a p_a(t)(a) - \frac{(0)}{t}(1 - t^{\mathbf{a}})(1 - t^{\mathbf{b}}) = \sum_a p_a(t)\,(a) \otimes [1]$$
$$= \sum_{a,b} p_b(t)\widehat{G}_{ab}(a) \,. \qquad (A.9)$$

Denote by G_{ab} the adjacency matrix of the *ordinary* Dynkin diagram, obtained from $\widehat{G}$ by removing the node called 0. Then identifying in (A.9) the coefficient of $(a) \neq (0)$ gives

$$a \neq 0 \qquad \left(t + \frac{1}{t}\right)p_a(t) = \sum_{b \neq 0} p_b(t)G_{ba} + p_0(t)\widehat{G}_{a0} \,. \qquad (A.10)$$

Following Kostant, write $p_a(t) = \sum_{n=0}^{h} p_{a\,n} t^n$, $p_{a\,n} = p_{a\,h-n}$, $p_0(t) = 1 + t^h$. Eq. (A.10) implies that $p_{a\,0} = p_{a\,h} = 0$ for $a \neq 0$, and thus

$$\left(t + \frac{1}{t}\right) p_a(t) = \sum_{n=1}^{h-1} p_{a\,n} \left(t^{n+1} + t^{n-1}\right)$$

$$= \sum_{n=0}^{h} (p_{a\,n+1} + p_{a\,n-1}) t^n \quad \text{with } p_{a\,-1} = p_{a\,h+1} \equiv 0 \quad (A.11)$$

$$= \sum_{m,n=1}^{h-1} t^n A_{nm} p_{a\,m} + t^0 p_{a\,1} + t^h p_{a\,h-1}$$

where A_{nm} is the adjacency matrix of type A. Compare the r.h.s. of equations (A.10) and (A.11). Since degree(p_a) $< h$ for $a \neq 0$, the two extra terms in (A.11) have to be identified with $p_0(t)\widehat{G}_{a0} = (1+t^h)\widehat{G}_{a0}$ in (A.10), hence

$$p_{a\,1} = p_{a\,h-1} = \widehat{G}_{a0}, \quad (A.12)$$

while the identification of the coefficient of the terms t^n, $1 \leq n \leq h-1$ yields

$$\sum_{m=1}^{h-1} A_{nm} p_{a\,m} = \sum_{b \neq 0} p_{b\,n} G_{ba}. \quad (A.13)$$

This may be read as a recursion formula determining uniquely $p_{a\,n+1} = p_{a\,n-1} + \sum_b G_{ba} p_{b\,n}$, starting from $p_{a\,1} = \widehat{G}_{a0}$. But this also proves that $p_{a\,m}$ is an intertwiner between the A and the G Dynkin diagrams. Its expression in terms of the matrices of sect. 3.6

$$p_{a\,m} = \sum_b \widehat{G}_{0b} n_{mb}{}^a \quad (A.14)$$

follows from the observation that both satisfy the boundary conditions (A.12).

This completes our proof of the statement (3) in sect 3.6, namely that Kostant polynomials are given by the intertwiners n:

$$p_a(t) = \sum_{m=1}^{h-1} \sum_b \widehat{G}_{0b} n_{m\,b}{}^a t^m, \quad a \neq 0. \quad (A.15)$$

Appendix B. The counting of essential paths

The following gives a slight variant of a proof given by Ocneanu [39], that the entry $n_{na}{}^b$ of the intertwiners of ADE type gives the number of "essential paths" on the diagram under consideration.

Given a graph of ADE type, consider the set of paths of length $n \geq 0$ starting from the vertex a and ending at vertex b, $(a_0 = a, a_1, \cdots, a_n = b)$. Consider the linear span $\mathcal{P}_{ab}^{(n)}$ of these paths. Define the operator Δ_i, the *contraction operator at step i*, by

$$\Delta_i (a_0 = 1, a_1, \cdots, a_n) = (a_0 = 1, a_1, \cdots, a_{i-1}, a_{i+2}, \cdots a_n) \delta_{a_{i-1}, a_{i+1}} \quad (B.1)$$

i.e. Δ_i gives zero if the path doesn't backtrack. Each Δ_i maps $\mathcal{P}_{ab}^{(n)}$ into $\mathcal{P}_{ab}^{(n-2)}$. Then define the subspace $\mathcal{E}_{ab}^{(n)}$ of essential paths from a to b of length n as those that are in the kernel of all Δ_i in $\mathcal{P}_{ab}^{(n)}$, for $i = 1, \cdots n-1$. It is important that this concept is defined in the vector space, because for two paths that both backtrack and would yield the same contracted path, their *difference* is in the kernel. This is typically what happens at the "fork" of a graph D: the two paths that "bounce" off the two end points have a difference that is essential. (As a side remark, it is more suitable to normalise differently the contraction operator [39]. This does not affect the dimension of its kernel but the explicit form of its null vectors is changed.)

As proved by Ocneanu [39], the counting of essential paths of length n with fixed ends a, b (i.e. the dimension of $\mathcal{E}_{ab}^{(n)}$) is given by the intertwiner $n_{n+1\,a}{}^b$. In particular the length of essential paths is bounded by the Coxeter number -1. I give here a proof slightly different from that of Ocneanu, in which I show that the recursion formula for essential paths is just the same as that for the intertwiners, namely (3.21).

Denoting by $a : b$ the property of a and b to be neighbours on the graph, one has

$$\begin{aligned}
\mathcal{E}_{ab}^{(n)} &= \cap_{i=1}^{n-1}(\ker \Delta_i)_{\mathcal{P}_{ab}^{(n)}} \\
&= \left(\oplus_{c:b} \cap_{i=1}^{n-2}(\ker \Delta_i)_{\mathcal{P}_{ac}^{(n-1)}} \right) \cap (\ker \Delta_{n-1})_{\mathcal{P}_{ab}^{(n)}} \\
&= (\ker \Delta_{n-1})_{\oplus_{c:b}\cap_{i=1}^{n-2}(\ker \Delta_i)_{\mathcal{P}_{ac}^{(n-1)}}} \,.
\end{aligned} \qquad (B.2)$$

Then consider the *image of* Δ_{n-1} in the space $\oplus_{c:b} \cap_{i=1}^{n-2} (\ker \Delta_i)_{\mathcal{P}_{ac}^{(n-1)}}$ i.e. in the subspace of $\mathcal{P}_{ab}^{(n)}$ of paths essential up to site $n-2$ (the paths of those kernels are continued from length $n-1$ to length n by" adding" the last step $c-b$). This subspace is $\mathcal{E}_{ab}^{(n-2)}$, since for a linear combination of paths p

$$\Delta_{n-1} \sum_p c_p [p = (a_0, \cdots, a_{n-1}(p), a_n = b)] = \sum_p c_p (a_0, \cdots, a_{n-2}(p)) \delta_{a_{n-2}, b} \quad (B.3)$$

which is a generic element of $\mathcal{E}_{ab}^{(n-2)}$. Hence the dimension of its kernel in that space i.e. the dimension of the r.h.s. of (B.2) is

$$\mathcal{N}_{ab}^{(n)} = \dim \mathcal{E}_{ab}^{(n)} = \sum_{c:b} \dim \mathcal{E}_{ac}^{(n-1)} - \mathcal{N}_{ab}^{(n-2)} \,. \qquad (B.4)$$

Thus the numbers $\mathcal{N}$ satisfy the same recursion formula (3.21) as the n's, and the same boundary conditions, hence are identical. QED

References

[1] A.A. Belavin, A.M. Polyakov and A.B. Zamolodchikov, *Nucl. Phys.* **B 241** (1984) 333-380.

[2] V.G. Kac, *Lect. Notes in Phys.* **94** (1979) 441-445;
B.L. Feigin and D.B. Fuchs, *Funct. Anal. and Appl.* **16** (1982) 114-126; *ibid.* **17** (1983) 241-242.

[3] V. Kac, *Infinite dimensional algebras*, Cambridge University Press; V.G. Kac and D.H. Peterson, *Adv. Math.* **53** (1984) 125-264;
J. Fuchs, *Affine Lie Algebras and Quantum Groups*, Cambridge University Press.

[4] P. Goddard, A. Kent and D. Olive, *Comm. Math. Phys.* **103** (1986) 105-119.

[5] P. Di Francesco, P. Mathieu and D. Sénéchal, *Conformal Field Theory*, Springer Verlag 1997.

[6] E. Verlinde, *Nucl. Phys.* **B 300** [FS22] (1988) 360-376.

[7] J. Cardy, *Nucl. Phys.* **B 270** (1986) 186-204.

[8] H. Sonoda, *Nucl. Phys.* **B 281** (1987) 546-572; **B 284** (1987) 157-192.

[9] G. Moore and N. Seiberg, *Nucl. Phys.* **B 313** (1989) 16-40; *Comm. Math. Phys.* **123** (1989) 177-254.

[10] A. Cappelli, C. Itzykson and J.-B. Zuber, *Nucl. Phys.* **B 280** [FS18] (1987) 445-465; *Comm. Math. Phys.* **113** (1987) 1-26;
A. Kato, *Mod. Phys. Lett.* **A 2** (1987) 585-600.

[11] T. Gannon, *Comm. Math. Phys.* **161** (1994) 233-263; *Ann. Inst. Henri Poincaré: Phys. Théor.* **161** (1994) 233-264, hep-th 9404185.

[12] C. Itzykson, *Nucl. Phys.* (Proc. Suppl.) **5 B** (1988) 150-165;
P. Degiovanni, *Comm. Math. Phys.* **127** (1990) 71-99.

[13] T. Gannon, *The monstruous moonshine and the classification of CFT*, hep-th 9906167.

[14] R.E. Behrend, P.A. Pearce, V.B. Petkova and J.-B. Zuber, *Nucl. Phys.* **B 579** [FS] (2000) 707-773, hep-th 9908036.

[15] M. Hazewinkel, W. Hesselink, D. Siersma and F.D. Veldkamp, *Nieuw Archief voor Wiskunde*, **3** XXV (1977) 257-307.

[16] N. Bourbaki, *Groupes et Algèbres de Lie*, chap. 4–6, Masson 1981; J.E. Humphreys, *Introduction to Lie Algebras and Representation Theory*, Springer Verlag, New York, 1972.

[17] H.S.M. Coxeter, *Ann. Math.* **35** (1934) 588-621;
J.E. Humphreys, *Reflection Groups and Coxeter Groups*, Cambridge Univ. Pr. 1990.

[18] For a review and references, see P. Slodowy, *Lect. Notes in Math.* **1008** (1983) 102-138, Springer; *Algebraic groups and resolutions of Kleinian singularities*, RIMS-1086, Kyoto 1996.

[19] V.I. Arnold, S.M. Gusein-Zaide and A.N. Varchenko, *Singularities of differential maps*, Birkhäuser, Basel 1985.
[20] P. Gabriel, *Manusc. Math.* **6** (1972) 71-103.
[21] E. Hille, *Ordinary Differential Equations in the Complex Domain*, J. Wiley, 1976.
[22] V.F.R. Jones, *Invent. Math.* **72** (1983) 1-25.
[23] F.M. Goodman, P. de la Harpe and V.F.R. Jones, *Coxeter Graphs and Towers of Algebras*, Springer-Verlag, Berlin 1989.
[24] J. McKay, *Proc. Symp. Pure Math.* **37** (1980) 183-186.
[25] J.-L. Brylinski, *A correspondance dual to McKay's*, `alg-geom 9612003`.
[26] T. Eguchi and S.K. Yang, *Mod. Phys. Lett.* **A 5** (1990) 1693-1701 ;
R. Dijkgraaf, E. Verlinde and H. Verlinde, Nucl. Phys. **B 352** 59-86 (1991) ; in *String Theory and Quantum Gravity*, proceedings of the 1990 Trieste Spring School, M. Green et al. eds., World Sc. 1991.
[27] V. Pasquier, *Nucl. Phys.* **B 285** [FS19] (1987) 162-172 ;
V. Pasquier, *J. Phys.* **A 20** (1987) 5707-5717.
[28] W. Lerche, C. Vafa and N. Warner, *Nucl. Phys.* **B 324** (1989) 427-474 ;
E. Martinec, in *Physics and mathematics of strings*, V.G. Knizhnik memorial volume, L. Brink, D. Friedan and A.M. Polyakov eds., World Scientific 1990 ;
P. Howe and P. West, *Phys. Lett.* **B 223** (1989) 377-385 ; *ibid.* **B 227** (1989) 397-405.
[29] P. Di Francesco and J.-B. Zuber, *Nucl. Phys.* **B 338** (1990) 602-646.
[30] N. Sochen, *Nucl. Phys.* **B 360** (1991) 613-640.
[31] F. Xu, *Comm. Math. Phys.* **192** (1998) 349-403.
[32] A. Ocneanu, lectures at this school.
[33] J.-B. Zuber, *C-algebras and their applications to reflection groups and conformal field theories*, lectures at RIMS, Kyoto December 1996, `hep-th 9707034`.
[34] A. Ocneanu, in *Operator Algebras and Applications*, London Math. Soc. Lect. Series, vol. 136, pp. 119-172, D. Evans and M. Takesaki eds, 1988, Cambridge University Press ;
D. Evans and Y. Kawahigashi, *Publ. RIMS, Kyoto Univers.* **30** (1994) 151-166.
[35] V. Pasquier, *Modèles Exacts Invariants Conformes*, Thèse d'Etat, Orsay, 1988.
[36] P. Di Francesco and J.-B. Zuber, in *Recent Developments in Conformal Field Theories*, Trieste Conference 1989, S. Randjbar-Daemi, E. Sezgin and J.-B. Zuber eds., World Scientific 1990 ;
P. Di Francesco, *Int. J. Mod. Phys.* **A 7** (1992) 407-500.
[37] E. Verlinde and R. Dijkgraaf, *Nucl. Phys. (Proc. Suppl.)* **5 B** (1988) 87-97.
[38] V.B. Petkova and J.-B. Zuber, *Nucl. Phys.* **B 463** (1996) 161-193, `hep-th 9510175`; in *GROUP21 Physical Applications and Mathematical Aspects of Geometry, Groups, and Algebras*, vol. **2**, (1997), p.627, Goslar Conference 1996, eds. H.-D. Doebner et al, World Scientific, Singapore, `hep-th 9701103`.

[39] A. Ocneanu, in *Lectures on Operator Theory*, Fields Institute Monographies, Rajarama Bhat et al eds, AMS 1999.

[40] J. Böckenhauer and D.E. Evans, *Comm.Math. Phys.* **197** (1998) 361-386; *ibidem* **200** (1999) 57-103, hep-th/9805023; *ibidem* **205** (1999) 183-228, hep-th/9812110; J. Böckenhauer, D.E. Evans, and Y. Kawahigashi, *Comm.Math. Phys.* **208** (1999) 429-487, math-OA/9904109; *ibidem* **210** (2000) 733-784; see also D. Evans' lectures at this school.

[41] J.-B. Zuber, in *Topological Field Theory, Primitive Forms and Related Topics*, Proceedings of the Taniguchi meeting, Kyoto Dec 1996, M. Kashiwara, A. Matsuo, K. Saito and I. Satake eds, Birkhäuser.

[42] S.M. Gusein-Zaide and A.N. Varchenko, *Verlinde algebras and the intersection form on vanishing cycles*, hep-th 9610058.

[43] J. Fuchs and C. Schweigert, *Nucl. Phys.* **B 530** (1998) 99-136.

[44] A. Recknagel and V. Schomerus, *Nucl. Phys.* **B 531** (1998) 185-225.

[45] J.L. Cardy, *Nucl. Phys.* **B 324** (1989) 581-596.

[46] G. Pradisi, A. Sagnotti and Ya.S. Stanev *Phys. Lett.***B 381** (1996) 97-104.

[47] V. Pasquier and H. Saleur *Nucl. Phys.* **B 330** (1990) 523-556.

[48] B. Kostant, *Proc. Natl. Acad. Sci. USA* **81** (1984) 5275-5277; *Astérisque (Société Mathématique de France)* (1988) 209-255.

[49] P. Dorey, *Int. J. Mod. Phys.* **A 8** (1993) 193-208.

[50] H. Braden, E. Corrigan, P.E. Dorey and R. Sasaki, *Nucl. Phys.* **B 338** (1990) 689-746.

[51] B.M. McCoy and W. Orrick *Phys. Lett.* **A 230** (1997) 24-32 ;
M.T. Batchelor and K.A. Seaton, K.A., *Excitations in the diluate A_L lattice model: E_6, E_7 and E_8 mass spectra*, cond-mat 9803206 ;
J. Suzuki, *Quantum Jacobi-Trudi Formula and E_8 Structure in the Ising Model in a Field*, cond-mat 9805241.

[52] E. Bannai, T. Ito, *Algebraic Combinatorics I: Association Schemes*, Benjamin-Cummings 1984.

[53] R.E. Behrend and P.A. Pearce, *J. Phys. A* **29** (1996) 7827-7835; *Int. J. Mod. Phys.* **11** (1997) 2833-2847; *Integrable and Conformal Boundary Conditions for sl(2) A-D-E Lattice Models and Unitary Minimal Conformal Field Theories*, hep-th/0006094.

Selected Titles in This Series

(Continued from the front of this publication)

270 Jan Denef, Leonard Lipschitz, Thanases Pheidas, and Jan Van Geel, Editors, Hilbert's tenth problem: Relations with arithmetic and algebraic geometry, 2000

269 Mikhail Lyubich, John W. Milnor, and Yair N. Minsky, Editors, Laminations and foliations in dynamics, geometry and topology, 2001

268 Robert Gulliver, Walter Littman, and Roberto Triggiani, Editors, Differential geometric methods in the control of partial differential equations, 2000

267 Nicolás Andruskiewitsch, Walter Ricardo Ferrer Santos, and Hans-Jürgen Schneider, Editors, New trends in Hopf algebra theory, 2000

266 Caroline Grant Melles and Ruth I. Michler, Editors, Singularities in algebraic and analytic geometry, 2000

265 Dominique Arlettaz and Kathryn Hess, Editors, Une dégustation topologique: Homotopy theory in the Swiss Alps, 2000

264 Kai Yuen Chan, Alexander A. Mikhalev, Man-Keung Siu, Jie-Tai Yu, and Efim I. Zelmanov, Editors, Combinatorial and computational algebra, 2000

263 Yan Guo, Editor, Nonlinear wave equations, 2000

262 Paul Igodt, Herbert Abels, Yves Félix, and Fritz Grunewald, Editors, Crystallographic groups and their generalizations, 2000

261 Gregory Budzban, Philip Feinsilver, and Arun Mukherjea, Editors, Probability on algebraic structures, 2000

260 Salvador Pérez-Esteva and Carlos Villegas-Blas, Editors, First summer school in analysis and mathematical physics: Quantization, the Segal-Bargmann transform and semiclassical analysis, 2000

259 D. V. Huynh, S. K. Jain, and S. R. López-Permouth, Editors, Algebra and its applications, 2000

258 Karsten Grove, Ib Henning Madsen, and Erik Kjær Pedersen, Editors, Geometry and topology: Aarhus, 2000

257 Peter A. Cholak, Steffen Lempp, Manuel Lerman, and Richard A. Shore, Editors, Computability theory and its applications: Current trends and open problems, 2000

256 Irwin Kra and Bernard Maskit, Editors, In the tradition of Ahlfors and Bers: Proceedings of the first Ahlfors-Bers colloquium, 2000

255 Jerry Bona, Katarzyna Saxton, and Ralph Saxton, Editors, Nonlinear PDE's, dynamics and continuum physics, 2000

254 Mourad E. H. Ismail and Dennis W. Stanton, Editors, q-series from a contemporary perspective, 2000

253 Charles N. Delzell and James J. Madden, Editors, Real algebraic geometry and ordered structures, 2000

252 Nathaniel Dean, Cassandra M. McZeal, and Pamela J. Williams, Editors, African Americans in Mathematics II, 1999

251 Eric L. Grinberg, Shiferaw Berhanu, Marvin I. Knopp, Gerardo A. Mendoza, and Eric Todd Quinto, Editors, Analysis, geometry, number theory: The Mathematics of Leon Ehrenpreis, 2000

250 Robert H. Gilman, Editor, Groups, languages and geometry, 1999

249 Myung-Hwan Kim, John S. Hsia, Yoshiyuki Kitaoka, and Rainer Schulze-Pillot, Editors, Integral quadratic forms and lattices, 1999

248 Naihuan Jing and Kailash C. Misra, Editors, Recent developments in quantum affine algebras and related topics, 1999

For a complete list of titles in this series, visit the
AMS Bookstore at **www.ams.org/bookstore/**.